Rescue of Sturgeon Species in the Ural River Basin

NATO Science for Peace and Security Series

This Series presents the results of scientific meetings supported under the NATO Programme: Science for Peace and Security (SPS).

The NATO SPS Programme supports meetings in the following Key Priority areas: (1) Defence Against Terrorism; (2) Countering other Threats to Security and (3) NATO, Partner and Mediterranean Dialogue Country Priorities. The types of meeting supported are generally "Advanced Study Institutes" and "Advanced Research Workshops". The NATO SPS Series collects together the results of these meetings. The meetings are co-organized by scientists from NATO countries and scientists from NATO's "Partner" or "Mediterranean Dialogue" countries. The observations and recommendations made at the meetings, as well as the contents of the volumes in the Series, reflect those of participants and contributors only; they should not necessarily be regarded as reflecting NATO views or policy.

Advanced Study Institutes (ASI) are high-level tutorial courses intended to convey the latest developments in a subject to an advanced-level audience

Advanced Research Workshops (ARW) are expert meetings where an intense but informal exchange of views at the frontiers of a subject aims at identifying directions for future action

Following a transformation of the programme in 2006 the Series has been re-named and re-organised. Recent volumes on topics not related to security, which result from meetings supported under the programme earlier, may be found in the NATO Science Series.

The Series is published by IOS Press, Amsterdam, and Springer, Dordrecht, in conjunction with the NATO Public Diplomacy Division.

Sub-Series

A.	Chemistry and Biology	Springer
B.	Physics and Biophysics	Springer
C.	Environmental Security	Springer
D.	Information and Communication Security	IOS Press
E.	Human and Societal Dynamics	IOS Press

http://www.nato.int/science
http://www.springer.com
http://www.iospress.nl

Series C: Environmental Security

Rescue of Sturgeon Species in the Ural River Basin

Edited by

Viktor Lagutov
Department of Environmental Sciences and Policy
Central European University
Budapest, Hungary

Springer

Published in cooperation with NATO Public Diplomacy Division

Proceedings of the NATO Advanced Research Workshop on
Rescue on Sturgeon Species by means of Transboundary Integrated Water
Management of the Ural River Basin
Orenburg, Russia
13–16 June 2007

Library of Congress Control Number: 2008934040

ISBN 978-1-4020-8922-0 (HB)
ISBN 978-1-4020-8924-4 (eBook)

Published by Springer,
P.O. Box 17, 3300 AA Dordrecht, The Netherlands.

www.springer.com

Printed on acid-free paper

PREFACE

While almost every aspect of society-nature interactions can be treated as an environmental security issue, the threats to human societies originating from inadequate freshwater management constitute one of the most widespread and pressing problems. For thousands of years rivers and river valleys have been the cradle of human civilizations. Rivers have provided not only food and freshwater, but also shelter and means of transportation, and they are still an essential component in every national and regional economy. In turn, growing needs of human societies, accompanied by growing abilities, have caused significant river alterations and ecosystem changes that have resulted in river contamination, biodiversity loss and general riverine ecosystem degradation.

The extinction of sturgeon species is one of the most eloquent examples of the negative and irreversible influence of human society on river ecosystems. The sturgeon, sometimes called the "living fossil" or living "dinosaur" of the fish world, is known to have lived since the time of the dinosaurs, for at least 250 million years, and is currently on the verge of extinction solely due to anthropogenic impacts.

There is no need to describe the importance of sturgeon conservation and worldwide concern over its fate. Apart from its high economic value (*black caviar*) and flagship function, sturgeon is an indicator (umbrella) species for the river basin it inhabits. Being the perfect natural bioindicator, an ecological endpoint, the well being of the sturgeon population in a river network allows us not only to determine the river ecosystems' health, but also to assess the sustainability of human activities in the basin. Restoring the sturgeon in the region would not only be of environmental benefit, it would also greatly contribute to economic and social stability in the region, as well as to food and water security.

Out of 15 sturgeon species known, most are considered critically endangered or vulnerable to extinction worldwide. Of the six different sturgeon species inhabiting the Ural river basin, five are indicated in the IUCN Red Book as endangered or critically endangered. Many authors consider even these conclusions as too optimistic and believe that the "point of no return" towards extinction for most sturgeon populations has already been reached.

The drastic decrease in the sturgeon population of the Caspian Basin and its extinction in European rivers is believed to have been caused by a variety of factors (e.g. overfishing, pollution, etc.), but the main ones are habitat degradation and blockage of the spawning places and migration routes by dams on the main basin rivers. Constructed in the lower river streams, the dams break the fish migration routes, cut off the spawning grounds, or destroy them by submerging them. Deprived of the access to the spawning grounds and the ability to reproduce, the long-living sturgeons were gradually harvested by constantly increasing fishing efforts over the course of several decades.

The Caspian Sea sturgeon stock, which according to some estimates once contributed up to 90% of the world sturgeon stock, was the last wild and actively harvested source of sturgeon. However, despite the sad experience and obvious reasons for sturgeon disappearance in other river basins, the Caspian stock was also brought to the verge of extinction by the very same reasons, namely barrier complexes and overfishing.

From this perspective the Ural river is unique since it contains the only self-sustaining, viable sturgeon population capable of natural reproduction.

The Ural River, the third longest river in Europe, is the only free-flowing river in the Caspian basin with a preserved natural hydrological regime and essential floodplain ecosystems. Generally speaking, this feature is unique not only for the Caspian basin, but also for most of the major water streams in the Northern hemisphere, most of which have undergone severe anthropogenic alterations (i.e. damming, channelization, etc.). Thanks to its natural hydrological regime, the Ural riverine biodiversity has not deteriorated as much as that of other big rivers.

On the other hand, the Ural river is also unique due to the traditions of rational water resources usage in the region. This area is historically populated by communities of Ural Cossacks, a self-governing paramilitary ethnic Russian group, enjoying exclusive rights to control natural resources for centuries. The economic activities were focused mostly on the river and sturgeon and were characterized by high appreciation and rational usage of both. Due to the regional specificities and mentality a high level of public participation and active cooperation of local communities with regional authorities can easily be achieved in the Ural watershed to help secure protected status for the sturgeon.

The Ural river contains the habitats for every Caspian sturgeon species. The future of the whole Caspian sturgeon stock and worldwide restoration programs depends on the Ural river's spawning and nursing habitats. The only available Caspian sturgeon spawning grounds are located in the Ural's upper branches on the territory of Russia, while the migration routes, nursing and feeding habitats are in Kazakhstan. Thus, the sturgeon can be preserved only by joint efforts and transboundary cooperation in river basin management. Taking into account the high economic value and worldwide demand for sturgeon, maintaining its natural reproduction and sustainable extraction is a genuine interest of the basin countries.

Though the importance of the Ural river basin sturgeon habitats for the conservation of the whole Caspian Sturgeon population is well understood, the practical measures which have been undertaken so far in this area are not satisfactory. For instance, the Russian National Action Plan developed within the framework of the Caspian Environmental Programme does not mention the river Ural even once, even though the restoration of the spawning habitats is one of the Caspian Strategic Action Programme's objectives.

Sturgeons utilize different habitat types throughout their life cycles: rivers for spawning; rivers, estuaries, or the sea for feeding and wintering. Depending on its life stages, sturgeon habitats are spread through the whole river network, estuaries and adjacent marine areas. The environmental conditions affecting the well-being of the population are defined by the river conditions: water quality, quantity, temperature, etc. In order to secure the sturgeon's well-being, integrated sustainable management of water resources in the basin should be assured.

At the end of the 20th century the river basins have finally been recognized as the most appropriate territorial units for integrated water resources management and sustainable regional development, which would secure current and future society needs as well as environmental flow requirements. The variety of river rehabilitation programs has increased lately, since modern society realized that access to fresh clean water is gradually becoming one of the limiting factors to development and a source of conflicts.

Environmental needs are often neglected in today's practice of water resource management. The currently prevailing concept of sustainable development and the corresponding approach to water resource management suggests assigning certain monetary values to each water-related product or service. This is a not an easy task even for the evaluation of the energy produced by the hydro power stations or an increase in the crops due to irrigation. With regards to the existence of a low importance aquatic species, which may become extinct due to inadequate water management, it is not possible to specify its economic monetary value, if any. In addition to that, the existing IWRM approach relies on stakeholder involvement in participatory decision-making process. However, as a rule, the defenders (if any) of the river ecological needs have much less power and influence on the decision makers than industry lobbies.

If sturgeon were to be used as an incentive for IWRM, its high economic and social values would allow the combination of both ecological and socio-economic aspects of sustainable development. Investment in IWRM and sturgeon conservation can be largely repaid later by "sustainable extraction" of sturgeon.

While this manuscript was being compiled, it was announced that more than 750 freshwater-related scientists in Brisbane, Australia in September 2007 adopted the Declaration on Environmental Flows. This introduces *Environmental Flows* as the "quantity, timing, and quality of water flows required to sustain freshwater and estuarine ecosystems and the human livehoods and well-being that depend on these ecosystems". The Declaration warns that freshwater ecosystems continue to degrade at alarming rates and demands that environmental flow assessment and management be made a basic requirement of IWRM. Under careful consideration this encouraging initiative reveals the fact that until now ecosystem needs were not paid sufficient attention in river management plans.

From this point of view, the sturgeon, or other anadromous species, can play the important role of an integrated indicator of how environmental the flow is. The well being of such species depends on various river ecosystem characteristics and the less natural (environmental) they are, the fewer sturgeons there will be.

The terrible situation with regards to sturgeon stock and the galloping price of caviar have already led to numerous national and international programs aimed at restoration of the Caspian sturgeon. The most commonly used approach is based on using hatchery-based sturgeon re-stocking in parallel with commercial exploitation of the fish stock. The large-scale sturgeon hatcheries began their activities in the Caspian Basin in 1955, releasing millions of sturgeon fingerlings per year. However, the efficiency of this approach in sustaining the wild stock and compensating the loss of natural reproduction has been challenged by many researchers and was to some extent confirmed by the sturgeon fishery collapse in the region. Instead, as is argued in this manuscript, sturgeon conservation and rehabilitation activities should focus on securing natural reproduction to make sturgeon populations self-sustaining.

Historically, sturgeon species inhabited almost every river of the Northern hemisphere and if successful river rehabilitation programs are established they could return to all these habitats as an indicator of river health and the sustainability of society development in the region. Until that time the gene pool of wild sturgeon species should be preserved. In the current situation, when most rivers are not capable of supporting a natural sturgeon population, it is important to conserve the last self-sustaining sturgeon population and to preserve the last river with undisturbed sturgeon habitats, the Ural river. Moreover, the Ural sturgeon conservation and rehabilitation program can serve as a basis for a sustainable regional development strategy, which can incorporate and link together economic, environmental and social aspects and greatly contribute to regional environmental security.

The Ural River Basin Project was launched in 2007 to facilitate sturgeon restoration and sustainable watershed management in the Ural River Basin. The Project is a joint initiative of the Central European University, the Russian Environmental NGO "Green Don" and DonEco Research and Consulting. The underlying idea of the Project is the concept of sustainable basin development by securing natural reproduction of migratory sturgeon species. In order to assure the implementation of this idea an international Ural Sturgeon Park, with features of a Biosphere Reserve and an Ethno-Natural Protected Area, should be established. In this way the Project aims not only to preserve this flagship species, but also to solve social and economic problems by restoration of the traditional life style of local communities.

The First Ural River Basin International Workshop (NATOARW) "Rescue of Sturgeon Species by means of Transboundary Integrated Water Management in the Ural River Basin" was held in Orenburg (Russia) on June 13–16, 2007 within the framework of the Ural Basin Project. Organized by the Research and Consulting Center DonEco and Central European University, the Workshop was also conducted with the active involvement and assistance of the Russian Federal Agency for Environmental Inspections. The workshop, co-financed by the NATO Science for Peace Program and the Caspian Environmental Program, was attended by more than 60 experts, researchers and practitioners from Governmental Environmental Agencies, NGO and business representatives from both basin countries (Russia and Kazakhstan), and representatives from relevant international organizations such as the Food and Agriculture Organization of the United Nations, the Secretariat of Wetland Convention (RAMSAR), the International Association on Danube Research, and many others covering the whole spectrum of Ural Basin management stakeholders.

The current volume is based on the contributions made by the workshop participants.

More information on the Ural river problems and the Ural Basin Project activities can be found on the Project's website at http://uralbas.ru.

Volume content

The problems of migratory species conservation and watershed management are highly interdisciplinary. Though the materials presented in this volume cover a variety of the water- and society-related scientific disciplines, they are only an attempt to grapple with some aspects of the processes and disciplines involved.

Successful integrated watershed management as well as sturgeon conservation is a policy/economy or implementation-related problem rather than a purely scientific challenge. The best available technologies and approaches in water management and fishery areas are often hampered by political or economic constraints. Therefore, to secure the holistic analysis of the environmental security issues in water management and sturgeon restocking the consideration of purely scientific aspects should always be accompanied by institutional, political and legislative analysis.

This volume, based on the materials submitted by the participants to the First International Ural Basin Workshop, consists of two parts.

The *first part* presents the international experience in integrated transboundary water management and sturgeon species conservation and rehabilitation programs.

Integrated Water Resources Management (IWRM) is high on the political and scientific agenda, and a number of national and international projects aim at integrated water management development and implementation worldwide. A review of such IWRM projects and case studies as well as its implications for environmental security is provided by Amar Čolakhodžić, Jerome Simpson and Trahel Vardanian.

This chapter also gives a deeper insight into the available institutional framework for transboundary cooperation on water resources and watershed management in Eastern European and Central Asian countries, where the Ural basin belongs. This analysis is provided by Sonja Koeppel, a representative of the Water Convention Secretariat (UN Economic Commission for Europe). The challenges and prospects of transboundary water management in the region are also discussed in details.

In this situation the conservation and sustainable use of migratory species (i.e. sturgeon), having their habitats along the transboundary shared river networks are challenging tasks. The management of the migratory species stock is often complicated by the transboundary nature of watersheds and species habitats. The implications of migratory fish stocks in transboundary basins for governance, management, and research are considered in the paper by UN FAO fishery experts John Valbo-Jørgensen, Gerd Marmulla and Robin L. Welcomme. The cascades of large dams and weirs were erected on every European river resulting in substantial modification of the environmental conditions, dooming sturgeons to extinction. The influence of the dams on the fisheries is also addressed in this paper.

The factors for sturgeon decline are mostly the same through the entire historical sturgeon habitats. The analysis of the factors causing the decline in sturgeon stock of the Danube river basin and possible rehabilitation activities is given in the next paper of the volume (Mirjana Lenhardt et al.).

Though the restocking of sturgeon population is a long term and complicated process some success was achieved by certain restoration programs in the USA. For instance, a conservation aquaculture program was developed by the Kootenai Tribe of Idaho and has been operating since 1991 (Mohammed Zaidi and Susan Ireland). Though the program uses hatcheries to prevent sturgeon extinction, the main rescue activities focus on re-establishing natural recruitment and mitigating biological and habitat alterations that have harmed the population.

Though the negative influence of dams on migratory fish stocks by blocking the migratory routes and spawning grounds is a well recognized problem, no effective means of sturgeon transfer through the dams has been implemented yet in any dams. Vladimir Lagutov in his paper analyzes the functioning of the fish passing facilities constructed on the South Russia rivers, indicating their inefficiency, and suggesting new inexpensive and efficient technology. The suggested "know-how" is based on "non-forced" nature-like principles, unlike the widely spread approach to fish transportation through the dams. Using this technology would allow the development of river rehabilitation programs and help secure the environmental flow needs.

The *second part* of the volume is devoted to the Ural river, with analysis of the situation regarding basin water and fisheries as well as the applicability of international experience to the regional problems.

The introduction to the Ural watershed and the river's hydrological regime is given by the first paper (Viktor Lagutov). Special attention is paid to the water availability and the environmental conditions affecting sturgeon populations through the 20th century.

Climate change is one of the potential factors which might have implications for regional environmental security. It intensifies the urgency of the improvements in watershed management and sturgeon rescue. A fragile balance or slow degradation can turn into accelerating and inevitable ecosystem changes. According to the results of an international modeling project, presented by Nikolaj Dronin and Andrei Kirilenko, the Ural River Basin will be subject to climate change impact in the near future. The possibilities of climate change in the region and its possible impact on water resources and regional agriculture are discussed in this paper.

The river flow depends not only on climate change or annual environmental conditions but also on human activities in the watershed. So, the influence of land use in the Ural catchment on the formation of the Ural flow is discussed by Yury Nesterenko and Maxim Nesterenko.

Due to the availability of the natural restocking in the Ural river the sturgeon fishery in the Ural has lasted longer than in other Caspian rivers. The next paper by Viktor Lagutov is devoted to the sturgeon species of the

Ural river, regional fishery analysis and the discussion of the restoration activities. The need to prioritize and secure natural sturgeon reproduction is emphasized. The sturgeon population conservation and rehabilitation programs in the Ural river should be aimed at assuring sturgeon's safe arrival in spawning grounds and natural spawning.

To develop such a scientifically sound restoration program extensive research is needed. One of the research techniques which can provide powerful insight into the biology and management of sturgeons is the combining of fish tagging and genetic sturgeon analysis. A study of the Ural sturgeon utilizing high-tech (satellite, acoustic) tagging methods was carried out by a joint team of US and Republic of Kazakhstan researchers (Phaedra Doukakis et al.).

Following the collapse of sturgeon fishery in the region some plans on reviving the protected areas established under the USSR and the development of new protected areas in the Kazakhstan portion of the Ural basin have been announced. However, such unilateral restoration activities are not sufficient in a transboundary watershed with shared valuable resources. Only joint transboundary efforts can be effective. The willingness to cooperate on this matter and readiness for the practical steps were expressed during the First Ural Basin Workshop discussions and consultations not only by scientists but also by officials from both basin countries. Moreover, active involvement of local communities is needed to secure the rational utilization of aquatic resources. The last paper discusses the plans for the creation of an International Ural Sturgeon park with the close involvement of local Cossack communities.

The volume concludes with the Resolution developed by the International Ural Basin Working Group (including both national and international experts) and adopted by the workshop participants. The Resolution was compiled on the basis of expert opinions and the presentations by the Russian and Kazakhstan fishery and water management authorities.

ACKNOWLEDGEMENTS

The authors acknowledge and highly appreciate the financial support provided by the NATO Science for Peace Programme and the Caspian Environmental Program, the technical and expert assistance by UN FAO and the Secretariat of the RAMSAR Convention as well as many other specialists who participated in and contributed to the First International Ural Basin Workshop.

The successful outcome of the workshop would not have been possible without the technical assistance and support of the Russian Federal Agency for Environmental Inspections in the Orenburg Oblast and the active contribution of the Agency Head, Anatolij Zhukov.

Some of the maps in this volume were created using materials kindly provided by Igor Lysenko (UNEP – WCMC).

The help in the preparation of this volume provided by Alan Watt (CEU) is highly appreciated.

Finally, the authors are grateful to Springer Publishing House for the opportunity to publish these materials and to contribute towards rescue of the sturgeon species worldwide and towards sustainable development of the Ural River basin.

CONTENTS

PART II THE URAL RIVER BASIN

2.1 WATER RESOURCES

2.2 FISH RESOURCES

2.3 RESTORATION ACTIVITIES

ABBREVIATIONS

AzNIRH – Azovskij Nauchno-Issledovatelskij Institut Rybnogo Khoziaystva (Azov Sea Fisheries Research Institute)

CITES – Convention on International Trade in Endangered Species of Wild Fauna and Flora

DSS – Decision Support Systems

EECCA – Eastern Europe, the Caucasus and Central Asia

FPF – Fish Pass Facilitiy

IWRM – Integrated Water Resources Management

KamUralRybVod – The Branch of Federal Fishery Inspections Agency in the Kama and Ural basins

KaspNIRH – Kaspijskij Nauchno-Issledovatelskij Institut Rybnogo Khoziaystva (Caspian Fisheries Research Institute)

RRVP – Radial Regulator of Variable Perforation

TAC – Total Allowable Catch

UNECE – United Nations Economic Commission for Europe

Uralbas – Ural River Basin Project

PART I
INTERNATIONAL EXPERIENCE IN TRANSBOUNDARY WATERSHED MANAGEMENT AND STURGEON RESTORATION

1.1 WATER

ENVIRONMENTAL SECURITY AND THE ROLE OF RIVER REGIMES IN FOSTERING (ENVIRONMENTAL) COOPERATION: CASE OF THE INTERNATIONAL SAVA RIVER BASIN COMMISSION

AMAR ČOLAKHODŽIĆ *

Central European University, Budapest, Hungary

Abstract This paper examines the potential and extent to which international river regimes can serve as a platform for encouraging basin wide environmental cooperation sustainable river basin management and in doing so complement regional stability. Contemporary literature indicates that river water resources are particularly volatile; as such they serve as sources of social instability and posses the power to promote regional stability by strengthening political and fostering environmental cooperation. The research complementing this paper was part of the Master's dissertation written by the author under the title – "*Comparative Approach to Environmental Cooperative Assessment of River Regimes and the Case of the International Sava River Basin Commission.*" The results obtained from analyzing four cases of river conflicts – conflicts on the Rhine, Danube, Nile and Jordan Rivers – indicate that even though river conflicts are complex and basin specific, the development of a relationship between stability and river regimes is possible. Accepting such a relationship, the analysis focuses on the recently established International Sava River Basin Commission and more so because the river was under a single jurisdiction for more than half of a century. It questions the regimes credibility to successfully act as a basin-wide administrational and institutional unit, one which practices sustainable water management, without acting as an environmental protection authority and promoting environmental cooperation under the principles of transparency and stakeholder involvement.

Keywords: International Sava Basin Commission, Sava River, environmental security, environmental cooperation, water security, water conflicts, River Commissions, Nile, Jordan Basin, Danube, Rhine, Integrated Water Resource Management, Sustainable Basin Management

* Amar Colakhodzic, Central European University, Nador ut.9, H1051, Budapest, Hungary; a.colakhodzic@gmail.com

Environmental security and water resources

Since the publication of Richard Ullman's paper – Redefining Security (1983) – a large variety of scholars have called for a redefinition and broadening of traditional concept of security to include, among other threats to social wellbeing and threats to environment. Studies carried out by researchers such as Thomas H. Dixon (1991) with an aim to provide linkages and causal relationships between environmental degradation, social stress and conflict led to the birth of a new field of study – Environmental Security. At present, the concept of environmental security is not grounded in a single definition but a number of definitions which support the claim that environmental degradation represents a threat to human species and the ecosystem and that due to its nature this threat "transcends beyond particular states and conceptions of national security" (Krause and Williams 1996). Understanding environmental security and the need for environmental cooperation is today not regarded only as a domain of international and intergovernmental organizations but also nation states. The problem of environmental change is anthropogenic and faced by nations worldwide, where preventative measures build on cooperation as a reaction to existing or potential conflicts. In understanding environmental threats to national security, the center role is being more often played by the state and in large depends on the ability of a state to adapt its national agenda to include and prioritize environmental degradation as a threat to social well being, national security and create an adequate response.

The concept of environmental security in terms of potential for conflict and necessary measures for cooperation is clearly illustrated by disputes over water resources. This stems from three central assumptions about water, namely, water is the foundation of life, "is a finite and scarce resource, and is a common and divided resource" (Haftendorn 2000). The assumption that water is a common and divided resource is critical in making a case that freshwater resources are subject to international conflict and political tensions when they cross national borders. According to Spector (2000), "water sharing inherently contains seeds of conflict as well as cooperation." In this context shared waters (rivers, lakes and aquifers) have become an overarching concern in international law and international relations/politics where efforts are directed to conflict resolution and prevention through constructing arrangements for "more" equitable distribution of water among the riparian sates. This problem is illustrated when observing the following two diametrically opposite principles: (1) the principle of sovereign ownership where the states have the right to control resources under their territory and (2) the principle of shared ownership and equitable joint use of rivers in particular (Dimitrov 2002). The first is granted to each state under the international law while the second principle is derived from international recognition of the need for equitable distribution of water and non-harm to other states. This is evident from the 1992 International Convention on the Protection and Use of Transboundary Watercourses and International Lakes and the 1997 UN Convention on the Law of the Non-Navigational Uses of International Watercourses. It appears difficult to reach a consensus on what equitable water utilization is and when another state is affected by such utilization. Thus, different interpretations of various uses of transboundary or international rivers have generated international conflicts in various parts of the world. These conflicts reach a peak in arid regions where the quantities of available water are insufficient to meet the needs of population within the basin, often resulting in violent conflict (Haftendorn 2000).

The above mentioned principles of sovereign and shared ownerships are one sided when "water" is observed from an ecological or environmentally friendly perspective. Both perspectives refer to the anthropogenic assumption that water resources are solely for human consumption while disregarding the ecological functions that water performs. Even if the "total quantity" of water was equitably divided for human utilization this would inevitably have a negative impact on the ecosystem, among others the loss of habitat and biodiversity. The different function of equitable water utilization will also have an impact on water quantity and quality. This would eventually "breach the security" of humans as they are naturally an integral part of the ecosystem (Dimitrov 2002).

International rivers as sources of conflicts

The word 'Rival' arrives from the Latin world 'Rivalis' meaning "one who uses the same river" (Dimitrov 2002). This certainly does not imply that every international river is a source for potential conflict or dispute, nor does it imply that every international river faces same management problems which can lead to such tensions. As Biswas (2004) notes, water problems are neither homogeneous nor constant or consistent with time. They vary from one region to another, from country to country, seasonally as well as annually. Most studied water conflicts include international rivers with a wide range of "situational" and geographical factors, such as the location of the river and the climatic impacts. Situational factors according to Spector (2000) include a wide array of indicators – developmental, political, institutional, and environmental – all of which influence the quality and quantity of water as well as the change in demand by each riparian state. Therefore, by analyzing causes and conditions of conflict in different river cases, and by identifying situational factors analysts aim to draw parallels and develop models for conflict assessment and promotion of cooperation, and thereby provide for environmental protection and equitable utilization of transboundary water resources.

Accepting that solutions to water conflicts as well as their causes and conditions, are often anthropogenic and depend on a myriad of basin specific factors it is necessary to draw on the following assumptions governing the study of water conflicts:

1. Water is a finite and scarce resource and is a common and divided resource. As such water sharing in transboundary context inherently contains seeds of conflict as-well-as cooperation.
2. River basins are studied on a case by case basis because the unique challenge in management depends on integration of geographical, physical, technological, environmental, social, economic, institutional and political factors.
3. All river conflicts in nature are asymmetrical, whereby there is a state or states controlling a river's source in the upper flow while placing lower lying states at a disadvantage.
4. All river conflicts can be sorted in two broad categories – conflicts through use and conflicts through pollution.
5. High levels of animosity or absence of transboundary institutions can exacerbate instability while positive international settings and existence of transboundary regimes mitigate potential negative effects.
6. Importance of institutional capacity within a basin defined as water management bodies or positive international relations often supersedes the physical aspects of a water system.
7. Existence of River regimes does not indicate that there is an absence of conflict but implies the existence of mutual will to prevent conflict through communication.

Considering the numerous factors defining the causes and effectiveness of treaties and particular intricacies present in each basin, four case studies were observed along the range of non-violent outcomes observed on the Danube and Rhine and violent outcomes observed on the Nile and Jordan, with consideration to water availability, origins of conflict, negotiated treaties, previous and present institutional mechanisms, roles of International Organizations (IO's), levels of development, transparency and the overall geo-political settings.

Observed violent and non-violent outcomes

The relationship between violent and non-violent outcomes is a relationship within basins characterized by underdevelopment and weak relationships and those characterized by development and regional stability. Considering the asymmetric nature of water conflicts, all potential conflicts in the observed basins were caused by an upper riparian attempting to manipulate the water resource at the consequence of a lower riparian state(s) – Sudanese plans to construct the Sennar reservoir on the Nile, joint plans of the Syrian and Jordan governments to construct the Mukheiba dam on the Yarmouk river, the original plans to construct the Gabickovo–Nagymaros on the border of Slovakia and Hungary, and the pollution of the Rhine by Alsatian potassium in France and the Sandoz Spill in Switzerland. The outcomes of these activities depended on the bargaining powers of basin countries, limited by their varying stages of development, regional relations and governing regimes.

Violent outcomes were observed in the Nile and Jordan Basins, otherwise characterized by underdevelopment, weak relationships, social and economic instability and low alternatives to water resources (Libiszewski 1997). The bargaining power in these two basins is vested in the military supremacy of Egypt and Israel. Both countries retaliated and prevented the plans of upper riparian state(s) to manipulate the otherwise dependant scarce water resources. In case of the river Jordan, the outcome impeded on the existing tensions in the Middle East, where the result was an Israeli initiated air-strike of the Mukheiba dam, and the beginning of the 6 Day war in which Israel secured the upper flow of the Jordan River. After almost four decades of international mediation, notably by the US Government, a bilateral agreement for Joint Utilization of the Jordan River was secured between Israel and Jordan in 1994 (Lonergan 1997; Libiszweski 1997). In contrast to crises in the Middle East where violent conflicts over the Jordan River are regarded as a side-effect of existing transboundary religious and ethnic tensions, the water conflicts observed in the Nile Basin are slightly milder. Egypt, Sudan and Ethiopia, which share the largest basin area, have experienced severe famine and in-country civil wars as result of ethnic differences, mismanagement of scarce water resources or both simultaneously. The severity is drastically reduced in Egypt which represents the basin "super power" and as the lowest riparian is highly dependent on the Nile waters (Haftendorn 2000; Murphy 1997). This superiority allowed the country to effectively use the military threat in preventing the construction of dams in Sudan (i.e. Sennar Dam) and Ethiopia (i.e.

series of water infrastructure development projects) and secure basin wide bilateral agreements which reflect the preferences of the dominant party, Egypt. Water utilization along the Nile Basin is under the mandate of the Joint Technical Committee (JTC) established between Egypt and Sudan (Lowi 2001; Murphy 1997).

Non-Violent outcomes were observed on the Rhine and Danube basins in which riparian and basin countries are presently characterized either by high development or in final stages of transition to democracy and free-market economy. The bargaining power in these countries is based on trade and commerce, while the democratic systems and access to alternative water resources and technology facilitate additional negotiation options and space for cooperative agreements. The former is clearly the case of the Rhine Basin which did not experience violent transnational tensions since WW II, and built basin-wide relations over utilization of the river, generally for economic purposes (Murphy 1997). Regardless, the utilization of the Rhine was not without consequences. As the region developed the demand for industrial water and irrigation increased, causing extensive pollution of the river. While its argued that this situation would likely have led to conflict in the Middle East and extreme tensions along the Nile, the Rhine Countries recognized the mutual stakes and initiated two clean-up projects. The second clean-up was a follow-up of constant pollution caused by Alsatian potassium works and the Sandoz Spill in 1986. The reached resolution – represented by the ratification of the Chemical Agreement in 1987, where each riparian agreed to accept a share of the clean-up costs – was greatly influenced by existing positive riparian relationships represented by the International Commission for the Protection of Rhine (ICPR) (i.e. established in 1963 following the first clean-up program) and the democratic governance system which allowed the stakeholder groups in Rotterdam, and transnational environmental NGO's to bring forward their concerns and exercises pressure on the polluting companies (Haftendorn 2000). In case of the Danube Basin, the bargaining power initially was not bestowed on trade and commerce but commanded for the benefit of the Soviet Union. Similar post WW II reconstruction processes as in the Rhine Basin increased transboundary industrial and agricultural pollution, resulting in the degradation of natural environment. The severe consequences of transboundary pollution were recognized in the 1960s, but the Danube cooperation remained limited to navigation overseen by the Danube Commission for Navigation (DCN) and under direct influence of the Soviet Union. This situation prevented any escalation of dispute or conflict otherwise more than likely to occur between "completely independent states" given the narrow mandate of DCN and exclusion of the western riparian basin states, Germany and Austria. Even if a potential escalation of conflict was prevented, the pollution of the Danube continued throughout the Cold War period (Murphy 1997). In the 1950s Hungary and Czechoslovakia initiated a large hydroelectric project at Gabickovo–Nagymaros, which in the early planning phases included three dams and two hydroelectric plants. The implementation of the project began in 1978 with a Czechoslovakian construction team breaking the ground. The objective was to provide electricity to these two countries, otherwise dependent on the Soviet Union, and demonstrate the efficiency of Soviet policies. The sheer size of the project raised international concern for impacts on wider scale, environmental and social. The Hungarian Academy of Sciences suggested that the construction should cease pending on the EIA. Following the evident break-up of the Soviet Union, the Hungarian government withdrew from the project in 1989 (Haftendorn 2000). At this point, the severity of environmental degradation on the Danube called for a collective response. A host of inter-governmental organizations, with the support of the EC, channeled aid to the Eastern Europe Basin countries and laid ground for establishment of the International Commission for the Protection of the Danube River (ICPDR) in 1994. The clean-up program initiated in 1992 started in relative obscurity. The opening of the Gabickovo Hydroelectric facility and the construction of 130 km canal, linking the Danube and the Rhine, received more attention as they were perceived as threats to basin resources. The projects limited water use of downstream countries and impacted their fragile economies in early stages of transition (Murphy 1997). Both projects resulted in varying degrees of non-violent outcomes. The canal provided clear economic benefits for all riparian states, by linking ports on the Black Sea with the Atlantic Ocean. The construction of Gabickovo energy, initiated by the Czechoslovakian government in 1978 and completed in 1992, was opposed by Hungary. As a result, the Gabickovo–Nagymaros dispute, as it is known today, was resolved at the ICJ where both parties were accused of breaking their contractual obligations, including the unilateral withdrawal of Hungary and the unilateral decision of the Slovakian government to divert the Danube (Haftendorn 2000).

Conflict outcomes described in these four basins illustrate how different factors, ranging from political, economic and social factors to hydrographic and environmental factors, are manifested in terms of water conflicts. Their relationship is extremely complicated and certainly varies in each case study. Still, certain parallels can be drawn. Developed countries and countries in transition seem to overcome violent outcomes through promotion of, primarily, economic interests while facilitating national and stakeholder cooperation through integrated river regimes. The same countries work on improving institutional capacities, as illustrated by ICPR and establishment of ICPDR, while acknowledging the importance of social well being and environmental protection in sustaining their river flows. The cases of the Nile and the Jordan demonstrate how these same, social, economic and environmental/hydrographic factors lead to violent or extremely tense outcomes. These semi arid regions are succumbed by underdevelopment, transnational tensions and weak state institutions, and degradation of scarce water resources exacerbates the potential for conflict. Institutional basin arrangements fail to act as preventative mechanisms, and processes for establishing cooperation are put to test. Therefore, establishing a relationship between water conflicts and water agreements is an evident step in understanding the roles of river regimes in facilitating regional stability.

Understanding the relationship between water conflicts and water agreements

Analysis of the conflict causes and outcomes in the previous four river basins points to one common denominator – the existing status of regional stability at the time of conflict and the incremental differences in stability changes after a resolution was reached. This conclusion is derived from five observed commonalities:

1. The intensity of conflict varied by: regional development, existing regional relations, governing regimes and water quantity.
   - Violent outcomes were observed in basins characterized by underdevelopment, weak transnational relationships and existing histories of social stress. These conflicts were further exacerbated by scarce water resources.
   - Non-violent outcomes were observed in basins characterized by developed countries and countries in transition where international assistance played a key role.
2. All treaties and cooperative agreements represent a dispute which has been solved. Basins with existing river regimes at the time of conflict experienced non-violent outcomes.
3. The level of reached resolutions varied by: regional development, existing regional relations, governing regimes and effectiveness of international pressure exerted through the processes of mediation or reconstruction by inter-governmental organizations and NGOs.
4. Present river regimes conforming to international principles tend to successfully mitigate adverse pressure on water systems that might otherwise present a source of conflict. Such river regimes play a crucial role in regional development and stability and are located in basins which experienced non-violent outcomes.

5. Environmental degradation and impacts on the well-being of the public, caused by extreme pollution of river waters, were considered only in basins occupied by developed and transitional countries, later being at the initiation of developed countries and inter-governmental organizations.

Therefore, the relationship between transboundary water conflicts and established or existing river regimes is directly related to degrees of regional stability in a given basin. This linear relationship is illustrated by Figure 1, under the section devoted to the Sava Commission. Accordingly, regions with a strong history of tensions and instability, as observed in the Nile and Jordan basins, lack the institutional capacity and/or political will to peacefully mitigate adverse water pressures. In such settings river/basin regimes are viewed as treaties, with low legitimacy and a limited scope of agreement, thus relying on the military supremacy of the dominating Party to enforce the treaty. With an increase in regional stability the probability of violent conflict decreases, allowing the basin countries to express their interests and negotiate river regimes for the wider benefits of water utilization, and lay ground for implementation of practices leading to sustainable water management. Developed regions and regions in transition seem to offer a wider array of possible arrangements as observed in the Rhine and Danube basins. Even if these basins have alternative access to water resources and water technologies, the degradation of abundant water resources caused by intense industrial development required an adequate response in form of basin administrative units. The services provided by the respective water management commissions include water quantity, water quality, protection from hazardous water impacts and environmental protection of perceived ecosystems. At present both commissions incorporate the developmental interests of respective basin countries and include provisions for transparency and social inclusion as key factors for securing an overall view of basin activities and providing legitimacy to decisions reached. These democratic principles governing the Rhine and Danube Basin Commissions are contrary to the non-transparent structures of the Nile and Jordan bilateral commissions which lack the necessary legitimacy and the enforcement of agreements strongly relies on the military powers of Egypt and Israel (Wolf and Hammer 2000).

Also evident in Figure 1 is the relationship between the international involvement in the process of (re)constructing regional stability and the development of river regimes. The presence of international organizations and foreign governments in the Jordan, Nile and Danube Basins often modified the severity of conflict outcomes and represented an important factor in negotiating water agreements The US Government led a number of initiatives for the settlement of conflict in the Middle East in which equitable water utilization represented a crucial objective. Eisenhower and, later, the Carter Administration believed that cooperation over management of scarce water resources would eventually pave the way for peace in the Middle East. Both administrations did not live to see such a solution. However, "water" played a crucial role in the signing of the Israeli and Jordanian Peace Agreement in 1994, in which the parties to the agreement declared to continue the joint construction of the Maquarin Dam on the Yarmouk River, and a joint bilateral water commission was established for regulation of the allocated water quantities. This represents the only legitimate bilateral agreement enforced in the region today (Lonergan 1997). The successful establishment of the Danube Commission is also largely attributed to the direct international and technical assistance provided by the international developmental organization and the EC after the break-up of the Soviet Union. The role of the international assistance was also inline with the political will of the Danube Countries (with the exception of ex-Yugoslav Republics) to cooperate. Similar international presence also urged the Hungarian government to reconsider the construction of their section of the Gabickovo–Nagymaros Hydro Project, thus limiting the severity of perceived environmental impacts (Murphy 1997). In all, the roles of international organizations and foreign interests differentiated on the basis of basin-wide stability and ranged from preventative, as in the Middle East and Nile Basin (notably Sudan and Ethiopia), cooperative and developmental, as in the early stages of transition in the Danube Basin, to participative, as in the case of the Rhine and the present functioning of the ICPDR. The sequential order of the nature of international involvement is illustrated further on in Figure 2, in relationship to regional stability and modes of cooperation and authority facilitated by river regimes.

Introducing the Sava Commission

The establishment of the International Sava River Basin Commission (ISRBC) in 2006 is today regarded as a strong indicator of stabilization of the Western Balkans. The Framework Agreement on the Sava River Basin (FASRB), the founding document of the ISRBC, at present stands as the only post-war *voluntary agreement* ever signed between Bosnia and Herzegovina (BiH, Bosnia), Croatia, Serbia and Montenegro (presently only Serbia) and Slovenia. The very nature of the agreement received international acclamation as a successful regional cooperative effort, indicating political will of the Sava Governments to go beyond their legacies of war.

From a historical perspective, the launching of the Sava Commission seemed unavoidable. The utilization of waterways as means of transport in Southeast Europe has been important for centuries and provides perhaps the earliest examples of interstate cooperation in this region. The same is valid for the Sava River, the second largest tributary of the Danube. Its unique location in the heart of the Balkans historically provided vital cultural and trading linkages within the region and beyond, bridging the gaps along its flow. However, the present task of the ISRBC goes beyond restoration of navigation, as post war economic stagnation and social poverty left environmental issues disregarded. Uncontrolled discharges of polluting substances and emissions and outdated industrial infrastructure contribute to pollution of the environment and increase pressure on the Sava Basin water systems. These issues are coupled by the present need for basin-wide cooperation and "water institutions" capacity building as the river was under as single jurisdiction for almost a century. According to Wolf (2003), this is the exact scenario where the potential for water conflicts is likely to exacerbate as the rate of change of water institutional capacities is outweighed by the rate of change of water systems. The question remains – is the Sava River still prone to regional instability or does it provide a platform for future basin-wide environmental cooperation?

Establishing the Sava Commission and its priorities

The ISRBC at present is regarded as a "single authority" overseeing the management of the Sava River. According to international acclamation and international principles of Integrated Water Resource Management, the Sava Commission is expected to provide a basis for basin wide environmental cooperative efforts and as such promote the reconciliation, peace building process, among other objectives. Theoretically, the extent to which the regime can be regarded as "cooperative" strongly depends on its ability to make legitimate decisions by providing for transparent decision making and the degree to which it perceives environmental protection as a priority. With an objective to understanding the level of cooperation facilitated by ISRBC it is crucial to understand the structure, mandates and degrees of exercised authority and their consistency with international principles and conventions dictating sustainable water management.

In the aftermath of the ratification of the EU WFD in 2000, the Sava Countries held a meeting in Sarajevo in 2001 and signed the "Letter of Intention" to initiate a series of activities in relation to the existing joint activities on Sava river and its tributaries. The basic idea complementing this letter was to ensure the utilization, protection and control of the Sava Basin in order to provide "better condition of social welfare and higher standards of the population in the region" (REC 2004). This marked the beginning of the five year process that was crowned by the establishment of the ISRBC in 2006. During the negotiation period, the Sava was governed by an Interim Commission for the Sava River Basin (ICSRB) that was established by the signing of the Framework Agreement on the Sava River Basin (FASRB) by the Sava Countries in 2002. The legal provisions in the FASRB form the present foundations of the ISRBC.

The interests expressed by the Sava Countries concerning the need to establish a river regime are initially characterized as economic. As mentioned by Djordje Stefanovic former president of the Interim Sava Commission, "a priority in the initial negotiations on defining the relations within the Sava Basin was the question of reestablishing navigation as the least costly means of transportation, as-well-as the ways of transport and the capacities for improving the infrastructure along the transport route(s)." This priority was supported by all Sava Countries, with exception of Slovenia remaining indifferent because the section of the river under its sovereignty is non-navigable. It is argued that the issue of navigation was also a major factor that stimulated the Stability Pact to facilitate the initial cooperative steps, since the environmental issues are addressed by the working table on economic development. The country initials were passed on to Southeast Europe Cooperative Initiative (SECI) which provided the initial political umbrella, together with the Regional Environmental Center for Southeast and Central Europe (REC) until the process became a self standing regime.

The international dimension of the establishment process represented a significant role in constructing the FASRB. It was the interest of the EU and well as NATO that led the Stability Pack to activate the second largest tributary of the Danube and provide means for acceleration of economic recovery and promote regional stability. The Danube Commission for Navigation (DCN), a remainder of the Soviet Union, was also directly involved in the negotiation process and exercised its power to place accent on navigation, trade and commerce. Environmental issues of concern were included in the agenda in the second round of the negotiations on establishment of the FASRB. The Sava Countries represented interests tangible with present environmental issues, but they were brought forward only after the indirect involvement of the International Commission for Protection of the Danube River and REC representatives. The ICPDR, otherwise responsible for environmental protection of the Danube strongly stressed the implications of IWRM and the EU water policy.

These modern principles governing international regimes strongly promote operational democratization, transparency, equality of represented parties and the need to reach decisions on the basis of consensus. This approach, together with support of international organizations (notably REC) and relevant river regimes (notably the ICPDR) provided for inclusion of all interests stipulated by the Sava Countries. The negotiation process brought to light other extremely relevant water management and environmental issues which were incorporated by the FASRB; Croatia and Bosnia were extremely concerned with flood prevention and the water capacities for potential energy production, Slovenia was concerned with environmental protection and development of responsible tourism, while Serbia raised the question water quality and quantity management. Protection of the aquatic eco-system was on one hand included as a side effect of mentioned interests, and on the other simply because the principles of sustainability, the EU WFD and the Helsinki *Convention on Protection and Use of Transboundary Watercourses and International Lakes* required this sensitive environmental issue.

The present legal provisions in the FASRB provide the ISRBC with international legal capacity to exercise the commissions' functions as an implementing body. The agreement also includes the Protocol on the Navigation Regime and defines the general principles and actions of the parties (i.e. the Sava Countries) and the realization of the mutually agreed goals:

1. Establishment of the international navigation regime on the Sava River and its navigable tributaries
2. Establishment of the sustainable water management and
3. Undertaking measures for prevention or restriction of danger, as well as elimination of the hazardous impacts of floods, ice, draught and accidents involving substances having negative impacts on waters

The FASRB document, besides defining the principles of conduct, also places emphasis on cooperation on the basis of sovereign equality, joint benefit and *good will.* In this, parties also accept to cooperate "by mutual respect of the national laws, institutions and organizations, and by acting in accordance with the European Directive 2000/607EC of the EU Parliament and Council dated October 23 2000" (ISRBC 2006c) The agreement also "envisages" necessary collaboration with ICPDR, the Danube Commission on Navigation, UNECE and "international organizations of the EU" under principles of "reasonable" and equitable utilization of water resources subject to the Sava River Basin. The parties further agreed to exchange data on water regime, navigation regime, regulations, organizational structures and administrative and technical practices.

The relationship between the Sava Commission and Basin stability

With application of a similar approach used to establish the relationship between river agreements and regional or basin stability, it is possible to comparatively establish a similar relationship between the Sava Basin and the Sava Agreement. Observing the regional stability side of the relationship the Sava Basin, at present, stands between the basins which experienced highly tense relations and/or acute conflict (Nile and Jordan Basins) and those which experienced varying degrees of non-violent conflicts (Danube and Rhine Basins), as illustrated in Figure 1. As discussed further, this position is a result of wars previously waged within the Sava Basin, present strong reconstruction efforts and the recognition of the need for trans-boundary water management.

The very fact in recent history of conflict differentiates the Sava Countries (with exception of Slovenia) from their European neighbors as the presence of international peace-building and developmental organizations indicates that the region is undergoing the process of reconciliation and is internationally still regarded as relatively unstable. In comparison to the Middle East, where religious and ethnic differences have led the region into turmoil for already more than half the century, the process of reconciliation in Western Balkans should be regarded as success. The recent efforts of the EU the accept Croatia as an accession country shows promise that eventually Bosnia and Serbia will also join the European Community. The burdens of the reconciliation process are increased by the process of transition experienced in BiH, Serbia and Croatia. The process towards parliamentary democracy certainly varies between the countries and its success is well indicated by the above mentioned varying prospects of EU accession. Even if the process is drastically set-back compared to most of the Danube Basin and post-communist countries, the degree of democratic rights granted significantly varies from countries in the Nile and Jordan Basins, with relative exceptions of Israel and Egypt. This fact is more attributed to the strong international pressures exerted on the Bosnian, Croatian and Serbian Governments at the end of the Balkan Wars than to the political leaders themselves. Still, the basic provisions for individual rights and free market economies are established and now it is for the governments to further incorporate this system of governance and inherent benefits.

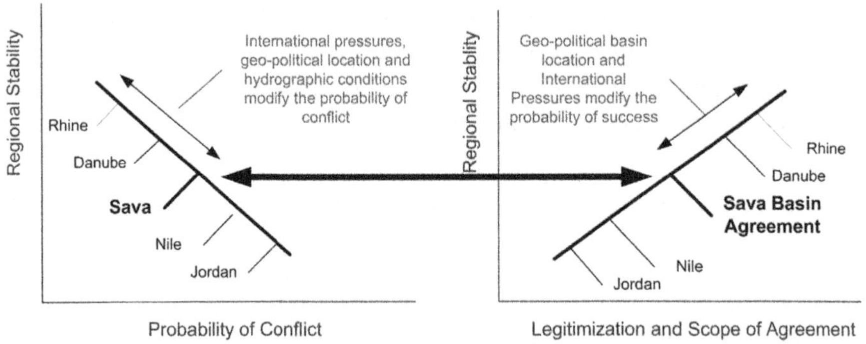

Figure 1. Relationship between water agreements, water conflicts and regional stability

Water scarcity is not an issue in the Sava Basin, as is the case in the Nile and Jordan Basins. The total renewable water resource (the sum of estimated internal and external water resources) in the region is estimated at the 341 billion cubic meters (BCM) and the annual average water availability on a per capita basis is approximated at 14,400 m^3/capita which is twice the average for the whole Europe. The issues of water quality and regulation of water quantity along the Sava River at present represent a growing problem. While the Sava River is subject to flooding, the annual flood damage is expected to increase "given the greater urbanization of flood plains [in Croatia and Serbia] and limited investment in flood mitigation." (World Bank 2003) During former Yugoslavia this issue was regarded as "national" but with its break-up and the regional path toward reconciliation, transboundary impacts of flooding may become greater than expected. The same is valid for dams, which were built during ex-Yugoslavia to regulate the river flow for navigation, generate water for irrigation or generate electricity. Most plans to construct additional dams for these purposes have been placed on hold during the Balkan Wars but their re-introduction is expected to be viewed with scrutiny given the growing concern of environmental and social, national and transboundary impacts. Water quality is another pressing issue, largely caused by deterioration of "discharging" industrial infrastructure and water piping and sewerage systems. The key issues present at the national level are also considered at the transboundary level given the prevalent interlinking water-system of the Sava Basin. The key challenges in the water sector include rehabilitation of water and wastewater treatment systems, enabling of mechanisms for evaluating and monitoring of water quality and further development of water-institutional capacities. On the overall, water management issues within the Sava Basin are similar to most countries in the SEE and other post-communist countries of the Danube Basin, with expectation that the whole process is set-back for obvious reasons. The present situation is further burdened by post-war regional relationships, lack of adequate funding, lack of technical capabilities and should be observed in the contextual process of transition and post war reconstruction (World Bank 2003).

The significantly improved state of stability, as the probability of conflict is drastically reduced by direct successful exercise of international pressure also indicates that, according to the established relationship with water agreements, the negotiated Sava river arrangement should pertain to certain elements of basin cooperation. The degree of river regime cooperation should roughly reflect a higher level of coordination than the JTC on the Nile, while still requiring significant improvements to achieve the degree of sustainable water management practiced by the ICPDR on the Danube, as illustrated by direct relationship between basin stability and regime development.

Shifting the perspective: river regimes as agents of cooperative efforts and regional (basin) stability

Analysis of relationships between water conflicts, water agreements, international involvement and regional stability suggest that relational factors dominating in one basin are often the same factor influencing the scope and legitimacy of the water agreements. Unstable basins negotiate "one sided" agreements or treaties which lack the legitimacy for peaceful resolutions of future disputes. Such agreements are influenced by interests of the dominating party, as observed in the Nile and Jordan Basins, where the regional power exercised by Egypt and Israel represents the means of enforcement. In stable regions the bargaining field is more equal, and perceived benefits of river sharing and cooperation, drive the demand for the establishment of basin regimes and override the potential for violent outcomes. As observed in the Rhine and Danube Basins, the perceived national benefits rely on the ability of basin regimes to sustain the water flows and provide legitimate means for equitable utilization of shared water resources. The degree of international involvement also varies with the perceived basin stability. In basins prone to conflict, the objectives include prevention of conflict escalation and resolution mediation, and shift to sustaining cooperation as relations stabilize, as observed in the Sava Basin. In stable basins, the involvement of international developmental organizations becomes negligent and representation of public interests becomes more important. In this developmental flow, from instability to stability, from the unequal bargaining field to more equal fields, established regimes represent a significant factor for constructing regional stability.

All negotiated agreements in the Rhine, Danube, Nile and Jordan basins represent a river dispute, violent or non-violent, which has been resolved while the Sava Basin represents the classical case of jurisdictional break-up of river governance, therefore – "it is safe to assume that an issue must arise for the parties to enter negotiations in the first place" (Wolf and Hamner 2000). While all rivers witnessed varying degrees of conflicts or political tensions, they also served as the basis of varying degrees of cooperation. After all, rivers *per-se* were not a direct cause of conflict but a side effect of dominating relationships between the basin countries at the time of conflict. Even if water scarcity exacerbated the potential for conflict in unstable regions the issue was not considered as an obstacle in negotiating agreements (Rogers 1997). However, the levels of cooperation reached were affected by situational factors and the existing regional relations which manifested themselves in the negotiation process and shaped the "forms of agreements." In this sense the treaties and established regimes should be regarded as *indicators* of regional stability and *facilitators* of regional stability.

A model for assessing the levels of facilitated (environmental) cooperation

It is accepted that Integrated Water Resource Management (IWRM) represents an ultimate goal for ensuring sustainability and productivity of a river system, which is ideally achieved by full integration of economic benefits and social needs under the framework of environmental protection (Biswas 2004). This approach represents a challenge in any setting since the priorities of "infinite" users must be reconciled. Considering this and the previously described obstacles surrounding the utilization of international rivers, achieving integrated management *requires international cooperation on levels of governance*. The complexities of achieving such degree of cooperation are often excruciating, especially in unstable basins such as the Jordan and Nile, and often must be achieved in the absence of an ultimate entity with a mandate and authority to impose a solution. On the other hand, international river management as well as river dispute resolutions rely on establishment of regimes, as demonstrated by the observed river basins, with various levels of delegated managerial responsibilities. On the contrary, international river management as well as river dispute resolutions rely on establishment of regimes with various levels of delegated managerial responsibilities. It is precisely the degree of mandates and authority of these established regimes which indicates the willingness of basin countries to cooperate. Therefore, the structures of the agreements, and the resulting commissions' as instruments for their implementation, tend to reflect regional stability to the degree of facilitated cooperation within a basin.

The relationship between regional stability, the modes of cooperation and structures of river regimes, with reference to the four river basins, is illustrated in Figure 2, by the River Regime Cooperation Continuum (RRCC). The RRCC is progressive model conceived of four modes of cooperation represented by the extent to which the different basin agendas converge. The focal points of converging agendas are the established forms of river regimes as observed in the Sava, Rhine, Danube, Nile and Jordan Basins. The correlation with regional stability is observed through two unlikely 'indicators' – *the degree to which environmental protection and water quality are perceived as a necessity* and *the degree to which transparent decision making is perceived to increase the regimes' legitimacy*. Their selection was based on the previous analysis of the basins which clearly indicated that the need for economic utilization of basin water resources, either for "production," agriculture or navigation represents the major factor driving the demand for international river agreements, which may or may not initially include a basin-wide administrative unit. In basins

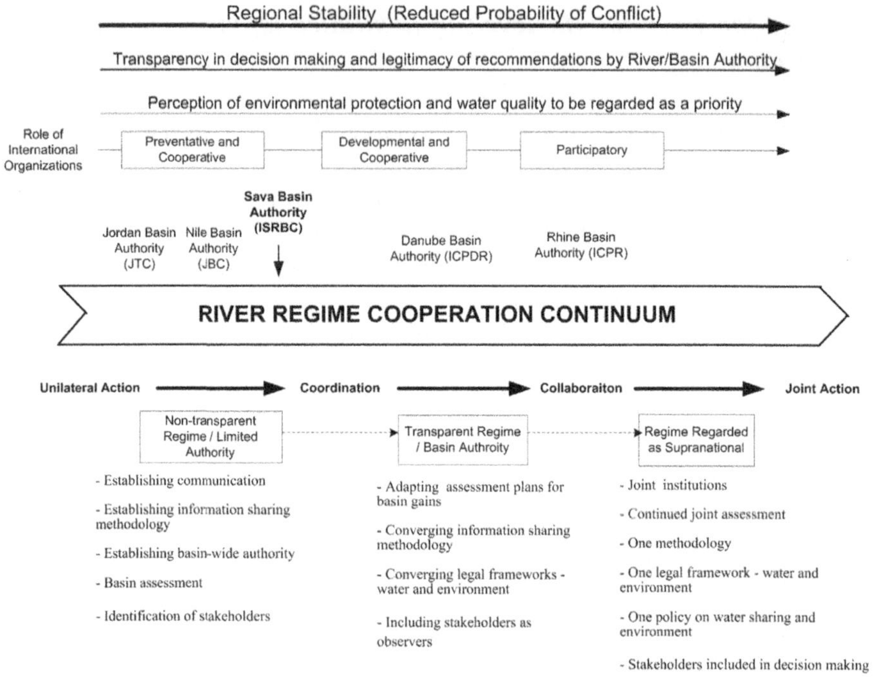

Figure 2. The River Regime Cooperation Continuum

where water resources are scarce the demand for a water utilization/allocation agreement is an absolute necessity to any future basin cooperation on water utilization and prevention of potential water conflict escalation, as was observed in the Nile and Jordan Basins. The need for environmental protection tends to arise after the "realization" that unregulated "economic activity" reduces water supply, water quality and impacts the dependant ecosystems, as was observed in the Rhine and Danube Basins. The ability of regimes to mitigate such adverse pressures on water systems depends on the regimes' structure, operational functioning and mandates. As the mandates of river regimes do not surpass national sovereignty, it is argued that their ability to mitigate adverse pressures also directly reflects the existing relationships between basin countries, economic development, environmental awareness of policy makers and the degree to which national basin regimes allow for the exercise of democratic rights. It is at this stage where the need for "transparency" receives its full meaning and importance in water management. In this progressive developmental and cooperative flow of a given regime environmental protection and transparency are often not considered as the initial subject of the basin agreements but rather gain importance as the regime matures and develops, and the overall basin stabilizes economically and socially.

Drawn implication on the present state of the International Sava Commission

According to the RRCC which describes and indicates a progressive flow of regime stages and modes of cooperation, the structure of the ISRBC and varying priority of objectives indicate that ISRBC is still an infant and non-transparent river regime with potential to establish basin-wide *coordination*, as previously illustrated in Figure 2. Since this model defines *limited authorities or* regimes as agreements limited either by: the number of stakeholder countries, number of objectives exercised, low priority/perception for environmental protection, water quality and low levels or by inexistence of transparency. While the ISRBC represents a basin-wide regime with a number of elements pertaining to sustainable water management, the regime is assessed as limited for the following three reasons:

1. Strong emphasis on navigation with a possibility to undermine the need for sustainable water management (i.e. limited objectives)
2. Initial reluctance to recognize the need for coordinated environmental protection and uncertain future progress (i.e. low perception for environmental protection)
3. Complete neglect for public participation and key-stakeholder involvement (i.e. non-transparency)

The first reason is derived from the fact that even though certain vital international principles of water management are included in the FASRB they are not the driving factor which launched the Sava Basin Initiative. Navigation as a driving factor is justifiable as long as it provides an initiative for enabling basin-wide cooperative efforts to jointly manage the water resources. This is derived from a widely accepted consideration that water transport is not an instrument for sustainable water management but another source of pressure exerted on water quality and aquatic biodiversity. The possibility that the Sava Countries and the ISRBC will politically prioritize restoration of navigation and undermine the purpose of a basin authority, which is sustainable water resource management, should be regarded with extreme caution. This is evident from the fact that the Stability Pact indicated navigation as the first issue to be resolved and that navigation represents the first common objective which all Sava Countries embraced. Prioritization of navigation will also place additional costs on the ISRBC budget, which is presently dependant on the limited national budgets of the Sava Countries, since restoration of navigation includes removal of war debris, unexploded ordinances, reconstruction of bridges and key ports. It is expected that this situation will remain unchanged in the near future, and will strain the otherwise limited funds predicted for other water management objectives – namely water quality and protection of aquatic ecosystems.

The second reason is derived from field research which indicates that even though the FASRB per-se was negotiated in a record six month period, the overall process lasted for seven years. In the early stage of negotiation this process was burdened by shifting national regimes in Serbia and later by the reluctance of the Sava Countries to expand the objectives of the early Sava Initiative beyond navigation and more specifically, embrace the need for protection of aquatic eco-systems. On the other hand, accepting the issues of hazardous water impacts and water quality was not viewed as an obstacle for two major reasons: (i) Sava River is subject to acute flooding and represents a major source of water and sanitation for Belgrade, northern Serbia and southeast Bosnia and (ii) both issues were subject to the First Sava River Management Plan introduced by the former Yugoslav government in the 1970s, thus their consideration required a logical revision and "adaptation to the present situation" – the transboundary context.

This "adaptation" also demanded the consideration and inclusion of current existing international and regional conventions and principles – namely the EUWFD, IWRM and the Danube Basin Agreement – which called for protection of aquatic ecosystems – numerous natural parks and wetlands dependant on the basin waterways. The negotiating parties demonstrated resilience toward this prospect without a clear explanation. However, this reluctance can be attributed to the inexperience of negotiating persons and advisors per-se which perceived the management of the Sava River through the narrow water management framework exercised by former Yugoslavia and DNC during the socialist period. This statement is supported by three facts: (i) a national institution devoted solely to environmental protection was non-existent during former Yugoslavia, (ii) water management was regarded as a sole responsibility of water and sanitation institutions – and as such the need for environmental protection was disregarded by the FSRBMP and (iii) DCN represented the only basin-wide regime during the socialist period, as Sava was considered an international fairway under a single national jurisdiction. It can also be argued that this lack of environmental practices under the Yugoslav regime is one of the major causes of low national environmental awareness and policy making observed in the Croatian, Bosnian and Serbian institutions. This lack of institutional environmental awareness seemingly transgressed into the negotiation process since most persons involved knew each other from before, while conducting functions in water management and political positions during former Yugoslavia as they knew the representatives of the DNC. This position changed after the direct involvement of the ICPDR and their emphasis on EU WFD as-well-as by partial coercion by REC which largely modeled the FASRB after the experiences of the Helsinki

Convention, and Danube and Rhine basin management commissions. The need for environmental protection of "dependant ecosystems" was brought to light and at present is considered as an objective of the ISRBC.

Finally, accepting that the initial reluctance by the Sava Country representatives to accept environmental protection has roots in low environmental practices and institutional awareness existing during the former regime; it is argued that the involvement of key stakeholders and more precisely the environmental advocacy groups, as observers would have certainly shed more light on the issue and probably accelerated the overall negotiation process. This possibility was never debated as the involvement of stakeholders was neglected by the Sava Country representatives in the last phase of negotiation. Key representatives of the international organizations involved regard this act as a complete neglect of the Aarhus Convention and basic democratic rights, and a violation of international obligations toward EU WFD and the Danube Basin Agreement (REC PCSIS 2005). As in the case of environmental protection, this decision remains to be clearly justified scientifically and politically. And as in the case of environmental protection it can be argued that this decision is attributed to low environmental institutional awareness and non-transparent Yugoslav water management practices which transcended into the present Croatian, Bosnian and Serbian Institutions, which however were not subject to this research. On technical and local level, this neglect is likely to decrease the legitimacy of commissions' future decisions and decrease operational effectiveness and efficiency. The question of legitimacy will surface shortly after the process of transition and stability enters its final stages and environmentally aware interest groups begin to question the decisions of ISRBC (as were the cases of ICPDR and ICPR), while the question of operational ineffectiveness was already posed during the debate on inclusion of environmental protection. Thus integrating "public opinion" and engaging interest groups is also a question of technical water management as the objective is to secure an overall view of basin activities and demands, and promote utilization of water for the benefit of all water dependants – the public, the environment as well as the industry.

Conclusions

In order to mitigate any adverse risk of security breach or to promote management and cooperation on basin scale the initial step is the establishment of a "Commission." The commission represents a medium for coordination of basin wide activities and ideally serves as a prevention mechanism. The success of the commission, again ideally, relies on its ability to make solid and ground decisions, balanced in terms of how much power they want to exercise over the parties and convert this into a gradual confidence and cooperation building process. The mandate of such a commission does not supersede the mandate of nation parties, i.e. the governments, thus any decision reached depends on the ability and willingness of "the government" to comply. This compliance at the formal level is regarded as "voluntary." Thus any indication of successful establishment of an international regime, such as is the case of the ISRBC represents a solid indicator of positive changes in regional stability.

From the perspective of water management and protection, the Sava Commission, an implementing mechanism of the Sava Agreement, officially expressed direct interest for sustainable water management in compliance with the EU WFD. This interest should not be misinterpreted in terms IWRM and the underlining transparent principles of water governance. The ISRBC at presents envisions sustainable water management more as a purely technical issue which should be a sole subject of regional water professionals and policy makers, and not a concern of the general public or key non-governmental stakeholders. Along this line of thinking the present 'commission' indisputably provides a platform for intergovernmental cooperation and includes provisions for containing escalation of international political tensions, but not their prevention. A full preventative approach, however, relies on the ability of a regime to secure basin wide flow of information on water needs and demands from local to national levels of governance by encouraging basin-wide environmental cooperative efforts. Thus, from the perspective of environmental cooperation and transparency the Sava Agreement represents a relative failure. This failure seems to stem from the misconception of the concept of stakeholder involvement or 'public inclusion,' also evident in institutional practices of Bosnia, Croatia and Serbia (the key implementing countries) and the benefits, environmental, social and economic, inherent to this transparent managerial approach.

However, considering the transitional national legislations of Bosnia, Croatia and Serbia, and their international obligations it is difficult to disprove that there is no ground for public participation. The logical foundation to the issue is in the difference between formal acceptance of obligations, the existing institutional capacities and political willingness of the Bosnian, Croatian and Serbian authorities to practically implement international agreements. The established relationship between basin stabilities and levels of regime development also indicates a similar observation. Therefore, the Sava Commission should be perceived as a direct outcome of existing basin-wide political relationships, governing regimes and national institutional practices, coupled by the lack of environmental policy making.

References

Biswas, A.K. 2004. Integrated Water Resources Management: A Reassessment. *Water International* 29(2): 248–256.

Dimitrov, R.S. 2002. Water, Conflict, and Security: A Conceptual Minefield. *Society and Natural Resources Natural Resources* 15(8): 678–690.

Haftendorn, H. 2000. Water and International Conflict. *Third World Quarterly* 21(1): 51–68.

Homer-Dixon, T.F. 1991. On the Threshold – Environmental Changes as Causes of Acute Conflict. *International Security* 16(2): 76–116.

International Sava River Basin Commission - Framework Agreement on the Sava River Basin (FASRB). 2006a. Official Web Site. URL: http://www.savacommission.org/history.php

International Sava River Basin Commission - Protocol on the Navigation Regime to the Framework Agreement on the Sava River Basin (FASRB). 2006b. Official Web Site. URL: http://www.savacommission.org/history.php

ISRBC. 2006c. International Sava River Basin Commission. Official Web Site. URL: http: //www.savacommission.org/history.php

Krause, K., and Williams, M.C. 1996. Broadening the Agenda of Security Studies: Politics and Methods. *Mershon International Studies Review* 40(2): 229–254.

Libiszewski, S. 1997. Integrating Political and Technical Approaches: Lessons from the Israeli – Jordanian Water Negotiations. *Conflict and the Environment*, ed. P. Nils, Chapter 23. Dordrecht: Kluwer.

Lonergan, S. 1997. Water Resources and Conflict: Examples from the Middle East. *Conflict and the Environment,* ed. N.P. Gleditsch, Chapter 22. Dordrecht: Kluwer.

Lowi, M.R. 2001. Rivers of Conflict, Rivers of Peace. *Journal of International Affairs* 49(1): 124–144.

Murphy, I.L. 1997. *The Danube: A River Basin in Transition.* Dordrecht: Kluwer.

REC PCSIS. 2005. *Public Communication and Stakeholder Involvement Strategy for the Implementation of the Framework Agreement on the Sava River Basin.* Final Draft as Submitted to the Sava. White Paper Document. Szentendre: Regional Environmental Center for Central and Eastern Europe.

Regional Environmental Center for Central and Eastern Europe (REC). 2004. *Integralno upravljanje vodnim resursima na podrucju Bosne i Hercegovine* [Integrated Water Resource Management in Bosnia and Herzegvoine]. Szentendre: Regional Environmental Center for Central and Eastern Europe.

Rogers, K.S. 1997. Pre-empting Violent Conflict: Learning from Environmental Cooperation. *Conflict and the Environment*, ed. N.P. Gleditsch, 503–534. Dordrecht: Kluwer.

Spector, B.I. 2000. Motivating Water Diplomacy: Finding the Situational Incentives to Negotiate. *International Negotiations* 5: 223–236.

Ullman R. 1983. Redefining Security. *International Security* 8 (summer): 129–153.

Wolf, A.T. 2003. The Present and Future of Transboundary Water Management. *Rethinking Water Management – Innovative Approaches to Contemporary Issue*, ed. C.M. Figueres et al., 165–179. London: Earthscan.

Wolf, A.T., and Hamner, J.H. 2000. Trends in Transboundary Water Disputes and Dispute Resolution. Environment and Security. *Environment and. Security: Discourses and Practices*, ed. M.R. Lowe et al., 123–148. Houndmills: Macmillan.

World Bank. 2003. *Water Resources Management in South Eastern Europe – Volume 1 Issues and Direction.* Washington, DC: International Bank for Reconstruction and Development.

CHALLENGES AND PROSPECTS OF TRANSBOUNDARY WATER MANAGEMENT IN EASTERN EUROPE, CAUCASUS AND CENTRAL ASIA

SONJA KOEPPEL*

United Nations Economic Commission for Europe (UNECE)
Convention on the Protection and Use of Transboundary Watercourses and International Lakes

Abstract Since the breakdown of the Soviet Union in 1990 transboundary water management has become very important in Eastern Europe, the Caucasus and Central Asia (EECCA) since many problems of water protection and use cannot be solved on the national level alone any more. Therefore, most states have taken measures for transboundary water cooperation. However, transboundary water management in the EECCA-region is complicated by many challenges, such as water pollution, lack of data, financial constraints, institutional and structural challenges and shortcomings of concluded agreements. On the other hand, many EECCA-governments are showing increased commitment and interest in transboundary cooperation by allocating additional financial resources, reforming their institutions etc. In addition, various International Organizations and national donors are supporting this trend through capacity-building, funding and other types of assistance.

Keywords: Transboundary water management, EECCA, Eastern Europe, the Caucasus, Central Asia, cooperation, water, Water Convention

Introduction

With about 150 transboundary rivers, more than 100 transboundary aquifers and 50 international lakes, water resources in Europe are characterized by their essentially transboundary nature (UNECE 2004). In the Eastern part of the region, after the breakdown of the communist block and the subsequent creation of numerous new states many problems related to

---

* Sonja Koeppel, United Nations Economic Commission for Europe, Convention on the Protection and Use of Transboundary Watercourses and International Lakes, 8-14 avenue de la Paix, 1211 Geneva 10, Switzerland; Sonja.Koeppel@unece.org

water protection, management and use have taken on a transboundary dimension. Therefore, since the early 1990s, all newly created countries established mechanisms for transboundary water cooperation to a greater or lesser extent. At the same time, the need for Integrated Water Resources Management (IWRM) and river basin management is increasingly recognized, which implies also the need to manage transboundary water resources jointly. International legal frameworks such as the United Nations Economic Commission for Europe's (UNECE) Water Convention require the creation of joint management institutions, oblige states to ensure sufficient water quality and to reduce pollution in their watercourses.

However, there are numerous challenges to this ideal state, especially in countries with economies in transition in Eastern Europe, the Caucasus and Central Asia (EECCA), among others. Therefore, this chapter analyses first the challenges, followed by the prospects of transboundary water management in the given region.

Challenges for transboundary water management

Growing competition for water

A very important challenge for transboundary water cooperation in general and also in the UNECE region is the competition between different use(r)s of water, agriculture (irrigation), industrial use, energy generation, households etc. (UNECE, UNEP et al. 2003). This competition exists even within national boundaries and becomes more problematic in transboundary settings, especially in situations of water stress. In Central Asia, upstream countries (Kyrgyzstan and Tajikistan) often need to use water for hydroelectricity generation, whereas downstream countries need water for irrigation in agriculture. For example, the need of the Kyrgyz government to use the Toktogul water reservoir for hydropower production in the winter causes serious problems and confrontations with agricultural production activities downland, where the main agricultural land of Uzbekistan and Kazakhstan is located. The problem is due to conflicting seasonal needs: Kyrgyzstan wants to store water in the summer in order to use it for hydropower generation in the winter, whereas the downstream countries need the water for irrigation exactly in the summer when it is stored for hydropower generation.

Competition for water is expected to grow due to population growth, economic growth, but also due to climate change. Although climate change impacts are still difficult to predict and insufficiently modeled on basin levels, it is certain that climate change will increase competition for water. Climate change impacts such as deteriorated water quality as well as an increase in extreme weather events such as floods, droughts are already visible in most of the river basins of the UNECE region (UNECE 2007a). Water availability will be especially reduced in Central, Eastern and Mediterranean Europe as well as in Central Asia (Alcamo et al. 2007). For example, in Central Asia, glaciers are melting faster than before, and most of the seasons are expected to be much drier than currently, which increases the risk for droughts and can reinforce existing competition for water. Adaptation measures are therefore necessary, but very little has been done so far in the region.

Pollution and degraded water quality

Water pollution remains a major problem in EECCA countries. Although the economic decline in the 1990s has led to an improvement of water quality due to the economic recession and the closing down of polluting industries, it also had as a consequence the deterioration of the maintenance and management of many sewage treatment facilities in the region (UNECE 2007a). For this reason, organic pollution is often a major problem in the region due to untreated wastewater. Further pollution sources include industrial facilities and especially mining, which often leads to a high concentration of heavy metals and other dangerous substances (acid mine drainage). In addition, water pollution is caused by illegal waste disposal along rivers as well as old and often uncontrolled waste disposal sites and abandoned contaminated military sites (UNECE 2007a). Deposit of armaments and munitions inherited from the Soviet Union and waste disposal sites belonging to the military, including toxic and radioactive material represent a significant threat to transboundary surface and groundwaters.

Insufficient water monitoring and data problems

These various pollution sources increase the need for appropriate and regular monitoring of water quality and quantity. However, lack of information and monitoring is a frequent problem in the EECCA region: since water quality is not always monitored or not at a sufficient level, data and information is partly missing between neighbouring countries despite its importance for good management. Various reasons explain the decline in monitoring activities such as insufficient and instable financing, insufficient replacement of monitoring stations and laboratory devices with up-to-date equipment, the worsening situation regarding sampling, and departures of qualified staff. Furthermore, EECCA countries sometimes do not want to share their data on water quality and quantity although this is a prerequisite for transboundary water management. Lacking harmonization of water data and methodologies for assessing water quality between neighboring countries represent an additional problem. In Central Asia, links between agencies responsible for water quality monitoring and river basin organizations are often missing (Hannan 2005).

Financial constraints

Monitoring as well as IWRM in general requires sufficient financial resources. However, financial problems significantly hamper appropriate water management, at the national, local and transboundary level especially in the EECCA countries. In addition, expenses for the improvement of water supply, sanitation and water protection measures are often not sufficiently prioritized by governments in the EECCA-region. This leads to degradation of water treatment plants and insufficient investments into water supply and sanitation as well as water pollution prevention activities. Almost 130 million persons still lack access to safe drinking water in the UNECE region, especially in EECCA-countries, 85 million do not have improved sanitation (WHO and UNICEF 2008). Thus, major investments are necessary to reach the Millennium Development Goals.

The elaboration and negotiation of regional conventions and transboundary water agreements as well as their implementation, especially the operation of joint bodies also require adequate funding (UNECE, UNEP 2003). However, joint commissions in the EECCA-region often only dispose of a very limited budget for implementation of joint programs, partially because of the lack of financial commitments of the Riparian Parties to cover these costs in the agreements which establish joint bodies.

Institutional and structural challenges at national level

Appropriate national water management is a precondition for the success of transboundary water management. However, lack of qualified specialized staff with water management expertise and experience in their water agencies, ministries etc. represents a significant barrier to effective transboundary water management in the region (UNECE, UNEP 2003). River basin organizations experience a similar problem regarding their staff members. The authorities in charge of transboundary water management are often not strong enough and their mandate is sometimes not clear. Even more problematic is the lack of cooperation between different government agencies and ministries, at the national as well as transboundary level, for example between the ministries responsible for water management and environment in Central Asia.

Insufficient legal frameworks

National provisions for transboundary water management are sometimes not sufficiently harmonized in the neighbouring countries which can cause problems, for example between EU member countries implementing the EU Water Framework Directive and non-EU member states. One example is lake Peipsi, shared by Estonia and the Russian Federation: these states have concluded an agreement in 1997, but as water quality standards diverge in the two countries it is difficult to reach a common conclusion on what is "good ecological quality" (Šumberová 2003). The Russian Federation has agreed to implement the EU Water Framework Directive in the lake Peipsi region, but implementation of the directive's provisions is much slower. This is due to divergences in objectives and interests which can lead to different management strategies.

Legal and regulatory frameworks for bilateral and multilateral cooperation in water management are partly lacking. A number of EECCA-countries have not yet ratified the UNECE Water Convention such as Tajikistan and Kyrgyzstan and numerous transboundary rivers still lack cooperation agreements such as the Psou river, shared by the Russian Federation and Georgia or the Zeravshan river, shared by Uzbekistan and Tajikistan. Transboundary water management is not considered as high priority by some governments in the region, especially upstream countries.

Shortcomings of existing agreements

The collapse of the Soviet Union in 1990 has triggered the creation of a number of transboundary agreements in EECCA-countries, even in Central Asia, such as the agreement between the Russian Federation and Estonia on the Lake Peipsi/Chudskoe – River Narva basin the Agreement between the Governments of Kazakhstan, Kyrgyzstan, Tajikistan, and Uzbekistan Concerning Use of Water and Energy Resources in Syrdarya River Basin (1998) (UNECE 2008a). However, many of these newly created joint bodies have a limited mandate, not covering all aspects of IWRM. Some are only responsible for boundary waters or for part of the watercourse, but not for the whole river basin. For example the 1994 agreement between Ukraine and Moldova covers only 120 km out of the 1,500 km long Dniester River, one of the transboundary watercourses concerned by the agreement.

Implementation of transboundary agreements and the related decisions is often problematic due to the lack of resources, insufficient motivation among national authorities, inadequate representation of national authorities in a joint body, as well as the lack of coordination at national level. Whereas in Western Europe transboundary water agreements are mostly implemented through joint Commissions, in EECCA-countries the institution of plenipotentiaries is still frequent, i.e. government officials are responsible for the implementation, but there is no permanent joint commission. The agreement between the Government of Ukraine and the Government of the Russian Federation concerning the Joint Use and Protection of Transboundary Waters (1992) represents an example for this. Since such a structure can lead to interruptions of activities and poor coordination in the case of certain agreements, it might be advisable for certain joint bodies to establish small but permanent secretariats.

Another related major problem is that the joint bodies often lack power to implement IWRM and have a narrow influence on issues important for the cooperation since water management authorities are usually the only governmental authorities represented in a joint body (UNECE 2008a). Environment, fishery, health, energy, hydrometeorology, economy and finance authorities rarely participate in the activities of a joint body although they play a very important role for ensuring successful and integrated transboundary water management. In addition, agreements frequently do not specify any mechanisms for dissemination of information, public participation and the involvement of stakeholders (such as NGOs, water user associations, business, local authorities, etc.) (UNECE 2008a). Finally, due to lack of reporting requirements and loss or degradation of previous facilities for water quality monitoring at the transboundary level, implementation and progress can often not be monitored. Thus, major barriers currently hamper the implementation of integrated water resources management in the region at national as well as at the transboundary level.

Prospects for transboundary water management

Despite of these numerous challenges, prospects for transboundary water management in the EECCA-region have significantly improved during the last few years: national governments increasingly understand the importance of transboundary water management, various International Organizations are providing financial and other support and, among other reasons, due to legal frameworks such as the UNECE Water Convention, several new transboundary water agreements have been signed since the 1990s (UNECE 2008a).

Reforms and increasing commitment on the national level

On the national level, reforms are ongoing to improve national water management in most EECCA-countries: these reform of ministerial environmental departments and water agencies provide an opportunity to harmonize responsibilities for water management and improve cooperation among entities involved in monitoring and assessment, including new partners (e.g. the research community and academia), and to designate appropriate institutions to supervise, guide and contribute to monitoring and assessment (UNECE 2007b).

National and regional policy-makers also seem to become increasingly aware of the need of transboundary water management. For example, in 1998, the presidents of Kazakhstan, Kyrgyzstan, Tajikistan, Turkmenistan and Uzbekistan jointly created the Special Programme for the Economies of Central Asia (SPECA). Azerbaijan joined it in 2002 and Afghanistan in 2005. It is supported jointly by the UNECE and the United Nations Economic and Social Commission for Asia and the Pacific (UNESCAP), but is governed by its member countries. Its Project Working Group on Water and Energy has for example elaborated a cooperation strategy on the rational use of water and energy resources in Central Asia and is currently focusing on the improvement of dam safety (UNECE 2008b).

The increasing commitment of policy-makers to transboundary water management is also reflected in growing financial commitments. Several EECCA-countries such as Armenia and Kazakhstan have significantly increased or even doubled their budget allocated for transboundary and national water management during the last years. In 1993, the Heads of States in Central Asia created the International Fund for Saving the Aral Sea (IFAS) which aims to fund joint environmental and research programmes and projects aimed at saving the Aral Sea, improving the environmental situation, and addressing common social and environmental challenges in the region (UNECE 2008a). The fund was founded by Kazakhstan, Kyrgyzstan, Tajikistan, Turkmenistan, and Uzbekistan, but is supported by numerous national and international donors.

International support

International Organizations and national donors are increasingly providing financial and other types of support for transboundary water management projects in the EECCA-region, for instance the European Union, the UNECE Water Convention, the Global Environmental Facility, national governments, development cooperation agencies and environmental protection agencies. Since the prevention of water conflicts through transboundary management is more and more considered as catalyst for peace and cooperation in other areas (UNESCO 2006) even security organizations such as the NATO are dealing with transboundary water cooperation. The "*Environment and Security Initiative*" *(ENVSEC)* recognizes the important role of water for human security. ENVSEC, NATO and OSCE have helped the previously hostile neighbours Armenia, Georgia and Azerbaijan to create mechanisms for data-sharing which could be a basis for joint IWRM-institutions (SIWI 2007).

As another example, the creation of the Chu-Talas Commission was supported by several International Organizations (UNECE 2008a). Since 2003, UNECE, UNESCAP and the Organization for Security and Cooperation in Europe (OSCE) assisted in the establishment of a permanent commission. In 2004, an EU-TACIS project resulted in a strategic document, "IWRM in Chu and Talas Basins". Since 2005, the Commission's secretariat is financially supported by the Asian Development Bank (ADB).

The 1992 UNECE Water Convention

The United Nations Economic Commission for Europe's (UNECE) Convention on the Protection and Use of Transboundary Watercourses and International Lakes has served as a basis for many transboundary water agreements (UNECE, UNEP 2003) providing a legal framework and support to Parties. It was adopted in 1992 with the aim to protect transboundary waters by preventing, controlling and reducing pollution, through integrated management and reasonable and equitable use of transboundary waters as well as conservation and restoration of ecosystems. The Convention obliges all Parties to develop, adopt and implement relevant legal, administrative, economic, financial and technical measures related to pollution control at source, impact assessments, sustainable water resources management, contingency planning etc. Waste-water discharges should be licensed and best available technology should be applied. The Convention is based on the Polluter Pays principle as well as the precautionary principle (UNECE 2004).

In 2008, 35 UNECE-countries and the European Community were Parties of the Convention, among them for example Azerbaijan, Belarus, Kazakhstan, Moldova as well as Ukraine. The Convention as well as the associated Protocol on Water and Health support parties through capacity-building, e.g. on water and health, elaboration of guidelines on different aspects of IWRM, several transboundary pilot projects, development of assistance programmes (e.g. for the establishment of legal frameworks and joint bodies) etc. Since 2007, one of the main activities of the Convention is supporting Parties in their efforts to adapt water management to climate change.

Creation of new transboundary water agreements

The UNECE Water Convention requires the establishment of agreements for transboundary water resources as well as the creation of joint water management bodies to implement these agreements. In 2000, app. 150 such agreements were concluded in the UNECE-region (UNECE 2008a). However, they vary widely in their implementation, existence and type of joint bodies, and effectiveness. For example, the Danube river commission as well as the Sava river commission have been established based on and supported by the UNECE Water Convention. In addition, some EECCA-countries are reforming or updating old agreements such as Moldova and Ukraine in the case of the Dniester river (UNECE 2008a).

Even in Central Asia where cooperation on transboundary water issues is difficult due to the general political tensions, the riparian states have set up several joint institutions such as the *Interstate Commission for Water Coordination* or ICWC, the *International Council for the Aral Sea* (ICAS) and other institutions (Ecologic 2005). The ICWC was established in 1992 by Kazakhstan, Kyrgyzstan, Tajikistan, Uzbekistan and Turkmenistan for implementation of their 1992 Agreement. It is composed of heads of national water management authorities and has a Scientific Information Center, a training Center and a coordination meteorological center (UNECE 2008a). The executive bodies of the ICWC, the Basin Water Organizations "Amudarya" and "Syrdarya" are in charge of exploitation of water management installations, interstate canals and other facilities in the respective river basins. The ICWC also has a secretariat responsible for the implementation of ICWC orders, preparation of draft documents, accounting and reporting, as well as international relations.

Conclusion

As shown, considerable progress has been made towards sustainable transboundary water management in the UNECE and especially the EECCA-region during the last few years, but many challenges still exist. Some of the challenges are being addressed through international programmes and projects, mainly through capacity-building, funding of projects or advisory services. Legal frameworks have been improved and several new agreements and conventions have been signed which provide a basis for cooperation, but will only be effective if implemented properly. However, implementation of the agreements is still a major challenge due to lack of funding, lack of information, of interest and of qualified staff. Non-harmonized institutional structures and diverging goals and interests between neighbouring riparian states as well as internal water management problems make effective transboundary water management difficult. Therefore, the prospects for transboundary water management in the UNECE region are rather positive, but numerous challenges still have to be addressed.

References

Alcamo, J., J.M. Moreno, B. Nováky, M. Bindi, R. Corobov, R.J.N. Devoy, C. Giannakopoulos, E. Martin, J.E. Olesen, and A. Shvidenko, 2007. Europe. Climate Change 2007: Impacts, Adaptation and Vulnerability. Contribution of Working Group II to the Fourth Assessment Report of the Intergovernmental Panel on Climate Change, M.L. Parry, O.F. Canziani, J.P. Palutikof, P.J. van der Linden and C.E. Hanson, Eds., Cambridge University Press, Cambridge, 541–580.

Ecologic/Newater 2005. Transboundary Regimes in the Amudarya Basin. Berlin.

Hannan, T. 2005. Kazakhstan and Central Asia. Information in Transboundary Water Cooperation & the Value of National IWRM & Water Efficiency Plans. Presentation at the Workshop on Information Management and Public Participation in Transboundary Water Cooperation 8–10 June 2005, St. Petersburg, Russian Federation.

SIWI (Stockholm International Water Institute) 2007. Strong Messages as 2007 World Water Week Ends: It Is Time to Do Better on Global Poverty, Sanitation, Water Scarcity and Climate Change. Stockholm.

Šumberová, B. 2003. *Transboundary water management on the future border of the European Union – Lake Peipsi.* Master of Science thesis, Department of Environmental Sciences and Policy, Central European University, Budapest.

UNECE 2004. What Is It - Why It Matters? The 1992 UNECE Convention on the Protection and Use of Transboundary Watercourses and International Lakes. New York/Geneva.

UNECE 2007a. Our Waters: Joining Hands Across Borders. First Assessment of Transboundary Rivers, Lakes and Groundwaters. United Nations, New York/Geneva.

UNECE 2007b. Implementation of UNECE Multilateral Environmental Agreements. Official Document ECE/BELGRADE.CONF/2007/12 for the Sixth Ministerial Conference "Environment for Europe" Belgrade, 10–12 October 2007.

UNECE 2008a. River Basin Commissions and Other Institutions for Transboundary Water Cooperation. Prepared mostly by Yulia Trombitcaia and produced for the workshop Capacity for Water Cooperation in Eastern Europe, Caucasus and Central Asia. UNECE, Geneva (in print).

UNECE 2008b. online: http://www.unece.org/speca/ accessed on 10 June 2008.

UNECE, UNEP 2003. Regional Office for Europe, Ministry for Natural Resources of the Russian Federation, Swedish Environmental Protection Agency, Agency for Environmental Assessments "Ecoterra". 2003. Transboundary Water Cooperation in the Newly Independent States. Moscow/Geneva.

UNESCO, World Water Assessment Programme, UN WATER 2006. Water - a shared responsibility. The United Nations World Water Development Report 2. UNESCO, Paris.

WHO (World Health Organization) and UNICEF 2008. Joint Monitoring Programme for Water and Sanitation 2008. Online: http://www.wssinfo.org/en/welcome.html. Accessed on 10 June 2008.

SUCCESSFUL TRANSBOUNDARY RIVER BASIN AND ESTUARY COOPERATION: BENCHMARKS FOR THE URAL RIVER BASIN?

JEROME SIMPSON*

Information Programme
Regional Environmental Center for Central and Eastern Europe, Szentendre, Hungary
based on separate papers by

STEPHEN STEC

Environmental Law Programme,
Regional Environmental Center for Central and Eastern Europe, Szentendre, Hungary

WIM COFINO, HELLE PEETERS, BERT VAN HOVE, AND ANNEMIEK VERHALLEN

Wageningen University, The Netherlands

Abstract Russia's new Water Code means its water resources remain under federal jurisdiction. However, this shouldn't rule out transboundary water cooperation. After acknowledging the current situation regarding water resources management in Russia, this paper presents three instances of transboundary cooperation, two case studies concerning nations within the Danube basin: the Sava and Tisza sub-basins, and a third concerning the Scheldt Estuary between the Netherlands and Belgium. Some of the challenges and obstacles encountered in ensuring successful transboundary basin cooperation are also presented. The paper concludes by summarizing domestic priorities concerning integrated water management in Russia, in the context of its Volga river and the new Water Code, as agreed by representatives of Russia's State Duma and envoys of leading international organizations in February 2007. It is hoped this political momentum will provide a foundation for advancing the more effective management of the Ural river basin.

Keywords: Transboundary cooperation, Water Code, Sava basin, Tisza basin

* Jerome Simpson, REC, Ady E. u. 9-11, H2000, Szentendre, Hungary; JSimpson@rec.org

Water Code, water resources, federal jurisdiction… opportunities for transboundary cooperation

Water Code No. 174-ФЗ stresses that all water bodies remain federal property, owned by the Russian Federation. Ponds or flooded pits are exempted.

Under the new law, Russia's 86 regional authorities gain new powers such as the right to issue water-use agreements (specifically for abstraction, recreation and power generation) and/or "grant" water-body use (e.g. for the discharge of effluents or drainage water or abstraction for agricultural purposes). In addition, these authorities are charged with overseeing water-body protection and prevention of adverse impacts.

Decisions to issue agreements and grants will be given within 30 days of the receipt of application. However, it is clear the central government still retains considerable power regarding water resources management. The permitting process remains under the close control and supervision of authorised federal executive bodies and their inspectors. And in issuing grants, the Russian Federation government's approval must be obtained.

Among the code's innovations are its river basin approach and the introduction of integrated water basin management schemes. This gives leeway for keeping in mind the importance of transboundary water basin management schemes. Diverse stakeholders should be part of the new basin councils for instance which will be invited on a consultative basis to make recommendations. Monitoring of basin districts' water bodies will be undertaken by federal executive bodies, in collaboration with regional authorities. Why not work toward involving those from neighbouring countries also?

Rustem Khamitov, director of the Federal Water Resources Agency, says "It is abundantly clear that it is impossible to manage a river partially, say a lower part of the Volga separately from the upstream parts. A region cannot do whatever it wants on the river flowing through its territory: build a dam, dispose whatever it wants into the river, construct something on it and so forth." Thus collaboration is key.

Examples of transboundary cooperation: the Tisza, Sava and Scheldt estuary

Both the Tisza and Sava basins are sub-basins of the Danube River.

The Tisza Basin includes the countries of Romania, Ukraine, Slovakia, Hungary, Serbia and Montenegro. It is the largest sub-basin in the Danube River Basin. Its total extent is 157,186 km^2, making it the largest sub-basin in the Danube River Basin, and the longest tributary of the Danube (966 km), and second largest by flow, after the Sava River.

In the Tisza, a major cooperative effort was the Tisza River Basin Initiative which was developed under the auspices of the Council of Europe. Originally, the Council and a part of the international community had an idea that the Tisza countries could negotiate a binding treaty on the Tisza river, but partly due to the ongoing dispute between the countries as a result of the Baia Mare incident, there was not the level of agreement necessary to give rise to such a treaty. As events moved forward, the initial idea of a kind of framework agreement on the Tisza River Basin became "reduced" to the level of what was labeled an initiative with an action plan attached.

In the mean time, the Tisza River Basin Sustainable Development Programme was brokered by UNDP and REC. (REC is an international organisation head-quartered in Hungary and with offices in 17 countries of Central and Eastern Europe. More info: www.rec.org) It started with developing a participatory framework for cooperation between the countries, sectors, communities and stakeholders in the river basin that would:

- Secure prosperity for the people living in the river basin
- Ensure sustainable use of natural resources
- Minimize environmental risks
- Preserve natural and cultural values

Activities got underway with a Diagnostic Audit that assessed regionally and at country level the current status of legal, policy and institutional frameworks related to sustainable water management.

Today, external factors like the EUs Water Framework Directive are driving further progress and international cooperation. The Directive seeks to ensure that all water meets 'good status' by 2015. The first step towards the objective is to create a River Basin Management Plan by 2009. The Member states pursuant to the Article 13 (5) of the Directive may supplement the River Basin Management Plan by the production of more detailed programmes and management planes for sub-basin.

On this basis, the Tisza countries signed a Memorandum of Understanding in December 2004 to work together in the frame of the ICPDR aiming at to produce a sub-basin level Tisza River Basin Management Plan by 2009, which will also integrate issues on flood management. The first step towards the objective is the preparation of the Tisza Analysis Report by 2007, which includes an overall characterisation of the basin, an analysis of anthropogenic pressures and impacts as well as flood risk mapping. The ICPDR helps to ensure consistency of the project with its own work and integration of the results in the preparation of a wider Danube River

Basin Management Plan (More info: http://icpdr01.danubeday.org/icpdr-pages/tisar_2007.htm).

The Sava River Basin includes the countries of Slovenia, Croatia, Bosnia-Herzegovina, Serbia and Montenegro. The Sava basin has a size of 95,419 km^2, which makes it the second largest after the Tisza basin.

A Framework Agreement (see insert) on the Sava River Basin (SRB) was established in 2002. It seeks to promote regional co-operation throughout the Sava River Basin on issues related to navigation, economic development, comprehensive water management and environmental protection (see: http://www.icpdr.org/icpdr-pages/sava_basin.htm).

An International Commission was established June 2005 with its Secretariat in Zagreb (Croatia). This is a kind of international body which has real powers and real authority. It seeks to fully implement the Agreement and is working to facilitate opportunities for economic development and to attract foreign investors and contribute to enhancement of relations and co-operation between the Parties to the agreement. It also issues decisions which are binding on the four Sava countries on important issues such as navigation (www.savacommission.org).

Sava Framework Agreement

- Navigation
- Transboundary impacts
- Protection against flood, excessive groundwater, erosion, ice hazards, drought and water shortages
- Water use/utilization
- Exploitation of stone, sand, gravel and clay
- Protection and improvement of water quality and quantity
- Protection of aquatic eco-systems
- Prevention of the water pollution caused by navigation, and
- Emergency situations

A UNDP-GEF Danube Regional Project supported the development of a pilot plan for the Sava River Basin, as a model for river basin management planning in line with the EU Water Framework Directive. In 2003, the Project completed a first situation analysis; examined the availability of information and data, and assessed institutional capacities in the four countries to carry out the development of the Sava River Basin Management Plan (RBM). A second phase has been recently completed (see: http://www.undp-drp.org/drp/activities_1-1_sava_river_basin_ management_ plan.html) in which various technical support has been provided to Sava countries to strengthen their institutional capacity to prepare a RBM Plan for the Sava River Basin (encompassing transboundary water management issues and the development of the Sava investment programme).

Next steps will be discussed 14–15 June, 2007 with EC support: (www.savariver.net).

The main characteristics of these countries' cooperation are: (1) that it is based on a legally binding, enforceable process aligned with the EU and international law; and (2) they have developed an innovative process which enables the countries to choose flexible tools (assessments, action plans, investment programmes, agreements and protocols etc.) to achieve commonly agreed objectives. Moreover, the overall aim is to have the countries work for themselves and for the international community to withdraw after a short time.

The Scheldt estuary is situated in the northwest of Flanders (Belgium) and the southwest of the Netherlands is the downstream part of the Scheldt river basin. The total basin area amounts to 21,863 km^2 and is divided over France, Belgium and the Netherlands. From its source in Northern France to its mouth in the North Sea, the river has a length of 355 km. The Scheldt estuary region is both an important agricultural and industrial area. It is of a high ecological importance. So, conflicting interests exist in the region with respect to water control and management. Moreover, since the 16th century the Scheldt estuary has been a source for political conflict between Flanders and the Netherlands (Figure 1).

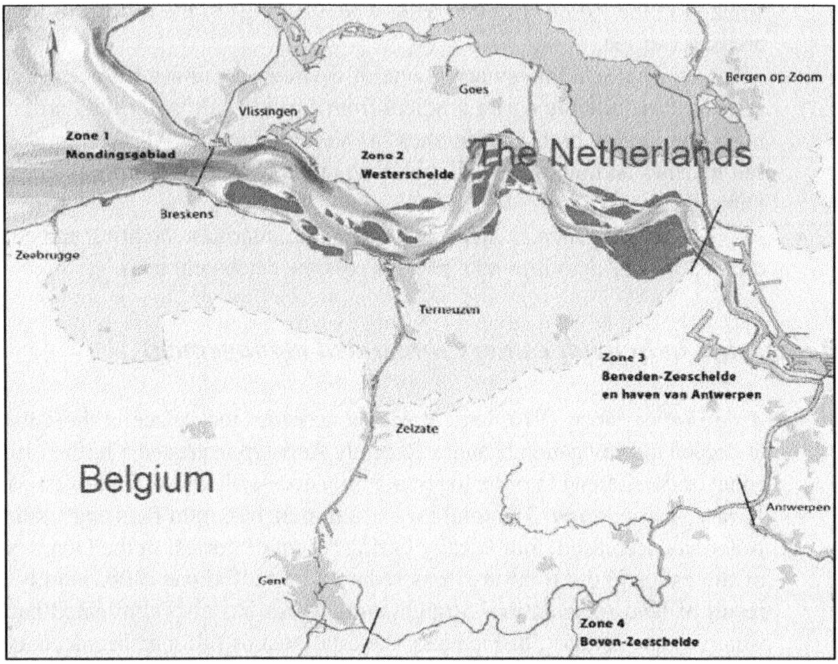

Figure 1. The Zeeschelde (zones 3 and 4) and the Scheldt estuary (zones 1 and 2) region

Coordination of policies, measures and approaches is essential. Recently, a closer cooperation between the Flemish and Dutch governments developed and a joint initiative was started, The Scheldt Estuary Development Project (ProSes). Its main purpose is to make a solid, broadly supported Development Plan taking into account the different interests of participating parties. This plan is the starting-point for a joint policy-making by the Flemish and Dutch government, aiming at a more sustainable development in the Scheldt estuary.

This chapter presents a short overview of the different functions of the river, the main issues and the institutional framework established to underlie joint policy-making.

Main functions

The main functions of the Scheldt estuary are navigation, ecology, recreation and fishery. Because the estuary contains salt or brackish water, it is not used for drinking water.

- *Navigation.* The estuary forms the maritime access to the port of Antwerp that is one of the largest ports in the world. Together with the port of Ghent (B), Vlissingen (NL) and Terneuzen (NL) the port of Antwerp is situated in the Rhine-Scheldt basin, which belongs to the most prosperous areas in Europe.
- *Ecology.* The Scheldt estuary is one of the few remaining European estuaries that include the entire gradient from fresh to salt water tidal areas. It hosts the largest brackish marshes of Western Europe. All the remaining salt marshes and mud flats fall under the protection of the European Habitat Directive.
- *Recreation and fisheries.* Riverside recreation, marinas, yachting and commercial fishery activities take place across the catchment area.

Main issues in Scheldt estuary integrated management

- *Accessibility.* Since 1970, large dredging activities took place in the estuary to deepen the navigation channel. Recently Antwerp requested a further deepening of the channel in order to remain fully accessible to larger sea vessels.
- *Nature conservation.* The total area of salt marshes, mud flats and shallow water has decreased dramatically during recent centuries. In the Dutch part of the estuary, the total area has reduced by half since 1800, mainly as result of land reclamation. Straightening dykes has also eliminated backwaters in the estuary. In Flanders, the total area of salt marshes decreased by nearly 25% since 1900, from 700 to 550 ha. Unrestricted deepening of the channel will lead to further serious decline in biodiversity.

- *Water quality.* The water quality is moderate to bad, although improvements have been made over recent years. About three million households drain off their untreated domestic wastewater in the Scheldt or its tributaries. Industry has made important efforts, but pollution with heavy metals and organic micropollutants is still significant. Agriculture is mainly responsible for the nutrient load, particularly of nitrogen, into the Scheldt estuary.
- Safety. In 1953, there was disastrous flooding in SW-Netherlands, at which more than 1,800 people drowned. This disaster formed the stimulus for a large-scale flood protection project called the Delta Plan. The region was again struck by a storm tide in 1976. Major floods occurred in Flanders. Flanders instigated the Sigma Plan to reinforce all dykes. Implementation is not yet complete, there is still a risk for flooding. Moreover, sea level rise due to climate change will eventually influence the safety against flooding in both states.

Institutional framework

Historical background

The Scheldt estuary has long been a source of conflict between the southern Netherlands (Belgium) and the northern Netherlands, which mainly had to do with the competition between the port of Antwerp and the ports of Amsterdam and Rotterdam. A number of treaties have been made and joint bodies have been established to deal with this conflict. The table below presents an overview of main historical events (Table 1).

Table 1. Overview of conflict and cooperation between Belgium and the Netherlands since 1948

Year	Event	Action
1948	Installation of Technical Scheldt Committee (TSC)	Permanent consultation on technical Scheldt issues
1994	Treaty of Charleville-Mezieres concerning the protection of the Scheldt	Establishment of International Commission for the Protection of the Scheldt (ICPS)
2001	Treaty of Liege	Appointment of an international basin according to the EU framework directive
2001	Memorandum of Kallo	
2002	Memorandum of Vlissingen	Agreement on objectives long-term vision for 2030 establishment of project organization ProSes
2004	Development Outline Scheldt Estuary 2010	
2005	Outline approved by both countries start of implementation	

The Technical Scheldt Committee (TSC)

In 1948 the Technical Scheldt Committee (TSC) was established. It is directed by Flemish and Dutch chairpersons. Its primary task is to advice Flemish and Dutch politicians on technical issues such as water infrastructure and general management. The present tasks result from the treaty concerning the deepening and widening of the navigation route that was drafted in 1995. In 2001, Flanders and the Netherlands reached agreement on the development of a long-term vision for the Scheldt estuary (respectively the memorandum of Kallo 2001 and that of Vlissingen 2002). For the elaboration of this plan TSC established the project organization ProSes (www.proses.nl), which operates in a 'triangle' with TSC, and the multi stakeholders' platform OAP ('Consultative Committee of Advisory Parties) which includes representatives of the participating governments, official bodies and interested parties. The figure below illustrates the institutional framework and relationships between the different actors (Figure 2).

ProSes and ProSes2010

The first task of ProSes was to make a solid, broadly supported Draft Development Outline aiming at a sustainable development in the Scheldt estuary till 2030. Several studies were carried out during recent years: a strategic environmental impact study, social cost/benefit analysis, a study on measures for developing the natural environment. During the preparation of the Development Outline, interested parties made contributions during e.g. workshops. They were regularly informed on the state of affairs via brochures, newsletters and the website. Furthermore, public hearings were held, in which draft versions of the Development Outline were presented. The responses were compiled and published, and used in formulating the final version. The 'Scheldt Estuary Development Outline 2010' (ProSes2010) was presented by the end of 2004 and approved by both governments in March 2005. It has three main foci:

- Safety: maximum protection against flooding in the region
- Accessibility: optimum accessibility to the harbors on the Scheldt estuary
- Natural environment: a dynamic, healthy natural environment

The Development Outline does not deal with all of the problems in the Scheldt estuary. For instance, it does not address the issue of improving water quality. This issue is already being dealt with jointly by Flanders and the Netherlands, along with other Belgian regions and France, in the International Commission for the Protection of the Scheldt (www.isc-cie.com).

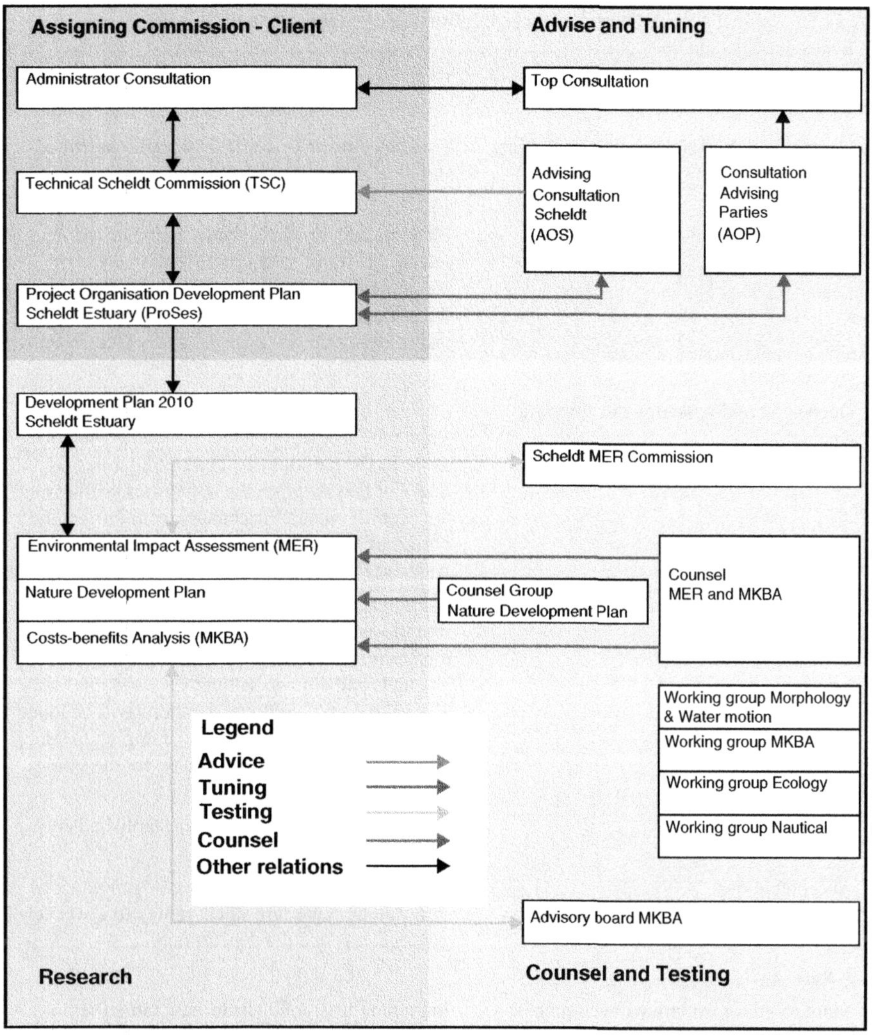

Figure 2. The institutional framework for the development of the Scheldt Estuary Development Outline 2010

Implementation

At present, the first steps are made for the implementation of the resolutions made by ProSes2010. Table 2 gives an overview of the different project plans. A new joint project management team has been established to coordinate this process. In 2006 the governments are to sign the new treaties

Table 2. Overview of resolutions and project plans until 2030

Resolutions	Projects
Safety	
Increasing dyke heights and establishing controlled flooding areas along the Zeeschelde	Flanders aims to establish 280 ha of controlled flooding areas by 2010.
Common approach to safety	Flanders and the Netherlands calculate the required level of safety in different ways. The Netherlands is presently examining whether the risk approach such as that used in Flanders is also desirable and possible in the Netherlands.
Accessibility	
Deepening and widening the shipping channel	Flanders and the Netherlands have decided that ships with a draught of 13.1 m must be able to sail as far as the harbour of Antwerp regardless of the tide. For this purpose, the authorities will lower the level of sills in the channel by 1.4 m. At the border of both countries, the Scheldt will be widened from 250 to 370 m over a length of 5 km.
Flexible dumping locations	All maintenance dredgings will be dumped back into the estuary. Careful selection of dumping locations is necessary to avoid silting-up of side channels and erosion salt marches and mud flats. The selection of dumping locations will be made more flexible in order to allow dumping to take place where it is most favourable for the vitality of the estuary.
Monitoring	A monitoring program will be established during and after deepening.
Acceptable risks	The governments will improve the provision of information regarding safety policy to lower-level governments and the general public.
Natural environment	
More room for estuarine developments	In total, at least 1,000 ha of new estuarine environment will be added to the Scheldt.
Increased vitality	To restore natural vitality where possible. For example, by alternative dredging and dumping strategies, constructing or removing breakwaters, excavating old marshes, and increasing or decreasing the depths of the channels.
Multifunctional environment	Combining natural environments with other objectives such as safety, agriculture, marine aquaculture, recreation and residential/ employment initiatives.

on financing of the resolutions, the order of significance of the resolutions to be implemented and on how they will further proceed to attain the target situation in 2030.

Challenges and obstacles

The establishment of the Technical Scheldt Committee was a first important step in the normalization in the relationship between Flanders and the Netherlands with respect to water control and management of the Scheldt estuary. From 1995, the decision making process developed slowly, from a situation of conflict, distrust and contra-productivity to a situation of interactive policy making by co-operation between different actors of both countries. The triangle formed by ProSes, the Technical Scheldt Committee and the multi-stakeholders' platform (OAP) proved to be a successful concept for process directed decision making. 'Joint fact finding' plays a prominent role in this. In this way commitment of the different actors can be obtained, that helps keeping the decision making process under way.

Of course, there were tensions, particularly with respect to participation and communication. The interests of the port of Anwerp are different from those of nature conservation organizations or those of agriculture. A number of representatives of the port of Antwerp threatened with juridical steps. Similar contrasts exist with respect to the safety measures to be taken in the area. In Flanders safety projects are conducted by another organization which operates separately from ProSes. Large areas of agricultural land are claimed by ProSes in order to be transformed into flooding areas. As a protest, agricultural organizations in the Netherlands refused to become a full member of the multi-stakeholders' platform. The Dutch province of Zeeland felt that the advantages of the Development Outline 2010 do not outweigh the disadvantages for its province and, therefore had large objections against the Outline. Flemish agricultural organizations and Dutch industry felt under represented. Groups of citizens were dissatisfied with the communication about The Development Outline because it was not quite clear who was responsible. However, despite all these difficulties the objectives of the ProSes2010 were obtained and the Proses organisation is seen as highly successful.

The importance of a sound legal basis

The Sava Commission process for cooperation is often mentioned as a good example for other rivers. One may think about the Tisza in this respect, where an initiative, despite having evolved into a Memorandum of Understanding, has not yet reached the level of agreement to develop a commission approach. Looking at both Tisza and Sava sub-basins together there are several interesting comparisons. They show in the first place that you can have a wide range of solutions for particular problems. One may have an example of an international agreement, as under the Sava, under which an international body like the Commission is established; on the other hand you can have another, looser cooperative structure called an initiative. For the Scheldt Estuary, the Technical Scheldt Committee became responsible for a long term vision, which became subsequently underlined by two memoranda of understanding (MoU). It depends, in all cases, on the respective will of the countries themselves to implement whatever has been agreed. So the fact that it might be an agreement, an initiative or an MoU doesn't make the most difference. The most difference is made by whether the countries are willing to implement it.

You can have a situation where you have possibilities to communicate in a common language where people know each other because they may have been integrated into one country, like in the case of the Sava or even the Northern and Southern Netherlands (Flanders). Or you can have a situation where you have sometimes quite difficult language or cultural barriers like the situation in the Tisza. Or you can have a situation where you may have a recent conflict or a low-level long term diplomatic situation. And the combination of countries is also different between Tisza and Sava. In two cases you have a clear EU context – besides the Netherlands and Belgium, there is the Sava, because all the countries are member states, candidates, or in the Stabilization and Association Process. Or you might have as you have in the Tisza. Until January 2007 two new member states, a candidate country, one in the Stabilization and Association Process, and finally you have one – Ukraine – which is completely outside this process. These different "constellations" of states and their relations to other states and international organizations are important because they determine the relationship to certain standards and instruments – for example the Water Framework Directive, which has been the driving force in all three examples.

In summary, there is no substitute for regional initiatives, but these initiatives must be flexible to the particular circumstances of the countries involved. There is no single formula, and countries may not be ready for highly developed mechanisms such as commissions, but the mechanism chosen will have an impact on the expectations and the results. As regional initiatives are increasing, the models looked at here may have increasing relevance as part of a wider debate about regional initiatives generally. Besides the Tisza, Sava and Scheldt contexts, another initiative is underway on the Caspian Sea. Plus there is the UNECE Convention on the Protection and use of Transboundary Watercourses and International Lakes.

Domestic priorities for water management in Russia

Finally I would like to put these three initiatives back into the Russian context by summarizing a recent demonstration of political will in the field of water management with particular emphasis on the Volga river.

- "Systematise hazard risk management;
- Effectively implement Russia's new Water Code; and
- Develop alternative financing solutions for tackling Volga related problems."

These were the three key recommendations heard and endorsed by representatives of Russia's state Duma and envoys of leading international organisations at a policy roundtable convened by the CABRI-Volga project (see www.cabri-volga.org) in Moscow on February 27, 2007.

"Systematising risk management is necessary to tackle the risk of accidents posed by unsafe industrial facilities, floods from the poor condition of dams, and health risks posed by sub-standard drinking water," reported Siegfried Rupprecht, CABRI-Volga project manager to roundtable participants.

To implement the new Water Code, administrative reform is necessary in order to effectively realise water basin district management. Although the Code may not directly address the issue of transboundary management, it is worth remembering that the development of secondary norms is still a process underway. Numerous procedures, impact limits, fee rates and indicators, alongside amendments to existing laws should take place. According to Natalia Davydova, director of the Moscow-based Ecological Projects Consulting Institute, some 24 statutory acts should be adopted to fully implement the provisions of the code. "However, only half of the necessary acts have been passed to date."

In overcoming cash-strapped budgets for dealing with water quality problems, implementing the polluter pays principle means "every rouble paid for polluting the water should be used for cleaning the water," concluded Rupprecht. A recent report published by the project called for enforceable water pollution standards, alongside the more effective enforcement of the "polluter pays" principle. This should generate funds that could be channelled back into improving water quality.

These recommendations, part of a broader set of ten, were drawn from a series of consultations organized by the CABRI-Volga project with experts in river basin management during its 27-month lifetime. The successful outcome of the project, whose aim was to support the sustainable management and development of the Volga basin, demonstrates there is political will to further advance river basin management, particularly in the context of Russia's new Water Code. I hope and trust the political momentum and international experiences described in this paper can be built upon and tapped into, in order to advance Ural river basin management in the coming future.

THE ISSUES OF TRANSBOUNDARY RIVERS IN SOUTH CAUCASUS AT THE END OF THE 20TH AND BEGINNING OF THE 21ST CENTURIES

TRAHEL VARDANIAN*

Yerevan State University, Armenia

Abstract This paper studies the present-day issues of water resources in South Caucasus. In particular, it discusses the change in the runoff of several transboundary rivers in Armenia in 20th century as well as providing forecasts for 21st century under the global warming of climate.

Keywords: Transboundary rivers, controlling and managing of water resources, anthropogenic load, quantitative water changes, global climate change

Twentieth century was distinguished by global changes in the public, political and socio-economic spheres, changes which left their trace on all the components of the landscape mantle, including water resources and, particularly, river runoff. What has the 21st century in store for the world?

The research of the Strategic Research Centres in a number of countries lead to the conclusion that water will become the major strategic resource in the 21st century, as was asserted in the resolution of the Interparliamentary Assembly in 1998. In particular, according to the experts of the Institute of the British Strategic Researches, in the near future a unit of volume of drinking water will cost much more than a unit of volume of oil; the desire to control water resources will cause the outbreak of numerous wars. Then, the regulation norms of international right in the sphere of water issues, which are already imperfect at present, will not function at all.

Then every state will strive to strictly control and manage the water resources which are self-restored in its territory, based on the principle "water is power".

* Trahel Vardanian, Department of Physical Geography, Yerevan State University, Faculty of Geography, Alek Manoukian Street 1, 0025 Yerevan, Armenia; tvardanian@ysu.am

From this point of view, the South Caucasus region, which may become a source of fresh water for the countries of the Middle East, the Persian Gulf and the Arabic world in general, has to carry out very serious tasks and activities. In this small region the water resources are distributed rather unequally: in the Kolchide and Lenkorane lowlands the water is abundant, while it is scarce in the Kura-Araks lowland and the Ararat depression – in other words, the region has its own problems to solve. Moreover, there is no common approach to the solution. Before the collapse of the Soviet Union there was a common management of water resources to some extent, with the countries of the region carrying out joint research and implementing a range of hydrological programmes; however, after the collapse of the Soviet Union almost all the links have been broken for various reasons. Rivers, the water of which freely flows into the countries of the South Caucasus, remained the only link. But the quality and the quantity of the water flow are still questioned.

Of the numerous issues in the Caucasus region, the issue of water use is one of the most acute. Water use is very inefficient, with major losses, and the culture of water use is very poor.

Agriculture, industry and utilities are the main spheres of water use in the region. A large portion of water use belongs to agriculture, particularly, irrigation, and most of the water losses are accounted for by the same sphere. The reasons can be found in the poor state of canals and their disparity with modern water requirements, due to which the efficiency coefficient is low.

There is major loss of water in utilities as well, due to the poor state of the water supply system, which has deteriorated since its installation some 50–60 years ago. For instance, about 60% of drinking water flowing to Yerevan is lost in the water supply system. The response to this problem has not been to repair the system and reduce losses, but rather to search for new sources of fresh water, and construct new water lines, resulting in an additional amount of water being taken from rivers. This approach if continued will bring about the quantitative exhaustion of water.

The issue of the present condition of water resources is also crucial. At first sight it may seem that in the countries with a transitional economy, like the ones in the South Caucasus region with little or no industry, the quantity of waste water must be low and of no threat to river waters. However, the reality is different. Whereas in the Soviet period industry was in full operation, and industrial sewage water was purified by about 60%, today the water purifying stations have almost stopped operating because there are run-down or outdated. As a result, a certain [small] amount of sewage water flows into the basin. Naturally, small rivers, which are very sensitive, cannot purify themselves, and are facing a serious ecological crisis. As a result, waters are qualitatively exhausted.

The anthropogenic load, i.e. the relation of the population size to the water resources of rivers is the simplest, but, at the same time, the most important index of evaluation of anthropogenic impact of water resources (Koronkevich et al. 1998). The figure for Armenia is 585,000 people/km^3, which is very high. This compares unfavorably with the other CIS republics, particularly with Russia, which was the richest in water resources among all the republics (Waters of Russia 1991), and where the anthropogenic load is only 40,000 people/km^3 (Koronkevich et al. 1998).

Obviously, the greater the anthropogenic load, the stronger the impact of human activity on water resources, both in quantitative and qualitative respects. It is worth mentioning that after 1990 the anthropogenic impact on river runoff has become minimal, due to the socio-economic state of the country.

Now, let us see what quantitative changes in runoffs were experienced by the relatively large transboundary rivers in Armenia in 20th century (Figure 1).

As can be seen from Figure 1, which displays the dynamics of changes of river runoff, there is a tendency of runoff increase in all rivers (except for the Debed-Ajrum). However, there is some disparity between the increase of amount of water use and rise of river runoff. If we take into account the circumstance that the observations concern the lower river basins, we can assume that water taken from rivers after use returns to the lower river basin in the form of sewage water (which is highly doubtful), in which case, anyway, no rise could take place. Perhaps, the increase is influenced by global climate change; this supposition, however, requires additional studies and explanations.

Such is the present condition of water use and conservation of water resources in the region.

What is expected in 21st century – what new problems may emerge and how can they be solved?

According to the specialists, in the next decade the countries of the region will not only revive their former industrial power, but under the conditions of new economic management they will also develop the economy unevenly, setting new requirements for water resources.

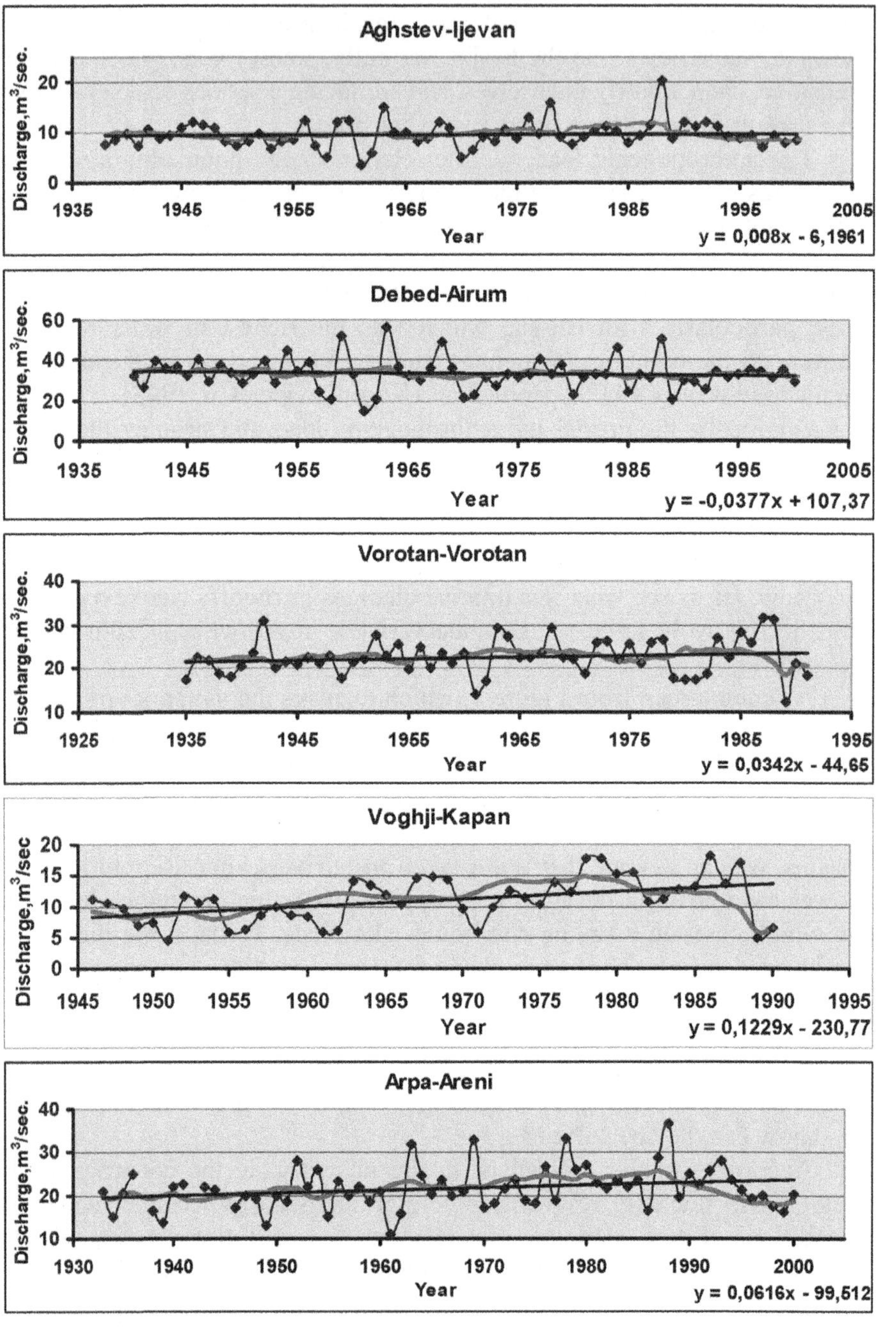

Figure 1. Dynamics of changes in average annual runoff in rivers of Armenia

Another, universally important problem is the global warming of climate. Various scenarios of probable climate change were used to forecast the level of river runoff vulnerability for different periods of time.

According to IPCC calculations, if greenhouse gas emissions continue at the current pace, then in Southern Europe, which includes Armenia, air temperature will grow by 1–3°C, and precipitation will drop by 5–15% by 2030 (IPCC European Regional ... 1997).

Based on the afore-mentioned considerations, as well as results of studies carried out in Armenia (Armenia: Climate Change ... 1999, 2003), and the peculiarities of the region, the following climatic scenarios were developed for evaluation of the level of river runoff vulnerability in Armenia:

(1) T + 1,0°C; 0,9P (2) T + 2,0°C; 1,1P (3) T + 2,0°C; 0,9P,

where T (°C) and P (mm) are perennial mean annual air temperature and precipitation.

Using the multifactor link (Q = f (P, T)) equations (Table 1), as well as applying the afore-mentioned scenarios, we calculated and evaluated the level of river runoff vulnerability in Armenia (Table 1).

The calculations of the studied rivers and all scenarios of climate change suggest that the river runoff in Armenia may decline by 12–13% on average, and in some basins by as much as 30–35% (Table 1), by the middle of the 21st century.

To adapt the problems of the use and conservation of water resources in the region to the conditions of new economic management and global changes, as well as to make certain decisions, it is necessary to:

1. Re-calculate and re-evaluate the volume of fresh waters.
2. Re-consider the direction of use of water resources in different spheres, undertake a new distribution, and lay out a new scheme of water use.
3. Organise monitoring observations of water resources in the region and implement joint scientific programmes.
4. Make an inventory of water resources in the region.

The problems mentioned in this article are very complicated, demanding large-scale scientific and organisational activities as well as funding.

In the implementation of the proposed changes the Geographic Society as well as other local and international NGOs can play an important role, they can cooperate with governmental bodies help solve these issues jointly with the support of donor organisations.

Table 1. Evaluation of river runoff change in case of different scenarios of climate change

River – observation post	Equation of multi-factor regression link	Correlation coefficient, R	Scenarios of climate change	River run-off, m^3/s	Runoff change	
					m^3/s	%
Aghstev-Ijevan	Q = 0.012P – 0.858T + 9.03	0.68	Basis	9.6	0	0
			T + 1.0°C; 0.9P	7.84	–1.76	–18.3
			T + 2. 0°C; 1.1P	8.45	–1.15	–12.0
			T + 2. 0°C; 0.9P	6.98	–2.62	–27.3
Debed-Ajrum	Q = 0.069P – 3.25T + 15.07	0.78	Basis	33.1	0	0
			T + 1.0°C; 0.9P	25.20	–7.90	–23.9
			T + 2.0°C; 1.1P	29.88	–3.22	–9.7
			T + 2.0°C; 0.9P	21.75	–11.3	–34.3
Arpa-Jermuk	Q = 0.004P – 0.053T + 2.27	0.60	Basis	5.22	0	0
			T + 1.0°C; 0.9P	4.65	–0.57	–10.9
			T + 2.0°C; 1.1P	5.19	–0.03	–0.57
			T + 2.0°C; 0.9P	4.59	–0.63	–12.1
Vorotan-Vorotan	Q = 0.025P + 0.04T + 13.02	0.56	Basis	22.6	0	0
			T + 1.0°C; 0.9P	21.96	–0.64	–2.8
			T + 2.0°C; 1.1P	23.90	+1.3	+5.8
			T + 2.0°C; 0.9P	22.00	–0.60	–2.6
Voghji-Kajaran	Q = 0.001P – 0.335T + 4.52	0.65	Basis	3.64	0	0
			T + 1.0°C; 0.9P	3.23	–0.41	–11.3
			T + 2.0°C; 1.1P	3.05	–0.59	–16.2
			T + 2.0°C; 0.9P	2.89	–0.75	–20.6

References

Armenia, Climate Change Problems/Collected Articles, 1999, I Issue. Editor Gabrielyan A.H. Yerevan, 373 p. (in Armenian)

Armenia, Climate Change Problems/Collected Articles, 2003, II Issue. Editor Gabrielyan A.H. Yerevan, 353 p. (in Armenian)

IPCC, 1997, European Regional Program 5; Regional Impacts of Climate Change, Geneva

Koronkevich N.I., Zajceva I.S., Dolgov S.V., Jasinskij S.V., 1998, Modern Anthropogenic Impacts on Water Resources. Proceedings of RAS, Geographic Series, N 5, pp. 55–68 Moscow (in Russian)

Waters of Russia, 1991, Sverdlovsk, 147 p. (in Russian)

1.2 STURGEON SPECIES

MIGRATORY FISH STOCKS IN TRANSBOUNDARY BASINS – IMPLICATIONS FOR GOVERNANCE, MANAGEMENT AND RESEARCH

JOHN VALBO-JØRGENSEN* AND GERD MARMULLA

Fisheries Management and Conservation Service (FIMF), FAO, Viale delle Terme di Caracalla, 00153 Rome, Italy

ROBIN L. WELCOMME

Long Barn, Stoke by Clare, CO10 8HJ, UK

Abstract Migratory fish species move between habitats in order to complete their life-cycles, and are therefore vulnerable not only to fisheries, but also to habitat degradation and loss of connectivity between habitats. In order to conserve migratory fish stocks fisheries must be properly managed, critical habitats maintained and rehabilitated where they have been degraded, and connectivity between habitats ensured. In river basins shared between various countries concerted efforts are needed as impacts in one part of the basin may affect fisheries elsewhere. In inland waters, fisheries compete with different stakeholders and powerful economic interests for access to water resources. Governments are responsible for setting goals for their river basins, elaborate river basins plans and follow up with appropriate legislation in compliance with international agreements e.g. FAO CCRF, Ramsar, CITES and CMS.

Keywords: Fish migration, Fisheries management, FAO, transboundary river basins, governance, river rehabilitation, environmental flows

Introduction

In all but the most northerly of rivers the number and complexity of the species making up river assemblages makes inland fisheries very resilient to fishing pressure in terms of absolute yield. However, individual species are highly vulnerable to fishing (Welcomme 2001) and to changes in aquatic environments caused by human developments. Migratory species are under

* John Valbo-Jørgensen, FIMF, FAO, Viale delle Terme di Caracalla, 00153 Rome, Italy; john.jorgensen@FAO.org.

particular pressure both from fishing and ecosystem changes as habitats crucial to their life histories have been destroyed and fragmented, and the flow regimes that ensure the functioning of the ecosystem have been altered. In most countries environmental impacts overshadow fishing as the major source of fish population decline and the most important threat to riverine fish stocks and fisheries comes from outside the fisheries sector.

Fish migrations

Migrations are systematic movements by animals between different habitats in order for them to meet their physiological needs. Many freshwater fish species require distinct habitats for feeding, reproduction, growth and refuge and as such movements constitute inherent elements in their life history. Distinction is made between obligatory migrants (which normally migrate over long-distances) and semi-migratory species that usually undertake shorter or local migrations and may be less dependent on migration for the completion of their life histories. Migrations may be between inland and marine habitats (diadromous), between lake and river and entirely within a river system (potamodromous). They normally occur during a particular time of the year, and may vary between different species as they are usually closely correlated with seasonal environmental changes (Smith 1985; Baran 2006). At the time of migration the fish may move as individual fishes or in large schools depending on the species.

The distances covered by species such as the sturgeons, many salmonids, pimelodid and pangasiid catfishes and anguillid eels are impressive, and may range from several hundreds to much more than a thousand kilometres (Welcomme 1985; Barthem and Goulding 1997; Lucas and Baras 2001; Carolsfeld et al. 2003). In many species adult fish move upstream to spawn and the eggs either hatch and develop in the upstream area or drift downstream as they develop so that the young reach their nursery areas in the main channel or adjacent floodplains (Welcomme 1985). The synchronisation of migration with the flood regime is often extremely delicate as adult fish must leave their downstream habitats so as to arrive at upstream spawning sites as water levels are rising in order for the young fish to benefit both from the increased flow to travel downstream and the maximum period of flooding of the downstream nursery site for growth. If the flow patterns are disturbed the young fish might not reach their nursery habitat in cases of low flow or may overshoot the favourable areas and die in excessive flows (Fuentes 1998). Some species such as eels migrate from inland waters to the sea for spawning and young eel may eventually return to rivers to move back upstream for ongrowing.

Depending on the species the adults may die, or return to their feeding grounds after spawning. Floodplains and estuaries are particularly important feeding areas for both adults and juveniles of many fish species because of the presence of decomposing organic material and nutrient-rich silt carried by the river giving rise to an important growth of microscopic plants and animals that form the basis of the food chain (Junk et al. 1989; Barthem and Goulding 1997). Migratory fish can also play an important role in nutrient inputs in upstream reaches. For example, in some North-American rivers the lack of carcasses of dead salmon following declines in that species, has led to the impoverishment of the upstream areas (Roni et al. 2005).

Fisheries

The importance of river fisheries

Migratory fishes are extremely important in freshwater fisheries throughout the world, and they often dominate fish landings. The 2.6 million tonnes of fish annually landed in the Mekong Basin, valued at US$1.7 billion, is thus dominated by migratory fish (Poulsen et al. 2003b; Van Zalinge et al. 2004). In Bolivia the migratory *Prochilodus lineatus* once made up 40% of annual landings from the Pilcomayo River (Payne and Harvey 1989), and in Bangladesh the most important freshwater fishery is aimed at the anadromous Hilsa shad (*Tenualosa ilisha*), and although the stocks have declined in recent years the species still accounts for about 13% of the fisheries (Payne et al. 2004). Large migratory species such as the Atlantic salmon (*Salmo salar*), and endangered species such as the Mekong giant catfish (*Pangasianodon gigas*), and the sturgeons (Acipenseridae) are among the most valuable and sought after species in inland fisheries around the world.

In temperate regions and, increasingly, in some prosperous tropical areas recreational fisheries are more important than the type of artisanal fishery common in the tropics. Recreational fisheries generate more financial benefit than food fisheries and integrate well with tourism (Shrestha et al. 2002). However, the allocation of benefits is often away from the people who were the traditional users of the resource thereby creating competition for access to fish, and since management objectives between food and recreational fisheries to a large extent are at odds, conflict between the two user-groups may be the result (Arlinghaus 2005; Cooke and Cowx 2006).

Fish may be very easy to catch during their migration, especially if they are forced to pass through narrow channels or localities with shallow water. The fishers know this and many fishing activities are therefore concentrated around the times of migration and focussed on such areas. This is for example well demonstrated by Weyden (1865) for the salmon of the River Sieg in Germany in the 19th century. It is also well illustrated by the gaff and spear fisheries for catfish in rapids of the Amazon (Barthem and Goulding 1997), and the wing trap fisheries at the Khone Falls in the Mekong (Singanouvong et al. 1996).

The annual migrations have big impacts on riparian human communities, where the fishers every year patiently wait for the fish to show up. When the fish arrive, everybody who owns fishing gear will be fishing, including people that only fish at this time of the year. Some fishers, who are specialized in this type of fishery, will even follow the fish for hundreds of kilometres (Buck in: Petrere and Agostinho 1993; Poulsen and Valbo-Jørgensen 2000).

The arrival of the fish is such an important event in some cultures that it is marked by ceremonies or festivals as an expression of gratitude for the bounty that will allow the riparian communities to survive another year (Hortle et al. 2004).

The fishery may be at different scales, i.e. subsistence, artisanal or it may even reach industrial dimensions. In many places fisheries of various scales and types occur side by side; as for example the Hilsa shad fishery in the Ganges, the *Henicorhynchus* spp. fishery in the Mekong, the catfish fisheries of the Amazon and the *Alestes* fisheries of the Yaeres floodplain of the Chari-Lake Chad system.

The enormous quantities of fish that suddenly flood the markets create a surplus of fish that needs to be preserved to allow the fish to be transported to markets further away and in order to distribute the supply of fish more evenly over the year (Hortle et al. 2004). Lacks of funds to invest in modern technology, and cultural preferences, frequently mean that people rely on traditional methods to preserve the fish such as drying, salting, pickling, fermentation or processing into fish sauce.

With the increasing shortfall in catches resulting from environmental changes, capture fisheries are quickly losing ground to the rapidly growing aquaculture industry. This development puts further pressure on limited water resources and has serious implications for access rights and policy development in the fisheries sector (FAO 2007).

Governance in inland fisheries

Large rivers and lakes may stretch across one or several international borders and activities in one region or country may thus affect fish stocks and fisheries in other regions or countries. Even if an impact on a certain species is confined to a particular area, the effects on the species may be felt by people or communities exploiting the fish stock in other regions or countries (Coates et al. 2000). While the need for international agreements and cooperation has been realized for marine fisheries (Munro et al. 2004), the recognition is only slowly emerging that a similar system of governance is indispensable at a basin scale in transboundary and international inland waters, where fish stocks are shared and may move between regions or cross borders during their migrations (FAO 2007).

Appropriate fisheries management in transboundary waters requires that suitable policies and strategies for sustaining shared fisheries resources and the water resources they depend on are developed at the regional level, and that these are incorporated in national legislation and implemented. A first step in this direction would be to identify the species and stocks that are shared and whether they are vulnerable, and to which threats. The countries would then move on to identify the particular management measures that are required. The FAO Code of Conduct for Responsible Fisheries (CCRF) emphasizes, *inter alia*, that "*States should ... co-operate at sub-regional, regional and global levels ... to promote conservation and management, ensure responsible fishing and ensure effective conservation and protection of living aquatic resources throughout their range of distribution, taking into account the need for compatible measures in areas within and beyond national jurisdiction*" and further "*For transboundary fish stocks, ... the States concerned ... should co-operate to ensure effective conservation and management of the resources. This should be achieved, where appropriate through the establishment of a bilateral, sub-regional or regional fisheries organisation or arrangement*" (FAO 1995).

There are already a range of regional frameworks, which provide advice on, or deal directly with management of inland waters and living aquatic resources. However, as only 44% of all the international basins have one or more agreements the governance system is incomplete (UNEP 2002).[1] Most agreements deal with a variety of issues and generally do not focus on fisheries, but on the water as a resource, i.e. the allocation of water for irrigation, flood protection, navigation or hydropower generation.

---

[1] The full texts of many international agreements on transboundary waters are available through the databases FAOLEX (http://faolex.fao.org/faolex), and the International Freshwater Treaties Database (http://ocid.nacse.org/cgi-bin/qml/tfdd/treaties.qml).

Nevertheless, many agreements do have a mandate in environmental matters, which, although fish are often not specifically mentioned, in many cases could be extended to comprehend fisheries (FAO 2007).

As part of many agreements, a basin committee or commission (RBO) has been established with the responsibility to oversee that the conditions of the agreement are respected by the parties, and to constitute a forum for discussion and consultation among them. In some of the best working and most active RBOs a permanent secretariat has been set up, which can institute a common vision for the development of the basin, and assist the countries in formulating and obtaining funding for programmes or projects. However, in some countries there is a tendency to measure the success of broadly mandated basin organisations by their ability to generate economic development rather than reaching softer objectives such as ensuring nature conservation and sustainable resource management. These organisations are therefore under increasing pressure to deliver tangible results to the detriment of the environment and fisheries.

However, there is some evidence that the international community is taking the issue of inland water resources governance more serious. A recent example is the adoption of resolution IX.4 of the Ramsar Convention (http://www.ramsar.org/res/key_res_ix_04_e.htm), on the conservation, production and sustainable use of fisheries resources where it is stressed *inter alia* that "*Local, national and international mechanisms should be established, as appropriate, whereby allocation of essential resources for the protection of aquatic resources and specifically fisheries resources are negotiated among all users of the resource.*"

In Europe, the European Water Framework Directive[2] calls for an for integrated and coordinated approach to the management of European rivers through comprehensive ecological assessment and classification on the basis of the composition and abundance of the aquatic fauna and flora taking into account the type-specific reference conditions of the waterbodies.

Managing transboundary fishery resources

There are two aspects of management of transboundary rivers for fisheries; management of the fishery itself and management of the environment both of which aim at sustaining the socio-economic benefits without destroying the underlying fish stock. The management requirements for a particular stock will depend on the specific biological demands of the species and the threats to its sustainability (Coates et al. 2000).

[2] See (http://ec.europa.eu/environment/water/water-framework/index_en.html).

Managing the fishery

Where large populations of fish move in the same direction within a discreet period of time they are very vulnerable to specialized and seasonally intensive fishing operations. Two major forms of fishing occur. The first: where fish concentrate at choke points in the system, such as channels draining floodplains, or at areas of restricted passages such as waterfalls and rapids. The second: where breeding populations gather at localised spawning sites. For such species the bulk of the annual catches are normally landed during the short period when the fish are concentrated and many countries have therefore implemented restrictions on the fisheries during the spawning season to ensure that sufficient numbers of spawners survive and get a chance to reproduce. Damaging forms of fishing such as cross river traps that prevent the fish from moving upstream are for example prohibited in many areas.

Co-management

Although it is necessary that the fishery continues to produce benefits year after year in order to sustain the livelihoods that depend on them, it does not necessarily mean that fishing practices must continue exactly as they are today. New fishing practices or alternative uses of the river system may be introduced as long as they do not threaten the sustainability of the resource base. However, many fish stocks are in decline so something will have to change in the way fisheries are currently managed. Traditional fisheries management frequently operates by regulating fishing activities and restricting access and can have positive benefits for fish stocks. Top-down, imposed regulations may result in management solutions that are locally inappropriate, but also in unwillingness of the public to comply with them. Solutions developed in partnership with local people, and building on and integrating their knowledge and traditions, are much more likely to be successful than those which simply aim to suppress activities labelled as undesirable by fisheries officers. In inland fisheries the fishers themselves are usually the *de-facto* managers, especially in countries that lack the resources to implement efficient enforcement mechanisms, and unless the fishers are actively involved in management decisions and implementation, these efforts are likely to fail. Thus even when the knowledge needed for management decisions is available any approach to management must be coupled with stakeholder involvement.

Local resource users are constantly present in the environment and consciously as well as subconsciously learn about system behaviour and function. Involving them in the entire management process, including defining management goals, as well as participating in data collection, analysis, management actions and monitoring can provide a win-win solution especially in the resource and data poor situations that characterizes many inland fisheries (Poulsen et al. 2003a). There are limitations to this approach, however, in that local fishers rarely see themselves in the context of the basin as a whole and may set goals that are damaging to the interest of fishers elsewhere in the basin. It is therefore important to remember that although management needs to be implemented at the local level, national initiatives are needed to link local and international management requirements. To be effective, national agencies must while promoting co-management for both local and transboundary stocks, simultaneously liaise with relevant agencies in other countries over joint-management requirements (Coates et al. 2000). However, when people are recognized as fully integrated components of the ecosystem, and their knowledge valued as a prerequisite for management, and when people know and understand the purpose, can see the benefits and have been involved throughout the process, they are much more motivated to comply with rules and regulations and collaborate in generating the necessary information.

Management must be based on relevant information, however, the kind of information and the level of detail needed depend on the exploited species, the types of fisheries of concern and the local situation. There is no single right answer for river fisheries management, which may be applied universally. However, while the traditional *one size fits all* approach to management operated with rigid management frameworks community led management approaches opens for more experimental adaptive approaches that seek to optimise benefits through gradual adjustments to the management strategy based on the outcomes.

Restocking and reintroduction

One of the traditional ways to improve the situation of threatened species has been breeding them in captivity and then stocking them into the waterbodies where they have become scarce or have disappeared.

It is technically possible to maintain a fishery through systematic stocking. However, in the case of transboundary fish stocks, the economic implications involved with such an approach would imply prior agreements between the relevant stakeholders including the riparian countries on ownership and rights to exploit the stocked fish. In river basins that have been severely impacted, stocking may be the only way to maintain fish stocks at a level where they can be exploited commercially, but if recurring stocking to support the fisheries is to be avoided, the reasons for rarity of the species must be identified and proper mechanisms to improve the situation put into place.

Large-scale translocations of species and genetic material pose risks for the receiving fauna irrespective whether the species in concern is already present there or not. Stocking, if done incorrectly, can endanger wild resources and should follow the precautionary approach and be regulated in accordance with internationally accepted guidelines (FAO 1996), such as those elaborated by ICES/EIFAC (Turner 1988; ICES 1995), and when stocking is to take place in transboundary river basins it should always be subject to mutual agreements among the riparian states involving also the stakeholders in concern.

Stocking material derived from too few breeders may result in a narrowing of the genetic base, which in turn will lead to poor adaptive capacity, higher mortality and reduced growth and reproductive potential (Mattson et al. 2002).

Where stocking is carried out with species that are already present in the receiving water body, there is a risk that genetic characters specific to sub-stocks adapted to the local conditions may be lost (Johnson 2000). This may for example cause problems with aspects of behaviour such as timing of reproduction and arriving at appropriate spawning grounds.

Genetic issues are a particular concern where natural fish stocks are already under pressure and the recruitment resulting from stocking is comparatively large (Allendorf et al. 2001). In situations where several stocks partially overlap during part of their lifecycle, such as in anadromous species, negative effects may spread to other neighbouring stocks.

In the case of reintroduction and in other situations where the stocking material comes from a foreign environment there is also a risk of introducing new diseases and parasites as well as the danger of co-introducing other fish species that may not be desirable (FAO 1996).

Managing the environment

The main obstacle to sustainable management of migratory fish stocks is that they are threatened not only from fishing but by a complex of changes resulting from other human activities in the river basin (Box 1). Their dependency on both habitat integrity and interconnectivity makes them highly vulnerable to environmental impacts. Interruptions to connectivity between sites by dam and weir construction can be especially damaging. Not only do dams and weirs act as physical obstacles to upstream migrating fish but they also damage downstream migrants, including larvae and juveniles, in the turbines, penstocks or discharge flumes. Impounding also changes the habitat from riverine (free-flowing) to stagnant water, often

Box 1. Non–fishery threats to migratory fish species

THE MOST SERIOUS HAZARDS TO MIGRATORY FISHES	
Dams and weirs (e.g. for hydropower production or irrigation)	
Block longitudinal migration routes	=> fish cannot complete their lifecycle
Diminish flood downstream	=> diminish nutrients and aquatic habitat available to fish
Changes in the timing of the flood	=> larvae of fish spawning in the river will not reach nursery areas
Decrease in silt load downstream	=> scouring and loss of habitats and nutrients
Release of anoxic water and water with different temperature from reservoir	=> change in species composition down stream and fish kills
Change from riverine to stagnant-water habitat	=> specialized species disappear and are replaced by common species that are often less appreciated in the fishery
	=> Reservoirs upstream of dams slow flow and impede the downstream drift and migration of fry and young fish.
Water abstractions for agriculture	
Alteration in timing, intensity and duration of flooding	=> failure of physiological cues
	=> failure of fish to colonise floodplain
	=> insufficient water for fish to negotiate shallows during migration
Dikes and polders	
Block lateral migration routes	=> reduced recruitment to the floodplain of fish spawning in the main river and vice versa
Modification of current	=> larvae of fish spawning in the main river will not reach nursery areas
Habitat conversions (e.g. conversion of flood forest into paddy fields)	
Less shelter available	=> reduction in average size
Fewer niches available	=> reduction in biodiversity
	=> reduction in productivity ?
	=> fewer feeding opportunities and reduced growth, reproduction etc
Dredging and removal of rapids	
Destruction of habitats and spawning	=> loss of biodiversity grounds
	=> reduction in fish production

accompanied by changes in the chemistry. The slowing of flow in large reservoirs upstream of dams prevents drifting larvae from upstream sites from reaching their destinations downstream, and may increase mortality due to predation (Jepsen et al. 1998). The association of several dams into cascades along a rivers course is especially damaging. For migratory fishes, the habitats that are home to different life stages of the targeted stock may be located far from one another and be under the influence of different user groups, and activities. This is most obvious for water resources management and water pollution. The construction of a dam may for example trigger, or contribute to, the disappearance of a species or a stock further downstream. The most serious problems arise where local activities have an impact upon a transboundary stock that is not exploited locally; for example, where the nursery areas of a certain species are located and threatened, in one area, but the fishery for the species takes place in a different area. If the resource managers are not fully aware of, or do not care about, such interactions between local disturbances and transboundary impacts, the prospects for good management are poor (Coates et al. 2000).

Effective management of transboundary stocks evidently requires co-operative management by the relevant parties that are either using the resource directly, or whose activities might have an indirect impact upon it. Management needs to be effective at all the different levels i.e. at the regional, national and local scale. The most obvious strategy to achieve this is to adopt a basin approach or, for large river basins, a sub-catchment approach where the basins are divided into more manageable units each of which can be managed at the appropriate level by the appropriate parties (e.g. village, intermediate, sub-catchment and catchment management areas (Hoggarth et al. 1999).

There are thus still important roles to play for the Government agencies, fisheries officers, NGOs, and development projects in raising awareness, improving capacity and providing technical information to the communities. The mentioned actors may also assist local people by facilitating coordination among communities that share a basin, and are fishing the same fish stock especially if these communities are located in different countries, and may further provide mechanisms for resolving conflicts that may arise. An externally sustained platform or catchment committee for intercommunity communications may not only serve the purpose of negotiations and consensus building, communities may also use it to share their experiences with different management approaches. By making management plans for entire sub-catchments across borders, it becomes easier to perceive the complexity of the resource situation and to structure research and management appropriately.

Whilst it may be difficult to formulate management plans for each and every species, progress may be made through the identification of critical habitat types where management efforts can be focussed, and by using species with stringent habitat requirements as indicators. If the species that are chosen as indicators at the same time can be promoted as "flagship species" (Box 2), it may be easier to create awareness about the need for implementing appropriate measures, and it may also be easier to attract national as well as international funding for management programmes.

Box 2. CITES, CMS and flagship species

The Convention on International Trade in Endangered Species of Wild Fauna and Flora (CITES, www.cites.org) and The Convention on the Conservation of Migratory Species (CMS, http://www.cms.int/) are legally binding international agreements between Governments that can be used to heighten the profile of threatened species of migratory fishes. Promoting high profile "flagship" species that are important in the fishery could make it easier to catch the interest among the public, as well as policy makers on issues relating to the preservation of biodiversity and fish production. Such species may provide good starting points for negotiating the conservation and management of aquatic resources between the countries, and may make it easier to reach consensus between the countries on the priorities and needs for action. While not many species of fishes currently are included in the appendices of these treaties, they do for example comprise all the sturgeon and paddle fish species and the Mekong giant catfish. In the case of sturgeon The Parties to the Agreement recognised that listing would not be enough to save the species, consequently regional agreements has been promoted and a number of activities to improve fishery management, legislation, control of illicit trade, marking systems, and aquaculture has been initiated since 1998 (http://www.cites.org/eng/prog/sturgeon.shtml).

It may be argued that the focus on individual species is inappropriate because other species may be equally important for preserving ecosystem functions. However large migratory species such as the sturgeons have strong requirements to habitat quality and ecosystem integrity, and it will not be possible to address the factors that make them vulnerable in isolation from the rest of the ecosystem (Poulsen et al. 2003a). It is therefore necessary to adopt an ecosystem approach, as advocated by FAO, if efforts to restore and maintain viable, self-sustaining, wild populations of such species shall be successful, and the conservation and management policies that are needed are thus likely to benefit a whole range of other species. Unless they are very carefully planned, single local initiatives such as setting up protected areas will by themselves do little to benefit migratory species that depend on access to habitats that may be widely dispersed. Closing part of a watershed for fishing may serve as a pretext for doing nothing else and result in habitat fragmentation. Instead the first priority should be to maintain key habitats and connectivity between them.

Rehabilitation

Principles

Human activities have impacted streams and rivers increasingly for several centuries, but the pressure on natural waterbodies has intensified and the degradation of aquatic habitats has accelerated during the last two hundred years. The consequences of changes to aquatic ecosystems brought about by industrialisation and population growth has been totally negative for aquatic species and therefore also for the fisheries. Currently, nearly all waterbodies in developed countries have been affected by development to various degrees, and most of the developing countries are following down the same path (Nilsson et al. 2005).

For most waterbodies it is neither realistic nor practical to aim for a complete restoration to a pristine condition. However, as upstream activities can counteract any effort at the local level it is a guiding principle that no rehabilitation project can be considered in isolation from its basin. Habitats supporting multiple species can only be restored by directing all rehabilitation actions at restoring ecosystem processes and functions and the goal of any rehabilitation programme should be to restore an ecosystem that favours whole communities of species, rather than specific fish populations. The habitat characteristics which need improvement must be identified accordingly, including all functional units used by fish and especially sensitive parts in the fishes' lifecycles, but the final rehabilitation strategy must be sufficiently flexible to allow new knowledge and tools to be incorporated. In Europe, rehabilitation measures are currently often guided by the principle of the "*potential natural species composition*" aiming not only at existing species but also species that lived there in the past and might return or be brought back in the future (FAO 2007). However, such a principle may be very difficult to apply in countries with more complex fish communities.

For rehabilitation to be sustainable a multidisciplinary basin-wide ecosystem approach including land and water management is needed. The ecological requirements of all riverine species and sizes, and particularly migrants, must be taken into consideration from the earliest stages in planning for the project to ensure maximum impact of remedial measures. To ensure this planning and subsequent actions are called for at three scales:

The landscape (basin) scale: Here major shifts in agriculture and land use patterns are controlled. For example, extensive deforestation and bad agricultural practice, particularly in the upper basin can change discharge patterns and increase siltation to the detriment of downstream habitats. Control of water quality should also be pursued at this level to prevent pollution by upstream activities being transmitted to downstream sites.

Similarly, provision for environmental flows should be made by general planning at the basin scale.

The habitat scale: where individual, critical habitats are rehabilitated. Here such rehabilitation should be part of a general, basin level plan as it is useless, for example, restoring downstream feeding habitats of a migratory species if the corresponding upstream spawning habitats are degraded or absent.

Connectivity: Connectivity should also be provided within the framework of a larger plan. Planning for the maintenance of longitudinal connectivity has to be included at the earliest possible stage in water development projects, i.e. ideally during the dam identification phase (Bernacsek 2001). However, restoring a migratory pathway should only be done if the upstream conditions are suitable for the target species, or might become suitable within a reasonable timeframe.

Social and economic considerations

The economic interests involved with power generation, navigation, agriculture, and industry are very difficult to counter because it is not easy to provide solid figures that demonstrate the true economic value of the intact aquatic habitat and its associated fish populations. Often social, economic and institutional issues and competing uses of inland waters impede application of the best approaches to rehabilitate rivers for fisheries. However, as FAO stresses: "*While the precautionary approach should be applied to fisheries ....there is an equal need to apply the approach to non-fisheries sectors whose capacity to damage the ecosystem is usually much greater than that of the fisheries themselves*" (FAO 1997a). Major interventions such as re-meandering, floodplain restoration, and the removal of dams are extremely costly and will require either the active cooperation of riparian landowners and other stakeholders, or the acquisition of the land by the state. Assessments of the cost-effectiveness of rehabilitation projects have, unfortunately, been greatly neglected, and this should be corrected soonest. However, it is clear, and now also more widely accepted, that habitat protection is often the most cost effective method to maintain riverine fisheries, as it is typically cheaper to protect habitat than trying to restore it later, and there is a great number of good studies on habitat rehabilitation and monitoring on which to base technical advice. The Code of Conduct also calls for increased efforts in capacity building as well as knowledge and technology transfer (FAO 1995, 1997a).

Utmost attention has to be laid on the fact that the decision-making process involves all the stakeholders and that the approach is multidisciplinary, i.e. involve for example engineers, biologists, socio-economists and the administration (FAO 1995, 1997a). To avoid tensions among the stakeholders resulting from the priority given to different objectives, politicians have to define an enabling framework so that net benefits can be derived from the many goods and services the aquatic ecosystems supply, including products for human consumption. In this process, it is the task of the fisheries managers and those responsible for the conservation of the environment to negotiate the best possible conditions for the maintenance of fish stocks and fisheries. Efforts should now focus on expanding the policy debate in order to find ways to resolve these issues so that inland fisheries can be better integrated into the whole field of natural resources management.

Restoring longitudinal connectivity

As a result of lost longitudinal connectivity in river systems, fisheries depending on migratory species have decreased or completely collapsed or were forced to target other species. Well documented examples are the salmon fisheries in Europe, and in North America. River rehabilitation for fish started there only a couple of decades ago and in some countries legislation has been amended to provide the basis for protection and rehabilitation of the aquatic environment, including construction of fish passes. However, despite its utmost importance to fisheries and biodiversity, and although it has been proved to enhance the reproduction and survival of fish species, the issue of preserving or restoring fish passage is not yet systematically perceived as a priority in terms of economical and social benefits. To some extent, this is due to the fact that the negative impact of dams on fisheries and livelihoods is not always noticed immediately because of some immediate benefits brought by the dams, the resilience of the concerned fish species in the short-term, or temporarily increasing catches as a result of stocking in reservoirs and blockage of migrants below dams (Marmulla et al. 2002).

It is important to emphasize that it is not only the upstream migration, which must be cared for. Also the downstream movements whether active or passive of eggs, larvae, juveniles and spent fish must be considered as mortality resulting from passage through hydraulic turbines or over spillways can be significant.

In promoting the implementation of the FAO Code of Conduct for Responsible Fisheries, FAO advocates restoring connectivity as an important and appropriate tool for sustainable management of inland waters and to maintain stocks at a level that can sustain fishing activities (FAO 1995, 1997a). Article 6.8 of the guidelines states *inter alia*:

> Elements for conservation: Certain basic elements are required of an aquatic system so that it can retain its functionality.

The maintenance and restoration of longitudinal and lateral connectivity in rivers in the interests of conserving fish migration patterns through removal of transversal (dams) or lateral (levees) obstructions or the provision of fish pass mechanisms (Box 3).

Box 3. Fish pass design

Dams on major river courses are among the most serious threats to migratory fish species. Any construction that prevents the fish from migrating will prevent them from completing their lifecycles, thus leading to the gradual disappearance of the affected stocks or in some cases even the species. The restoration of longitudinal connectivity is thus usually regarded as a priority measure in river rehabilitation, and the removal of barriers can restore both upstream and downstream passage as well as habitat.

Depending on the behavioural characteristics and swimming capacity of individual species, even low obstacles can interrupt upstream migration. Strong swimmers, such as some of the salmonids, can, to some extent, leap over obstacles. While less vigorous swimmers tend to use the slacker water close to the bottom to move against the current and such species are easily impeded even by relatively low obstructions.

Many different types of fish passes have been designed with various levels of technology. Most of them to facilitate upriver movements of salmonids in temporary rivers. There exist, however, types fish passes (e.g. vertical slot passes) that seem, more than others, to have the potential to accommodate also the needs of other species. However, the most effective mechanism for meaningful habitat restoration is the complete elimination of existing dams whereby the original condition of the river may be restored in the long term.

Removal of obstructions or retrofitting them with fish passes might be praiseworthy, but should be carefully assessed against the availability of essential habitat conditions upstream that these structures will reconnect and the wider planning at river basin level.

Environmental flows

Flow plays many roles in regulating fish populations and the environment in which they live. Natural regimes maintain the environment by regulating the erosion-deposition cycle and by such functions as cleaning spawning gravels. They also give physiological clues that trigger migration and spawning behaviour. They provide sufficient water for fish to migrate up the river and for the fry to drift downstream. They allow the river to overflow its banks and to flood the floodplains and riparian wetlands. Disturbances to natural flow regimes by damming and water abstractions diminish the viability of fish faunas by displacing or suppressing flood peaks thus removing the physiological triggers, by failing to maintain sensitive habitats in suitable condition and by failing to flood floodplains and riparian wetlands. Conversely, excessive flows at the wrong time of year can wash away young fish or wash drifting fry past the target floodplains resulting in the loss of most individuals.

Very often the activities that modify flows occur in other countries or administrative regions than those where the fish are affected. International agreements are, therefore, needed to maintain a flow pattern that is acceptable to the fish community at any point in the river. Generally, there has to be some form of assessment as a basis for planning and Tharme (2003) catalogues a number of methods that have been used to do this. The simplest approaches rely on specifying minimum flows (usually Q_{95}) during the period of low flows but are unsatisfactory as they fail to account for the many different roles played by flow in the fish's life history, the close synchronization in some species between migration and temporal flow signals, and the need for year-to-year variations in flow to satisfy different elements of the fish assemblage (Poff et al. 1997; Welcomme and Halls 2004).

As a result a variety of more flexible methods have been developed to accommodate the various needs of the fish. For example, the Building Block Methodology (BBM) developed and applied in South Africa (Tharme and King 1998; King et al. 2000) defines a series of critical points in the life history for a species and attempts to define desirable flows that need to be delivered (see Figure 1) especially for dams and impoundments, although similar considerations can be used for limiting abstractions. Other systems such as the DRIFT (Downstream Response to Imposed Flow Transformations) procedure developed in South Africa (Arthington et al. 2003; King et al. 2003) are capable of considering a range of uses of water and determining a series of scenarios of possible water releases.

Research priorities and tools

To manage the fish resources sustainably, the fish species, their behaviour and the important habitats in a catchment area must be known as well as the migrations in and out of the catchment, and where the fish go from there. A variety of techniques have been developed over the last three decades for studying spatial behaviour of fish (see Lucas and Baras (2000) for a review). However, the current level of knowledge on inland waters, their fish populations and fisheries are in most parts of the world still far from a technical level that is sufficient to allow for management of these resources according to the aspirations expressed in the CCRF and other international agreements such as the Convention on Biological Diversity.

Most research on fish ecology and fisheries has focused on individual habitat units or are at reach or country scale. Although the information generated is important, it leaves behind uncertainty about the situation at a population or basin level.

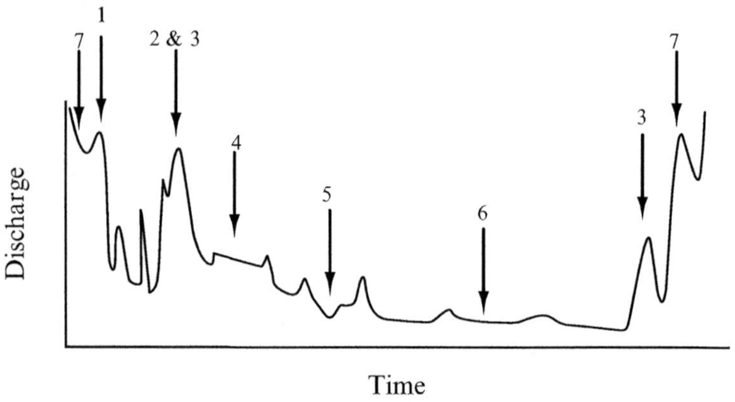

Figure 1. Theoretical flow curve (black line) and some elements that may affect fish that could be used as building blocks to designing flow releases. 1: channel maintenance flows – extreme high flows that modify channel structure, 2: habitat maintenance flow – flows that perform a specific function such as cleaning spawning gravels, 3: flows that provide longitudinal and lateral connectivity or serve as migration freshets, 4: water levels over spawning habitats, 5: flows influencing larval survival and drift, 6: low water flows to maintain sufficient water and water quality, 7: physiological trigger flows

In large river basins many important species may consist of several distinct populations or stocks with different migration and spawning patterns (Bonetto 1986; Quiros and Vidal 2000), and some of them may be transboundary others may not. The distribution range of some stocks may overlap, while others are more distinctly separated (Quiros and Vidal 2000). In order to develop management strategies for transboundary fish stocks, it is crucial to have these stocks identified, because each stock is a management unit and may have separate management requirements.

However, studies addressing these questions are often both difficult and costly to implement.

The financial resources allocated for the research and the time available for any particular study put restrictions on any research programme. There are obviously trade-offs between the accuracy of the information gathered, the area that can be covered, the duration of the study, the numbers of species for which information can be gathered, and the availability of resources for the study.

Before embarking on a study the specific needs must be carefully identified and compared with the existing level of knowledge. In data-poor situations it may be better initially to design studies in a way that requires only a low level of technical equipment and therefore low technical costs, but covering a large area to get an overview of the situation. At a later stage when more information is available, more specific questions that can be answered through more focussed studies can be asked, possibly applying advanced methodologies.

In many situations a considerable amount of often surprisingly detailed ecological information can be obtained collecting the knowledge of expert fishers. Such local knowledge may provide the basis, which can be used to formulate testable hypotheses, and it may later supplement and help interpret the data gathered through more conventional scientific approaches (Valbo-Jørgensen and Poulsen 2000; Silvano and Valbo-Jørgensen 2008). If local knowledge is collected systematically over large areas, large-scale ecosystem issues may emerge, and it is important to keep in mind that the knowledge possessed by fishers may cover a much longer time horizon than any research programme, sometimes even beyond their own lifetime (Poulsen and Valbo-Jørgensen 2000).

Local knowledge may for example indicate the existence of discrete stocks based on different migration patterns reported in different parts of a basin. More focussed studies using tagging, radio telemetry, or genetic markers may then be used to test the validity of this hypothesis.

GIS is a very powerful tool for the fisheries managers because it can incorporate a variety of information from different sources at the same time, thereby revealing patterns that may otherwise be difficult to see. It can for example be used to analyse and illustrate migration patterns, fish occurrences and spawning grounds in relation to physical data such as water quality, substrates, current, presence of physical obstacles etc. By combining environmental data with population statistics a GIS can also yield information about the status of fisheries, people's dependency on aquatic resources and their vulnerability to environmental change.

Summary of recommendations

Sustainable and effective management of transboundary fish stocks requires co-operation among the parties that are using the resource directly, or whose activities have an indirect impact upon it. Management needs to be effective at all the different levels i.e. at the regional, national and local scale. Thus close cooperation among all the countries sharing a basin is essential, and Governments need to develop clearly articulated goals or visions for their river basins, expressed as international agreements, river basin plans and appropriate legislation, and mechanisms for collaboration between basin states should be established so as to share information and collectively make decisions as to the management of the system for migratory fish species.

Governments are obliged to comply with international agreements such as the FAO CCRF, Ramsar, CITES and CMS. These frameworks can therefore be used actively to promote inland fisheries management and conservation. Developing joint management plans for entire sub-catchments across borders is one way of structuring research and management. As pointed out by the FAO CCRF, it is crucial that the precautionary approach be applied not only to fisheries but also to non-fisheries sectors whose capacity to damage the ecosystem is often much greater than the fisheries itself, and inland fisheries must become better integrated into overall natural resources management schemes. Management strategies should be developed through stakeholder participation and the chosen strategy will define data and information needs. Government agencies, fisheries officers, NGOs, and development projects are key-actors in raising awareness, improving capacity and providing technical information to the communities. When making plans for development governments should keep in mind that habitat protection is more cost efficient than rehabilitation. And river ecosystems that have already become degraded should be rehabilitated.

Rehabilitation should focus on restoring ecosystem processes and functions for the benefits of whole communities of species. The most important issue in relation to the ecological functioning of river ecosystems from the point of view of migratory fishes is that critical habitats are maintained in time and space. This includes the provision of upstream (spawning) and downstream (feeding and refuge) habitats and the connectivity between them, which should be agreed upon as part of any river basin plan. Appropriate annual hydrological patterns must also be conserved, including their role in the creation of seasonal floodplain habitats, triggers for migration, and distributor of fish larvae and juveniles through passive drift. Environmental flows based on scientific evidence of the requirements of the fish species should be negotiated with other users of water at basin level.

Stocking and reintroduction of endangered species should follow the precautionary approach, and before carrying out such a programme the reasons for rarity of the species must be identified and alleviated.

Where and how can FAO assist?

The FAO Code of Conduct for Responsible Fisheries (CCRF) and the related CCRF Technical Guidelines are instruments to develop good practices and policies for sustainable capture fisheries and aquaculture. Although the CCRF is a voluntary code, its guiding principles are internationally accepted in the management of fisheries. The code was developed to cover both inland and marine fisheries and is relevant to the management of most fish stocks, including transboundary ones. The CCRF is supplemented by a series of technical guidelines on how to implement specific provisions (e.g. #2 Precautionary Approach to Capture Fisheries and Species Introductions (FAO 1996), #4 sup 2 Fisheries Management: The Ecosystem Approach to Fisheries (FAO 2003), and #6 Inland Fisheries (FAO 1997a)).

However, translating international or regional agreements into workable national strategies will require assistance and support in many cases, FAO's Regional Fisheries Bodies could make a significant contribution as neutral fora for designing regional strategies and management plans, with FAO providing technical guidance and advice on how to develop the legal and institutional agreements and assisting with the execution of the necessary technical surveys. FAO's Regional Fisheries Bodies are open to all member countries in the appropriate region.

Depending on the availability of funds and the in-kind support that countries could provide, assistance could include a variety of subjects, ranging from technical support for the establishment of consultation structures, up to training at national and sub-regional levels in the management of international resources and in the negotiation and resolving of controversies. Assistance could even include the establishment of the appropriate bi-national or sub-regional structures through an international agreement or treaty where such institutions do not already exist.

Previous work by FAO in the Caspian region

Inland fisheries are very important in many of the countries around the Caspian Sea (inland fisheries of the Soviet Union were reviewed by Berka (1989). However, due to the Soviet Union not being a member of FAO and the Russian Federation not joining the Organisation until 2006, the amount of work in the Caspian Sea region which has been supported by FAO is very limited.

Water scarcity is a dominating feature in many of the countries in the Central Asian region. In Water Report 15, published by FAO's AQUASTAT programme, irrigation and drainage in rural areas in 15 countries of the former Soviet Union including the Azerbaijan, Kazakhstan, the Russian Federation and Turkmenistan in the Caspian Sea Catchment were presented and compared (FAO 1997b). Dams have been built and reservoirs created on most rivers in the region mainly for irrigation purposes. This development had severe impacts on fish migration and resulted in the installation of multiple fish passes and sparked considerable research into fish migratory behaviour in the former Soviet Union, these experiences were reviewed by FAO (Pavlov 1989).

However, reservoirs and irrigation canals also provided new opportunities for fishing (reservoir fisheries were reviewed by Karpova et al. 1996), and in 2001 FAO organised an expert consultation in Almaty, Kazakhstan on the use of irrigation systems for sustainable fish production in arid countries of Asia, with the participation of experts from ten countries including the two Caspian countries the Islamic Republic of Iran and Kazakhstan. The objectives of the workshop were to review and improve fisheries management in waterbodies used for irrigation in the arid part of Central Asia (Petr 2003).

With the dissolution of the Soviet Union the fisheries management system that was formerly in place collapsed and the independent riparian countries to a large extent failed to coordinate their management efforts resulting for instance in an enormous increase in the fishing pressure on the sturgeon stocks, that at the same time was under pressure from pollution and habitat loss. The situation became so critical that it prompted CITES to list all sturgeon species in Appendix II, as a condition to lift the ban on sturgeon products CITES Appendix II required that the countries seek the advice of FAO to improve their fisheries management capacity. The countries consequently approached FAO with a request for assistance for strengthening national and regional capacities in research and monitoring of the fisheries. A regional Technical Cooperation Programme "Capacity building for the recovery and management of the sturgeon fisheries of the Caspian Sea" was approved to be implemented from February 2006–September 2007, and is currently ongoing. In addition to enhancing national and regional capacities of the fisheries institutions, the project is expected to facilitate investment possibilities through the World Bank/Strategic Partnership Programme.

Acknowledgements

We wish to thank the following people for their contributions to this paper: Dr. Devin M. Bartley, FAO who commented on the first draft of the document. Dr. Tomi Petr, FAO (retired) and Dr. Ross Shotton, FAO provided information on respectively FAO's former and current activities in the Central Asian region.

References

Allendorf, F. W., Leary, R. F., Spruell, P., and Wenburg, J. K., 2001, The problems with hybrids: setting conservation guidelines, *Trends Ecol. Evol.* **16**(11): 613–622.

Arlinghaus, R., 2005, A conceptual framework to identify and understand conflicts in recreational fisheries systems, with implications for sustainable management, *Aquat. Resour. Cult. Dev.* **1**(2): 145–174.

Arthington, A. H., Rall, J. L., Kennard, M. J., and Pusey, B. J., 2003, Environmental flow requirements of fish in Lesotho rivers using the DRIFT methodology, *River Res. Appl.* **19**: 641–666.

Baran, E., 2006, *Fish Migration Triggers in the Lower Mekong Basin and Other Tropical Freshwater Systems*, MRC Tech. Pap. No. 14, Mekong River Commission, Vientiane, 56 pp; http://www.mrcmekong.org/download/free_download/Technical_paper14.pdf.

Barthem, R. B., and Goulding, M., 1997, *The Catfish Connection. Ecology, Migration and Conservation of Amazon Predators*, Biology and Resource Management in the Tropics, Columbia University Press, New York, 144 pp.

Berka, R., 1989, *Inland Capture Fisheries of the USSR*, FAO Fish. Tech. Pap. No. 311, FAO, Rome, 143 pp.

Bernacsek, G. M., 2001, Environmental issues, capacity and information base for management of fisheries affected by dams. In: *Dams, Fish and Fisheries. Opportunities, Challenges and Conflict Resolution*, FAO Fish. Tech. Pap. No. 419, G. Marmulla., FAO, Rome, pp. 139–166; ftp://ftp.fao.org/docrep/fao/004/Y2785E/y2785e.pdf.

Bonetto, A. A., 1986, Fish of the Paraná system. In: *The Ecology of River Systems*, Monographiae Biologicae 60, B. R. Davies and K. F. Walker (eds.), Dr. W. Junk, Dordrecht, The Netherlands, pp. 573–588.

Carolsfeld, J., Harvey, B., Ross, C., and Baer, A. (eds.), 2003, *Migratory Fishes of South America: Biology, Fisheries and Conservation Status*, World Fisheries Trust/The World Bank/International Development Research Centre, Washington, DC, 372 pp; http://www.idrc.ca/openebooks/114-0/

Coates, D., Poulsen, A. F., and Viravong, S., 2000, *Governance and transboundary migratory fish stocks in the Mekong River Basin*, paper presented at the MRC Third Fisheries Technical Symposium. Mekong River Commission, Phnom Penh, Cambodia, 8–9th December 2000.

Cooke, S. J., and Cowx, I. G., 2006, Contrasting recreational and commercial fishing: searching for common issues to promote unified conservation of fisheries resources and aquatic environments, *Biol. Conserv.* **128**: 93–108.

FAO, 1995, *FAO Code of Conduct for Responsible Fisheries*, FAO, Rome, 41 pp; ftp://frp.fao.org/docrep/fao/005/v9878e/v9878e00.pdf.

FAO, 1996, *Precautionary Approach to Capture Fisheries and Species Introductions*, FAO Technical Guidelines for Responsible Fisheries No. 2, FAO, Rome, 54 pp; ftp://ftp.fao.org/docrep/fao/003/W3592e/W3592e00.pdf.

FAO, 1997a, *Inland Fisheries*, FAO Technical Guidelines for Responsible Fisheries No. 6, FAO, Rome, 36 pp; ftp://ftp.fao.org/docrep/fao/003/W6930e/W6930e00.pdf.

FAO, 1997b, *Irrigation in the Countries of the Former Soviet Union in Figures*, Water Report 15, FAO, Rome; http://www.fao.org/docrep/W6240E/W6240E00.htm.

FAO, 2003, *The Ecosystem Approach to Fisheries*, FAO Technical Guidelines for Responsible Fisheries No. 4, Sup. 2. FAO, Rome, 112 pp; ftp://ftp.fao.org/docrep/fao/005/y4470e/y4470e00.pdf.

FAO, 2007, *The State of World Fisheries and Aquaculture*, FAO, Rome, 162 pp; ftp://ftp.fao.org/docrep/fao/005/v9878e/v9878e00.pdf.

Fuentes, C. M., 1998, Deriva de Larvas de Sábalo Prochilodus lineatus (Valenciennes, 1847) y Otras Especies de Peces de Interés Comercial en el Rio Paraná Inferior, PhD Thesis, Universidad de Buenos Aires, Buenos Aires, 138 pp.

Hoggarth, D. D., Cowan, V. J., Halls, A. S., Aeron-Thomas, M., McGregor, J. A., Garaway, C. A., Payne, A. I., and Welcomme, R. L., 1999, *Management Guidelines for Asian Floodplain River Fisheries. Part 1: A Spatial, Hierarchical and Integrated Strategy for Adaptive Co-Management*, FAO Fish. Tech. Pap. 384/1, 63 pp; ftp://ftp.fao.org/docrep/fao/006/x1357e/x1357e00.pdf.

Hortle, K. G., Lieng, S., and Valbo-Jørgensen, J., 2004, *An Introduction to Cambodia's Inland Fisheries*, Mekong Development Series 4, 41 pp; http://www.mrcmekong.org/download/free_download/Mekong_Development_No4.pdf.

ICES, 1995, *ICES Code of Practice on the Introductions and Transfers of Marine Organisms*, ICES Co-Operative Research Report No. 204.

Jepsen, N., Aarestrup, K., Økland, F., and Rasmussen, G., 1998, Survival of radio-tagged Atlantic salmon (*Salmo salar* L.) and trout (*Salmo trutta* L.) smolts passing a reservoir during seaward migration, *Hydrobiologia* **371/372**: 347–353.

Johnson, M. S., 2000, Measuring and interpreting genetic structure to minimize the genetic risks of translocations, *Aquac. Res.* **31**(1): 133–143.

Junk, W. J., Bayley, P. B., and Sparks, R. E., 1989, The flood pulse concept in river-floodplain systems, *Can. Spec. Publ. Fish. Aquat. Sci.* **106**: 110–127.

Karpova, E. I., Petr, T. and Isaev, A. I., 1996, *Reservoir Fisheries in the Countries of Commonwealth of Independent States*, FAO Fisheries Circular No. 915, FAO, Rome, 131 pp.

King, J. M., Tharme, R. E., and de Villiers, M. S. (eds.), 2000, *Environmental Flow Assessments for Rivers: Manual for the Building Block Methodology*, Water Research Commission Report TT 131/00, Pretoria, 339 pp.

King, J., Brown, C., and Sabet, H., 2003, A scenario based holistic approach to environmental flow assessments in rivers, *River Res. Appl.* **19**: 619–639.

Lucas, M. C., and Baras, E., 2000, Methods for studying the spatial behaviour of freshwater fishes in the natural environment, *Fish Fish.* **1**: 283–316.

Lucas, M. C., and Baras, E., 2001, *Migration of Freshwater Fishes*, Blackwell Science, Oxford, 420 pp.

Marmulla, G., Kairys, A., and Baras, E. (compilers), 2002, Fish passage in developing countries: issues and perspectives, in: *Proceedings of the Workshops Held in Conjunction with the Environmental Flows for River Systems*, Cape Town, South Africa, March 2002; ftp://ftp.southernwaters.co.za/downloads/sumws1.pdf.

Mattson, N. S., Buakhamvongsa, K., Sukumasavin, N., Nguyen Tuan, and Ouk Vibol, 2002, *Mekong Giant Fish Species: On Their Management and Biology*, MRC Tech. Pap. No. 3, Mekong River Commission, Phnom Penh, 29 pp; http://www.mrcmekong.org/download/free_download/Technical_paper3.pdf.

Munro, G. R., Van Houtte, A., and Willmann, R., 2004, *The Conservation and Management of Shared Fish Stocks: Legal and Economic Aspects*, FAO Fish. Tech. Pap. No. 465, FAO, Rome, 69 pp; ftp://ftp.fao.org/docrep/fao/007/y5438e/y5438e00.pdf.

Nilsson, C., Reidy, C. A., Dynesius, M., and Revenga, C., 2005, Fragmentation and flow regulation of the world's large river systems, *Science* **308**: 405–408.

Pavlov, D. S., 1989, *Structures Assisting the Migrations of Non-Salmonid Fish: USSR*, FAO Fish. Tech. Pap. No. 308, FAO, Rome, 97 pp; http://www.fao.org/docrep/008/t0138e/t0138e00.htm.

Payne, A. I. and Harvey, M. J., 1989, An assessment of the *Prochilodus platensis* Holmberg population in the Pilcomayo River fishery, Bolivia using scale based and computer assisted methods, *Aquacult. Fish. Manage.* **20**: 233–248.

Payne, A. I., Sinha, R., Singh H. R., and Huq, S., 2004, A review of the Ganges basin: its fish and fisheries, in: *Proceedings of the Second International Symposium on the Management of Large Rivers for Fisheries*, Vol. I, RAP publication 2004/17, R. J. Welcomme and T. Petr, FAO Regional Office for Asia and the Pacific, Bangkok, pp. 229–251; http://lars2.org/Proceedings/vol1/A_Review_of_the_Ganges_Basin.pdf.

Petr, T., 2003, Fish stocks and fisheries in irrigation systems in arid Asia. In: *Fisheries in Irrigation Systems of Arid Asia*, FAO Fish. Tech. Pap. No. 430, T. Petr, FAO, Rome, pp. 1–16; ftp://ftp.fao.org/docrep/fao/007/y5082e/y5082e00.pdf.

Petrere, M. and Agostinho, A. A., 1993, *La Pesca en el Tramo Brasileño del Río Paraná*, Informe de la sexta reunión del grupo de trabajo sobre recursos pesqueros, Montevideo Uruguay 10–13 mayo de 1993, Informe de Pesca N° 490, Comisión de Pesca Continental para América Latina, FAO, Rome; http://www.fao.org/docrep/008/t0849s/t0849s00.htm.

Poff, N. L., Allan, J. D., Bain, M. B., Karr, J. R., Prestegaard, K. L., Richter, B. D., Sparks, R. E., and Stromberg, J. C., 1997, The natural flow regime, *Bioscience* **47**: 769–784.

Poulsen, A. F. and Valbo-Jørgensen, J. (eds.), 2000, *Fish Migrations and Spawning Habits in the Mekong Mainstream – A Survey Using Local Knowledge*, AMFC Technical Report, Mekong River Commission, Phnom Penh.

Poulsen, A. F., Hartman, W., and Mattson, N., 2003a, *The role of local communities in the generation of fisheries information based on the ecosystem approach*, paper submitted to the Second International Symposium on the Management of Large Rivers for Fisheries, Phnom Penh, Cambodia; http://www.lars2.org/unedited_papers/unedited_paper/Poulsen%20local%20knowledge.pdf.

Poulsen, A. F., Ouch Poeu, Viravong, S., Suntornratana, U., and Nguyen Thanh Tung, 2003b, *Fish migrations of the lower Mekong Basin: implications for development, planning and environmental management*, MRC Tech. Pap. 8, Mekong River Commission, Phnom Penh, 62 pp; http://www.mrcmekong.org/download/free_download/Technical_paper8.pdf.

Quiros, R. and Vidal, J. C., 2000, The cyclic behaviour of potamodromous fishes in large rivers. In: *Management and Ecology of River Fisheries*, I. G. Cowx (ed.), Blackwell Science, Oxford, pp. 71–86.

Roni, P., Hanson, K., Beechie, T., Pess, G., Pollock, M., and Bartley, D. M., 2005, *Habitat Rehabilitation for Inland Fisheries. Global Review of Effectiveness and Guidance for Rehabilitation of Freshwater Ecosystems*, FAO Fish. Tech. Pap. No. 484, FAO, Rome, 116 pp; ftp://ftp.fao.org/docrep/fao/008/a0039e/a0039e00.pdf.

Shrestha, R. K., Seidl, A. F., and Moraes, A. S., 2002, Value of recreational fishing in the Brazilian Pantanal: a travel cost analysis using count data models, *Ecol. Econ.* **42**: 289–299.

Silvano, R. A. M. and Valbo-Jørgensen, J., 2008, Beyond fishermen's tales: contributions of fishers' local ecological knowledge to fish ecology and fisheries management, *Environ. Dev. Sustain.* (online version available at http://www.springerlink.com/content/y5n3u5p2675863h5)

Singanouvong, D., Soulignavong, C., Vonghachak, K., Saadsy, B., and Warren, T. J., 1996, *The Main Wet-season Migration Through Hoo Som Yai, a Steep-Gradient Channel at the Great Fault Line on the Mekong River, Champassack Province, Southern Lao PDR*, IDRC Fisheries Ecology Technical Report No. 4, 115 pp.

Smith, R. J. F., 1985, *The Control of Fish Migration*, Zoophysiol. 17, Springer, Berlin, 243 pp.

Tharme R. E., 2003, A global perspective on environmental flow assessment: emerging trends in the development and application of environmental flow methodologies for rivers, *River. Res. Appl.* **19**: 397–396.

Tharme, R. E. and King, J. M., 1998, *Development of the Building Block Methodology for Instream Flow Assessments and Supporting Research on the Effects of Different Magnitude Flows on Riverine Ecosystems*. Report to Water Research Commission, 576/1/98, Cape Town, South Africa.

Turner, G. (ed.), 1988, *Codes of Practice and Manual of Procedures for Consideration of Introductions and Transfers of Marine and Freshwater Organisms*, EIFAC Occasional Paper No. 23. European Inland Fisheries Advisory Commission, FAO, Rome, 49 pp; ftp://ftp.fao.org/docrep/fao/009/ae989e/ae989e.pdf.

UNEP, 2002, *Atlas of International Freshwater Agreements*, UNEP Press, Nairobi, 183 pp; http://www.transboundarywaters.orst.edu/publications/atlas/atlas_zipped.zip.

Valbo-Jørgensen, J. and Poulsen, A. F., 2000, Using local knowledge as a research tool in the study of river fish biology: experiences from the Mekong, *Environ. Dev. Sustain.* **2**(3–4): 253–276.

Van Zalinge, N., Nuov, S., Choulamany, X., Degen, P., Pongsri, C., Jensen, J., and Nguyen Van Hao, 2004, The Mekong river system, in: *Proceedings of the Second International Symposium on the Management of Large Rivers for Fisheries* Vol. I, RAP Publication 2004/16, R. L. Welcomme and T. Petr (eds.), FAO Regional Office for Asia and the Pacific, Bangkok, pp. 335–357; http://lars2.org/Proceedings/vol1/Mekong_River_system.pdf.

Welcomme, R. L., 1985, *River Fisheries*, FAO Fish. Tech. Pap. 262, FAO, Rome, 330 pp; ttp://www.fao.org/docrep/003/T0537E/T0537E00.HTM.

Welcomme, R. L., 2001, *Inland Fisheries: Ecology and Management*, Blackwell Science, Oxford, 358 pp.

Welcomme, R. L. and Halls, A., 2004, Dependence of tropical river fisheries on flow, in: *Proceedings of the Second International Symposium on the Management of Large Rivers for Fisheries* Vol. II, RAP Publication 2004/17, R. L., Welcomme, and T. Petr (eds.), FAO Regional Office for Asia and the Pacific, Bangkok, Thailand, pp. 267–284; http://lars2.org/Proceedings/vol2/dependence_p267-284.pdf.

Weyden, E., 1865, Das Siegthal – Ein Führer von der Mündung bis zur Quelle des Flusses und durch seine Seitenthäler. Leipzig, Germany.

THE KEY THREATS TO STURGEONS AND MEASURES FOR THEIR PROTECTION IN THE LOWER DANUBE REGION

MIRJANA LENHARDT*

Institute for Biological Research, Serbia

IVAN JARIĆ, GORČIN CVIJANOVIĆ AND MARIJA SMEDEREVAC-LALIĆ

Center for Multidisciplinary Studies, Serbia

Abstract The six native sturgeon species have been commercially harvested in the Danube Basin for more than 2,000 years, with rapid decrease in catch by mid 19th century. Additional negative effect on sturgeon populations in the Danube River was river regulation in Djerdap region, due to navigation in the late 19th century, as well as dam construction in the second half of 20th century that blocked sturgeon spawning migrations. Beside overfishing and habitat loss, illegal trade, life history characteristics of sturgeon, lack of effective management (due to lack of transboundary cooperation and change in political situation in Lower Danube Region countries) and pollution all pose serious threats on sturgeon populations in Lower Danube Region. International measures established by the Convention on International Trade in Endangered Species (CITES) in late 20th century, listing of beluga (*Huso huso*) as an endangered species under the U.S. Endangered Species Act, as well as development of Action plan for conservation of sturgeons in the Danube River Basin, had significant impact on activities related to sturgeon protection at beginning of 21st century. These actions were aimed towards diminishment of pressure on natural sturgeon populations and aquaculture development in countries of Lower Danube Region. The main goal of the Action Plan was to raise public awareness and to create a common framework for implementation of urgent measures. Black Sea Sturgeon Management Action Group (BSSMAG) was founded as a consultative body to improve transboundary cooperation among countries in Lower Danube Region. Nevertheless, more actions are necessary in the field of basic scientific investigation, NGO involvement and better connection among all stakeholders. Everyone involved in sturgeon protection must be aware that the sturgeon recovery, as well as their extinction, is a multi-decadal affair, especially for species with

---

* Mirjana Lenhardt, Institute for Biological Research, Despota Stefana 142, 11000 Belgrade, Serbia; lenhardt@ibiss.bg.ac.yu

V. Lagutov (ed.), *Rescue of Sturgeon Species in the Ural River Basin.*

long life and late maturing like beluga. Resource users and other stakeholders must be patient enough to support recovery plans, which will allow only small-scaled sturgeon fisheries, or even fisheries that will be performed by some future generations.

Keywords: Acipenseriformes, CITES, aquaculture, beluga, over-fishing

Introduction

For biologists, sturgeons are a group that has been extant for a quarter of a billion years, while for the public they are prized exhibits in aquaria and, of course, the source of caviar (Bemis and Findeis 1994). In spite the fact that many sturgeon species have outlived the dinosaurs and survived two Ice ages, nowadays they are on the verge of extinction, due to river regulation, dam building, pollution and market demand for their meat and well prized caviar. Sturgeons, with 27 existing species, have some characteristics that distinguish them from teleost fish and make them unique. While modern teleosts may have lost the ability to synthesize ascorbic acid since the late Triassic, sturgeon can produce ascorbic acid in the kidney (Moreau and Dabrowski 2000). Sturgeon fish sperm possess acrosin-like activity (loss in teleost fish) that shares many properties with mammalian acrosin and has some unique properties that may represent adaptations of this enzyme to the environment of external fertilization (Ciereszko et al. 1996). Sturgeons have a unique gastrointestinal tract, because the pyloric stomach wall is hypertrophied to a gizzard-like organ (Hung and Deng 2002). Bearing previous in mind, sturgeon protection is very important for biological and genetic diversity salvation.

Decline in sturgeon catch

Historically, five acipenserid species migrated from the Black Sea into the Danube River: beluga (*Huso huso*), Russian sturgeon (*Acipenser gueldenstaedtii*), stellate sturgeon (*A. stellatus*), ship sturgeon (*A. nudiventris*) and Atlantic sturgeon (*A. sturio*) (Bacalbasa-Dobrovici 1997). The freshwater sterlet (*A. ruthenus*) once thrived in the Danube and its tributaries. Presently, only three anadromous species occur in the Danube: *H. huso, A. gueldenstaedtii* and *A. stellatus*. Atlantic sturgeon nowadays occurs only along the Eastern coast of the Black Sea, in the area adjacent to the Inguri and Rhioni Rivers in Georgia (Bacalbasa-Dobrovici and Holcik

2000). Specimens of ship sturgeon are rarely recorded, with the latest finding at 1,401 rkm of Danube (Simonovic et al. 2005), while *A. ruthenus* is still present in the Danube River basin.

Inhabitants of the Greek colonies in the area noted the catch of sturgeon in the Lower Danube River, in 5th and 6th century B.C. By the beginning of the 16th century sturgeon catch decreased in the Middle Danube River, and in the 18th century fishing of migratory sturgeons in the Austrian stretch of the Danube River was completely abandoned. In 19th century, over-fishing and river modification became main reasons for decrease in sturgeon population, with water pollution becoming one of the most important factors in 20th century (Nikcevic et al. 2004). Sturgeon resources in the Danube River have been over-exploited since the early 1900s. Since ancient time, navigation through Djerdap sector was not safe during low water levels, due to underwater rocks that emerged from the water. Modifications of this part of river flow lasted since 1890 to 1896, and even then it became partial barrier for sturgeon upstream migrations (Petrovic 1998). But the most drastic decline of capture followed the construction of the dam (Patriche et al. 1999; Lenhardt et al. 2004a).

Huge anthropogenic influence, expressed through river bottom modification, over-fishing, dam construction and pollution, lead to decrease of fish catch in Lower Danube countries. The sturgeon catches have decreased dramatically in Romania, from 1,144 t in 1940 to less than 8 t in 1995, while Ukraine catch decreased from 114.2 t in 1952 to lack of catch record since 1994 (Navodaru et al. 1999). Romania was the second world major exporter of caviar, after the former Soviet Union (Patriche et al. 1999). According to data on total sturgeon catch in the Bulgarian part of the Danube River, for the period 1920–2002, it varied between 9.1 t and 72.4 t (Vassilev and Pehlivanov 2003). Yugoslavia catch declined from 38.5 t in 1975 to 5.2 t in 1986 (Stamenkovic 1991).

Based on catch statistics for Serbian part of the Danube River, during the period 1960–1997, predicted extinction of Russian sturgeon was estimated to fall around the middle of this century, and of beluga approximately at middle of this millennium (Lenhardt et al. 2006d).

Activities related to sturgeon protection

Principal reasons that triggered action for diminishment of pressure on sturgeon natural populations and shifted attention to aquaculture were significant decrease of sturgeon species catch, political changes that took place in Lower Danube countries during the last decade of 20th century, inclusion of all sturgeon species in CITES Annexes on April 1, 1998, as well as United States Department of the Interior, Fish and Wildlife Service declaration (U.S. Fish and Wildlife Service 2005) on trade suspensions for Black Sea basin beluga (*Huso huso*) caviar and meat.

CITES

Quotas for sturgeon species were established since 2001. Figure 1 shows quotas for beluga, Russian sturgeon and stellate sturgeon (total for Ukraine, Romania, Bulgaria and Serbia) for 2001–2006 period, with visible fall of quotas for all three species. In 2006 Romania declared ban on sturgeon fisheries for the following ten years. In 2007, there were no prescribed quotas for sturgeon fisheries in Black Sea basin.

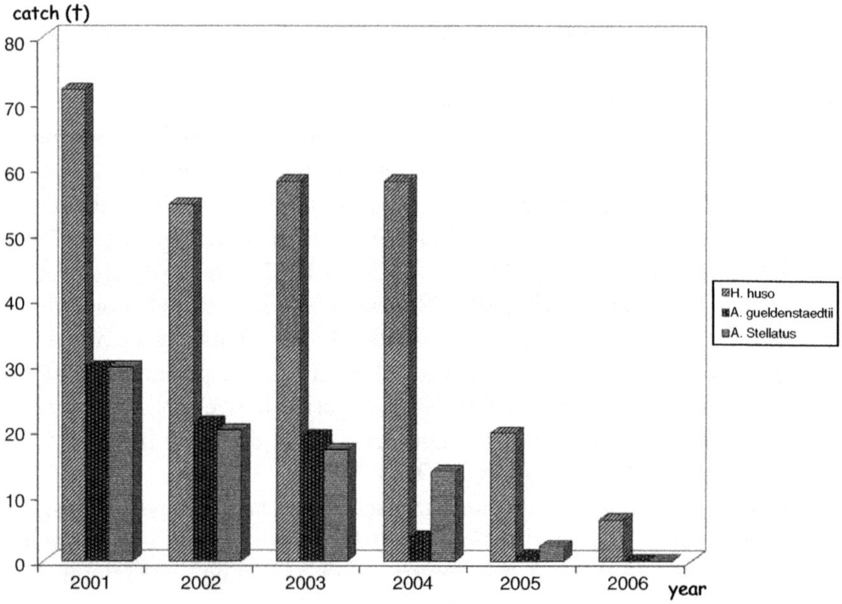

Figure 1. CITES catch quotas for beluga, Russian sturgeon and stellate sturgeon of the N-W Black Sea and the Lower Danube River (total quota for Ukraine, Romania, Bulgaria and Serbia)

CITES, in cooperation with member countries governments, faciltitates organization of meetings for both scientists and governmental representatives. One of the meetings resulted in establishment of Black Sea Sturgeon Management Action Group (BSSMAG) – consultative body of sturgeon range countries. It was formed in October 2001, during the First Regional CITES Meeting on Sustainable Management of Black Sea Sturgeons, held in Sofia, Bulgaria.

"Regional Strategy for the Conservation and Sustainable Management of Sturgeon Populations of the N-W Black Sea and Lower Danube River in accordance with CITES" (http://www.iucn.org/themes/ssc/sgs/sturgeon/activities-romania2.html) was established in accordance with Resolution Conf. 12.7. Representatives of the Fisheries and CITES Management Authorities of N-W Black Sea and Lower Danube River countries (Bulgaria, Romania, Serbia and Ukraine) met in Tulcea, 24–27 November 2003 and agreed on the Regional Strategy. The strategic goals are: improvement of knowledge on biological features of the sturgeon species, development and application of standardized assessment of all existing sturgeon species, protection of important habitats, insight in genetic structure and proper methods of artificial spawning for restocking and reintroduction, monitoring of fishery load compared to the potentials of sturgeon populations, development of stock assessment system for different sturgeon species, improvement of national legislation and ensuring enforcement mechanisms, adaptive management of sturgeon species, development of sturgeon aquaculture, development of cooperation and promotional programs for sturgeon protection and inclusion of economic and social components.

Scientific research

Since drastic decreases of stocks were observed in the Danube River, there is increasing demand for specific knowledge about biology and ecology of the various sturgeon species. It is surprising that there is still a significant lack of knowledge regarding species identification, migration behavior and natural reproduction (Reinartz et al. 2003). At the end of 20th and beginning of 21st century started international funding for sturgeon related projects and consequently publication of scientific papers (Finn et al. 2003; Kolarevic et al. 2004; Lenhardt et al. 2005a). International projects, related to location of essential sturgeon habitats in the Danube River and their genetic structure, were so far conducted in Romania. Research was also focused on specific species, such as stellate sturgeon (Ceapa et al. 2002a, b; Vecsei et al. 2007), beluga (Vecsei et al. 2002; Vassilev 2003), sterlet (Lahnsteiner et al. 2004; Lenhardt et al. 2004b).

The impact of industrial, agricultural and domestic wastewater discharge on sterlet populations was also investigated (Lenhardt et al. 2004c; Stanic et al. 2004). Development of Action plans was also initiated (Lenhardt et al. 2005b).

Aquaculture

Private enterprises in Romania, Bulgaria and Serbia began development of sturgeon hatcheries and rearing facilities. Quotas for export of sturgeon species produced in aquaculture have increased in last two years. Nevertheless, aquaculture development can also have negative effects. Hatcheries have long been proposed as a means to sustain and restore sturgeon population, but beside the risk of inbreeding, there is also a lack of adequate understanding of artificial rearing influence on behavoiur after their release (Secor et al. 2002).

However, the new trend of emerging private sturgeon hatcheries may add more problems than provide solutions, especially with respect to the intentional, incidental or accidental release or escape of exotic species, such as the accidental escape of North-American paddlefish (*Polyodon spathula*) from ponds into the Danube River (Lenhardt et al. 2006a, b; Simonovic et al. 2006).

NGO – International Association for Danube Research

The International Association for Danube Research (IAD), as a non-governmental organization that celebrated 50 years since foundation, had significant impact on Danube sturgeon species protection. As a result of its activities, IAD published a book that deals with biology and state of sturgeon species and their protection in the Danube River (Reinartz 2002). Also, former IAD president dr Jürg Bloesch initiated development of the "Action Plan for the Conservation of the Sturgeons (Acipenseridae) in the Danube River Basin" (Action plan 2006). Part of the IAD 2006 Congress was dedicated to sturgeon species in the Danube River (Guti 2006; Lenhardt et al. 2006c; Paraschiv and Suciu 2006; Bloesch 2006).

Internet presentations

In order to enable better availability of information related to sturgeon protection issues, a number of Internet presentations was developed: "Sturgeons of Romania and CITES" (http://www.indd.tim.ro/ rosturgeons/ index1E.htm) and "Sturgeons in Serbia" (http://www.sturgeons.info). However, there is a constant need for continuus input of new data.

Activities that could resolve problems

There is an emerging atmosphere for cooperation of all interested parties. On the other hand, there is still a lack of nongovernmental sector involvement and public awareness. Cross-sectoral cooperation is still undeveloped, especially regarding problem of fish passes which requires involvement of electrical industry and hydro-engineers.

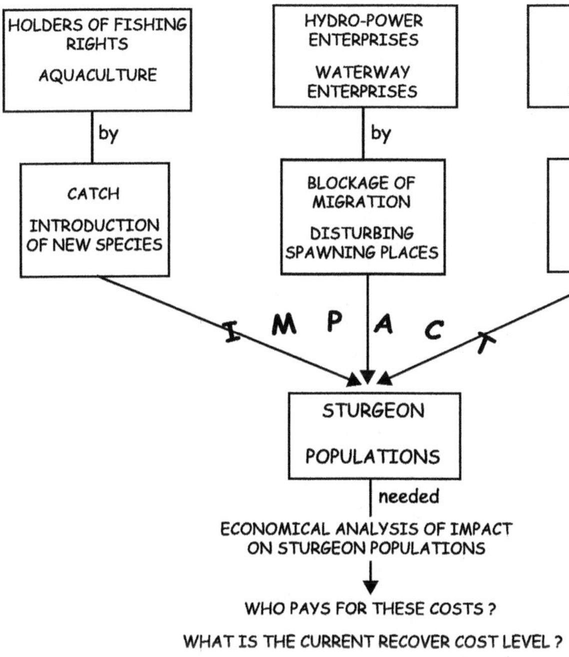

Figure 2. Flow chart presenting who makes impact on wild sturgeon populations

Mutual work of all interested parties would be best to organize through projects which would, beside biological component, also include socio-economic component. Application for project funding is currently available at the GEF (related to biodiversity), or for FP7 projects that refer to fishery, aquaculture or Water Framework Directive (WFD).

In every Lower Danube Region country it is also possible to work in accordance with ideas from WFD. One of the ideas in the WFD is that there should be full recovery of costs for water services. The water users should pay what the use of water costs (Berge and Dahl-Hansen 2005). In that sense, all who impact sturgeon populations (Figure 2) could pay for sturgeon recovery in Danube.

References

Action plan, 2006, Action plan for the conservation of sturgeons (Acipenseridae) in the Danube River basin, *Nat Environ* **144**, Council of Europe.

Bacalbasa-Dobrovici, N., 1997, Endangered migratory sturgeons of the lower Danube River and its delta, *Environ Biol Fish* **48**:201–207.

Bacalbasa-Dobrovici, N., and Holcik, J., 2000, Distribution of *Acipenser sturio* L., 1758 in the Black Sea and its watershad, *Bol Inst Esp Oceanogr* **16**(14):37–41.

Bemis, W. E., and Findeis, E. K., 1994, The sturgeon's plight, *Nature* **370**:602.

Berge, D., and Dahl-Hansen, G., 2005, Management of freshwater fisheries on bordering rivers; EU-regulations in the field of water and fish management, NIVA report SNO 5001-2005, ISBN 82-577-4699-1.

Bloesch, J., 2006, The ultimate need for implementation of sturgeon protection in the Danube River – a view of 2006 and call for actions with the sturgeon Action Plan under Bern Convention, Proceedings of the 36th International Conference of IAD, Austrian Committee Danube Research/IAD, Vienna, pp. 132–136.

Ceapa, C., Williot, P., and Bacalbasa-Dobrovici, N., 2002a, Present state and perspective of stellate sturgeon brood fish in the Romanian part of the Danube, *Int Rev Hydrobiol* **87**:507–513.

Ceapa, C., Williot, P., Le Menn, F., and Davail-Cuisset, B., 2002b, Plasma sex steroids and vitellogenin levels in stellate sturgeon (*Acipenser stellatus* Pallas) during spawning migration in the Danube River, *J Appl Ichthyol* **18**:391–396.

Ciereszko, A., Dabrowski, K., and Ochkur, S. I., 1996, Characetrization of acrosin-like activity of lake sturgeon (Acipenser fulvescens) spermatozoa, *Mol Reprod Dev* **45**:72–77.

Finn, R. N., Lenhardt, M., Ferrara, A., Isely, J. J., Dean, J., and Fyhn, H. J., 2003, Yolk proteins and free amino acids in Acipenseriformes, Proceedings of the 26th Annual Larval Fish Conference, Bergen, 22–26 July 2002, Norway.

Guti, G., 2006, Past and present status of sturgeons in Hungary, Proceedings of the 36th International Conference of IAD, Austrian Committee Danube Research/IAD, Vienna, pp. 143–147.

Hung, S. S. O., and Deng, D. F., 2002, Sturgeon, Acipenser spp. In: *Nutrient requirement and feeding of finfish for aquaculture*, C. D. Webster, and C. Lim, eds., CAB International, New York, NY, pp. 344–357.

Kolarevic, J., Lenhardt, M., Cakic, P., and Jaric, I., 2004, Population study of the Danube sterlet (*Acipenser ruthenus* L.). In: *Otterstad, releasing development potentials at the Eastern Adriatic*, O. Oddmund, ed., NTNU; http://www.easternadriatic.com.

Lahnsteiner, F., Berger, B., Horvath, A., and Urbanyi, B., 2004, Studies on the semen biology and sperm cryopreservation in the sterlet, *Acipenser ruthenus* L., *Aquac Res* **35**:519–528.

Lenhardt, M., Cakic, P., and Kolarevic, J., 2004a, Influence of the HEPS Djerdap I and Djerdap II dam construction on catch of economically important fish species in the Danube River, *Ecohydrol Hydrobiol* **4**:499–502.

Lenhardt, M., Cakic, P., Kolarevic, J., Mickovic, B., and Nikcevic, M., 2004b, Changes in sterlet (*Acipenser ruthenus* L.) catch and length frequency distribution in the Serbian part of the Danube River during the twentieth century, *Ecohydro Hydrobiol* **4**:193–197.

Lenhardt, M., Kolarevic, J., Jaric, I., Cvijanovic, G., Mickovic, B., Gacic, Z., Cakic, P., and Nikcevic, M., 2004c, Assessment concepts for river ecosystems characterization based on sterlet (*Acipenser ruthenus* L.) population research, Proceedings of the Fifth International Symposium on Ecohydraulics "Aquatic habitats: analysis & restoration", Madrid, 12–17 September, pp. 153–156.

Lenhardt, M., Finn, R. N., Cakic, P., Kolarevic, J., Krpo-Cetkovic, J., Radovic, I., and Fyhn, H. J., 2005a, Analysis of the post-vitellogenic oocytes of three species of Danubian Acipenseridae, *Belg J Zool* **135**(2):205–207.

Lenhardt, M., Hegedis, A., and Jaric, I., 2005b, Action plan for sturgeon management in fishery waters of Republic Serbia, Institute for Biological research, Belgrade, pp. 1–21.

Lenhardt, M., Bloesch, J., Reinartz, R., Suciu, R., Ivanova, P., Guti, G., Pannonhalmi, M., and Tsekov, A., 2006a, The actual situation of the endangered sturgeons in the Danube River Basin, *Danube News* **13–14**:18–22.

Lenhardt, M., Hegediš, A., Mićković, B., Višnjić-Jeftić , Ž., Smederevac, M., Jarić, I., Cvijanović, G., and Gačić, Z., 2006b, First record of the North American paddlefish (*Polyodon spathula* Walbaum, 1792) in the Serbian part of the Danube River. *Arch Biol Sci* **58**(3):27P–28P.

Lenhardt, M, Jaric, I., Bojovic, D., Cvijanovic, G., and Gacic, Z., 2006c, Past and current status of sturgeon in the serbian part of the Danube River, Proceedings of the 36th International Conference of IAD, Austrian Committee Danube Research/IAD, Vienna, pp. 148–151.

Lenhardt, M., Jaric, I., Kalauzi, A., and Cvijanovic, G., 2006d, Assessment of extinction risk and reasons for decline in sturgeon, *Biodivers Conserv* **15**:1967–1976.

Moreau, R., and Dabrowski, K., 2000, Biosynthesis of ascorbic acid by extant actinopterygians, *J Fish Biol* **57**:733–745.

Navodaru, I., Staras, M., and Banks, R., 1999, Management of the sturgeon stocks of the Lower Danube River system. In: *The Delta's: State-of-art protection and management*, Conference Proceedings, Tulcea, Romania 26–31 July 1999, pp. 229–237.

Nikcevic, M., Lenhardt, M., Cakic, P., Mickovic, B., Kolarevic, J., and Jaric, I., 2004, Historical review and new initiatives for sturgeon fisheries, aquaculture and caviar production in Serbia and Montenegro. In: *Otterstad, releasing development potentials at the Eastern Adriatic*, O. Oddmund, ed., NTNU; http://www.easternadriatic.com.

Paraschiv, M., and Suciu, R., 2006, Present state of sturgeon stocks in the Lower Danube River, Romania, a review, Proceedings of the 36th International Conference of IAD, Austrian Committee Danube Research/IAD, Vienna, pp. 152–158.

Patriche, N., Pecheanu, C., and Billard, R., 1999, Sturgeon fisheries, caviar production in Romania and possible measures of rehabilitation in the Lower Danube, *Rivista Italiana di Acquacultura* **34**:83–91.

Petrovic, M., 1998, Ribarstvo [Fisheries], Zavod za udzbenike i nastavna sredstva, Belgrade [In Serbian].

Reinartz, R., 2002, Sturgeons in the Danube River, International Association for Danube Research, Vienna.

Reinartz, R., Bloesch, J., Ring, T., and Stein, H., 2003, Sturgeon are more than caviar: A plea for the revival of sturgeons in the Danube River (Literature review), *Arch Hydrobiol* **14**(*Suppl Large Rivers* 3–4):387–403.

Secor, D. H., Anders, P. J., Winkle, W. V., and Dixon, D. A., 2002, Can we study sturgeons to extinction? What we do and don't know about the conservation of North American sturgeons, American Fisheries Society Symposium 28, pp. 3–10.

Simonovic, P., Budakov, L. J., Maric, S., and Nikolic, V., 2005, Recent record on ship sturgeon *Acipenser nudiventris* in the middle Danube (Serbia), *Biologia* **60**(2):231–233.

Simonovic, P., Maric, S., and Nikolic, V., 2006, Occurence of paddlefish *Polyodon spathula* (Walbaum, 1792) in the serbian part of the lower River Danube, *Aquatic Invasions* **1**(3):183–185.

Stanic, B., Andric, N., Zoric, S., Grubor-Lajsic, G., and Kovacevic, R., 2004, Biochemical response of sterlet *Acipenser ruthenus* collected from the River Danube, Proceedings of the 35th International Conference of IAD, IAD, Vienna, pp. 281–287.

Stamenkovic, S., 1991, Evolution des captures de troits especes d'esturgeons *Huso huso*, *Acipenser gueldenstaedtii*, et *Acipenser stellatus* dans la partie Yougoslave du Danube. In: *Acipenser*, P. Williot, ed., Cemagref Publications, Bordeaux, pp. 179–183.

U.S. Fish and Wildlife Service, 2005, Endangered and threatened wildlife and plants; special rule to control the trade of threatened beluga sturgeon (*Huso huso*), *Federal Register* **70**(42):10493–10507.

Vassilev, M., 2003, Spawning sites of beluga sturgeon (*Huso huso* L.) located along the Bulgarian-Romanian Danube stretch, *Acta Zool Bulg* **55**(2):91–94.

Vassilev, M., and Pehlivanov, L., 2003, Structural Changes of Sturgeon Catches in the Bulgarian Danube Section, *Acta Zool Bulg* **55**:97–102.

Vecsei, P., Sucui, R., and Peterson, D., 2002, Threatened fishes of the world: *Huso huso* (Linnaeus, 1758) (Acipenseridae), *Environ Biol Fish* **65**:363–365.

Vecsei, P., Peterson, D., Suciu, R., and Artyukhin, E., 2007, Threatened fishes of the world, *Acipenser stellatus*, Pallas, 1771 (Acipenseridae), *Environ Biol Fish* **78**(3):211–212.

RESCUE EFFORTS TO SAVE STURGEONS IN AMERICA

MOHAMMED K. ZAIDI*

Idaho State University, College of Engineering, USA

SUSAN IRELAND

Fish and Wildlife Program, Kootenai Tribe of Idaho, Bonners Ferry, ID 83805, USA

Abstract The sturgeons – the primary source of commercial caviar, have experienced severe population declines worldwide because of overexploitation, habitat alteration and excessive take for international trade. Some sturgeon species are at serious risk of extinction. The rescue effort to save a declining species of sturgeon in the United States of America (USA) is aimed at using a hatchery to prevent extinction while effective habitat measures are identified and implemented. Recovery is contingent upon re-establishing natural recruitment, minimizing additional loss of genetic variability and successfully mitigating biological and habitat alterations that have harmed the population. A conservation aquaculture program was developed by the Kootenai Tribe of Idaho (KTOI) and has been operating since 1991 to prevent extinction while efforts are made to re-establish suitable habitat conditions to increase natural production. A recovery program 2005–2010 will be presented.

Keywords: Kootenai River, protection, rare species, sturgeon hatchery

Introduction

Sturgeon is a term for a genus of fish (Acipenser) and is considered living fossils, appearing first about 136 million years ago, some refer it to an aquatic dinosaur. They have retained many primitive characteristics that have been lost or modified in other modern-day fishes. They have bony plates instead of scales, and a reptile-like tail. The fish, grayish-white on the back and sides, can weigh over 1,000 lbs and reach over 12 feet in length. Their mouth is toothless and positioned under the snout for sucking small fishes and invertebrates from the river bottom. Sturgeon pose no

* Mohammed K. Zaidi, Idaho State University, College of Engineering, Pocatello, ID 83209-8060, USA; zaidmk@gmail.com

threat to man. They have a long life span and some can live over 100 years but mature and reproduce slowly (Artyukhin et al. 1999; Artyukhin and Romanov 2000). The size and age of first sexual maturity is variable for white sturgeon. The youngest age at sexual maturity was estimated at age 22 for females and age 16 for males in the Kootenai sturgeon population (Paragamian and Wakkinen 2002). Although female white sturgeon have been reported to spawn every 2 to 11 years (Paragamian et al. 1997), empirical evidence suggests that female Kootenai sturgeon exhibited spawning periodicities of 4 or 5 years (Conte et al. 1988).

Sturgeons in the USA

American sturgeon populations are managed under federal, state and tribal jurisdictions, as well as interstate commissions. Acipenseriformes are a primitive group of approximately 27 species of fish, whose biological attributes make them vulnerable to intensive fishing pressure or other agents of elevated adult mortality. Although females produce large quantities of eggs, juvenile mortality is high; sturgeons are generally long-lived and slow to mature (reaching sexual maturity at 6–25 years); and depend upon large rivers to spawn. Sturgeons are fished for meat and caviar, with caviar being the most valuable product and in highest demand in international trade. Many species of sturgeons, the primary source of commercial caviar, have experienced severe population declines worldwide because of over-exploitation through historic fisheries, dam construction, blocking or inundating spawning and nursery habitat caused both habitat destruction and excessive take for international trade. Some sturgeon species are at serious risk of extinction (Smith 1986). In USA, the six different sturgeons listed as threatened or endangered are listed in Table 1.

Table 1. Different sturgeons found in the states of the USA

Name of the species		Listing status	Current distribution
Scientific	Common		
Acipenser brevirostrum	Shortnose	E	CT, DE, FL, GA, MA, MD, ME, NC, NJ, NY, RI, SC, VA
Acipenser oxyrinchus desotoi	Gulf	T	AL, FL, LA, MS
Acipenser transmontanus	White	E	ID, MT
Huso huso	Beluga	T	NA
Scaphirhynchus albus	Pallid	E	AR, IA, IL, KS, KY, LA, MO, MS, MT, ND, NE, SD, TN
Scaphirhynchus suttkusi	Alabama	E	AL, MS

USFWS Threatened and Endangered Species System (TESS). Report 02/14/2007

Sturgeons in Idaho: a case study of the Kootenai River white sturgeon population

The white sturgeon population in the Kootenai River was listed as endangered in 1994 by the US Fish and Wildlife Service (USFWS) (USFWS 1999). The Kootenai river basin covers 45,584 km^2 in watershed area and contributes the second largest runoff volume of all tributaries to the Columbia River (PWI 2000). The headwaters originate from Kootenay National Park, British Columbia (BC), Canada. The river flows south into Montana, USA and then turns west and flows into Idaho and turns north just west of Bonners Ferry returning to BC. The Kootenai River flows north into Kootenay Lake (Figure 1).

Over the past decades, native fish and wildlife populations have declined significantly due to large-scale habitat and ecosystem changes. The Kootenai River White Sturgeon Study and Conservation Aquaculture Project was initiated by the KTOI as a stopgap measure in 1989 to produce fish from wild Kootenai River adults until effective habitat restoration measures could be identified and implemented. Only the long life span of the sturgeon has forestalled extinction to date. Natural recruitment has been absent or limited for decades and the current population of large old fish is steadily dwindling. Continued failure of natural recruitment means that the next generation of Kootenai white sturgeon will come almost entirely from the hatchery. The Tribe, in cooperation with many agencies and stakeholders, is implementing a native fish restoration program in the Lower Kootenai River for sturgeon using conservation aquaculture techniques with wild broodstock.

Restore Natural Recruitment, Kootenai River white sturgeon, is based on a holistic and elastic, ecosystem-based management approach to sturgeon recovery.

Without immediate intervention, Kootenai River white sturgeon, are at extreme risk of extinction. KTOI in coordination with agency partners, is pursuing a multifaceted approach to recovery that includes habitat restoration actions designed to address a variety of limiting habitat conditions and a conservation aquaculture program designed to capture adequate genetic diversity and ensure sufficient numbers of Kootenai River white sturgeon for recovery. Through this project, KTOI will develop a comprehensive master plan (strategic plan) to ensure efficient coordination of these diverse recovery efforts.

The master plan development will incorporate input from steering and technical committees, and provide opportunities for public education and input.

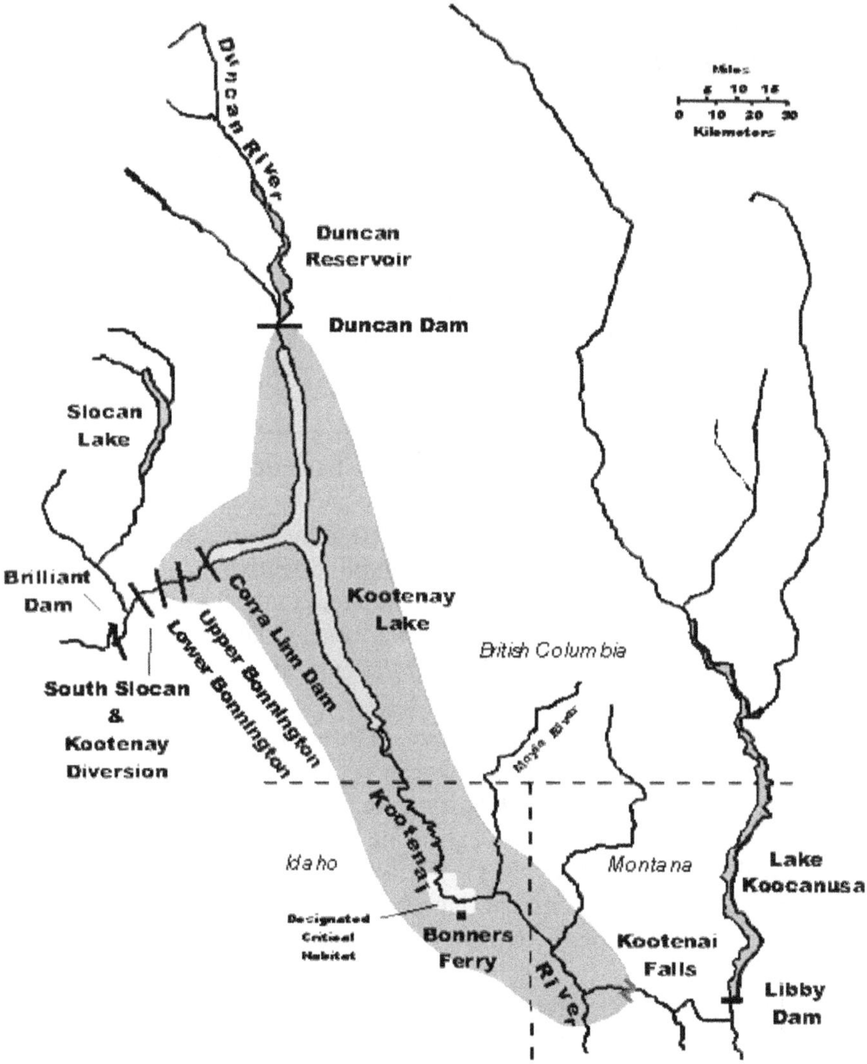

Figure 1. The Kootenai River basin in British Columbia, Montana, and Idaho

Kootenai sturgeon currently occupies the waters in the shaded area, although they are now very rare in Montana (Paragamian et al. 2005). Current critical habitat designated by the USFWS under the Endangered Species Act (ESA) listing is shown and includes known spawning areas. Critical habitat information is available on: http://www.fws.gov/policy/library/66fr20962.pdf.

The white sturgeon, the largest freshwater fish in North America, has not successfully reproduced in the Kootenai since before the completion of

Libby Dam in Montana in 1974. An estimated 500 wild sturgeon remain, a number expected to dwindle to 50 by 2030. Over the past decade, the hatchery has released over 80,000 juvenile sturgeons into the river. But since the fish do not reach sexual maturity until about age 30, the oldest of those hatchery-raised sturgeons are not expected to begin spawning until 2025 (Figure 2). Bred from captured wild sturgeon, the young sturgeon may represent the last hope of preventing the species' extinction.

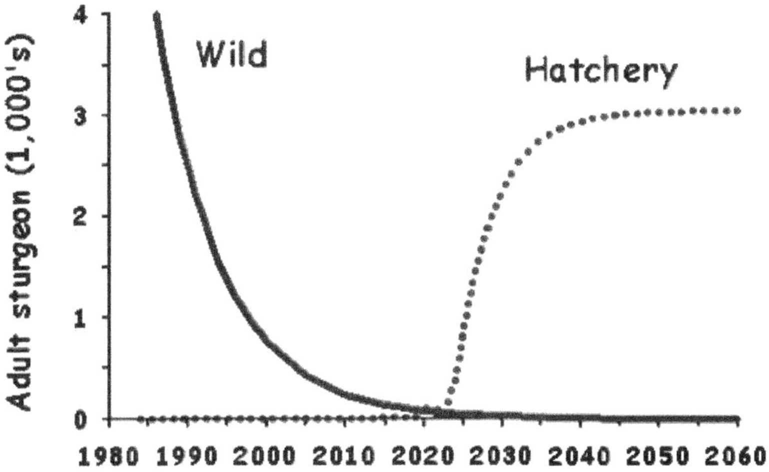

Figure 2. Projected trends in wild and hatchery-produced adult sturgeon in the Kootenai sturgeon population (Paragamian et al. 2000)

Experimental hatchery releases of age 1–4 juvenile sturgeon from 1992 through 2004 have included nearly 47,000 fish (Figure 3) (Paragamian et al. 2005). Subsequent recaptures of hatchery fish in an annual monitoring program indicate that significant numbers have survived introduction and grow well after an initial period of adjustment to the natural environment (KTOI 2005; Ireland et al. 2002). Survival rates of hatchery-produced juveniles averaged about 60% during the first year at large, and about 90% during all subsequent years. Updated growth, condition, and survival analyses are ongoing to track the effects of hatchery releases.

In a study of spawning behavior of the Kootenai white sturgeon performed by Idaho Department of Fish and Game, it was estimated that the number of white sturgeon spawning events ranged from 9 to 20, with spawning days ranging from 17 to 31 days. Average daily temperature during spawning ranged from 7.5°C to 14°C, with the highest probability of spawning (48%) at 9.5–9.9°C. Average daily flow for spawning events ranged from 141 to 1,265 m^3 s^{-1}, but most (63%) spawning took place

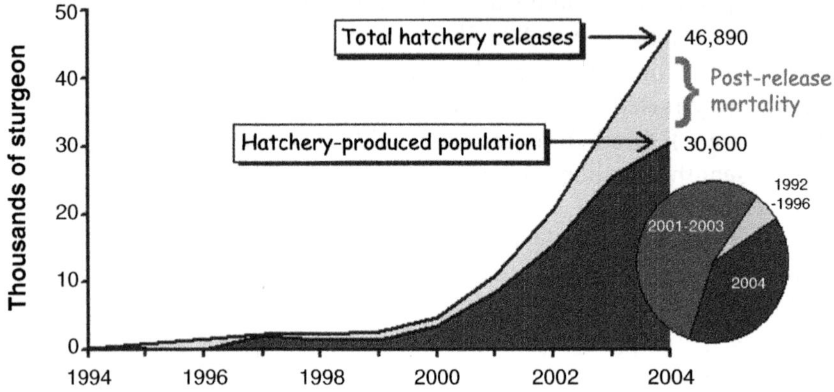

Figure 3. Estimated population surviving from hatchery-reared Kootenai sturgeon released into the Kootenai River from 1992 through 2004. The pie chart identifies the contributions from various release periods to the 2004 population (KTOI 2005)

above 630 m^3 s^{-1}. Initial spawning white sturgeon during spring may be synchronized with the arrival of females from downstream staging reaches. After the onset of spawning, the temporal distribution of spawning events appears to be dependent on the shape and stability of flow and temperature. It was observed that a water temperature decrease of ≥0.8°C could disrupt sturgeon spawning (DFG 2006).

Production

Annual broodstock numbers are established based on recovery goals, hatchery capacity, and the availability of ripe fish for spawning. Annual numbers have included 3 to 6 females and 8 to 14 males since 1999 when the KTOI hatchery was upgraded and the BC facility began to be used (Paragamian et al. 1997). A total of 148 wild broodstock have been spawned from 1990 through 2005, producing 109 full or half-sibling families. Almost four million eggs have been harvested from 51 of 77 mature females brought to the hatchery and spawned successfully (Figure 4).

Through 2003, the goal of the captive breeding program was to use 3–9 females and an equal number of males to produce 4–12 families per year. This goal was based on a plan designed to approximate a normal expanding natural population and to avoid exaggerated genetic contributions of a small fraction of the parent population from the hatchery to the natural

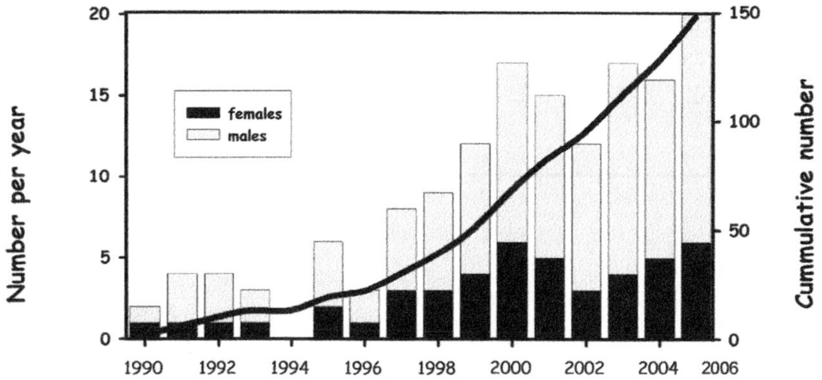

Figure 4. Broodstock numbers used in the Kootenai sturgeon hatchery program

population (Kincaid 1993). The Kincaid Plan (1993) also included very conservative rearing requirements for balanced family contributions at release which constrained the number of broodstock that could be used in the available hatchery space. Initial production goals assumed that significant natural production would be restored during the current sturgeon generation.

Current plans call for using up to 12 females per year where hatchery capacity allows. Increased numbers of males are also sought to mate with each female. This change accompanied the realization that significant natural recruitment had not been restored and the next sturgeon generation would be produced entirely in the hatchery. With this determination, the nature of the genetic risk changed from unbalanced contributions of hatchery and natural spawners to a founder bottleneck effect caused by too few spawners to preserve the natural diversity of the native wild population. Increased broodstock numbers were an attempt to propagate more of the existing genetic diversity and to front load hatchery production as a hedge for uncertainty in future broodstock availability. It will become increasing difficult to obtain ripe females and males as the wild population continues to decline but hatchery-produced fish are not yet mature.

Numbers released

A total of 86,523 endangered Kootenai white sturgeon juveniles have been reared and released from hatcheries from 1992 through 2005 (Figure 5). Release numbers are determined by recovery goals, hatchery capacity, and survival rates in the hatchery. Hatchery releases prior to 1997 were largely experimental. Significant releases began in 1997 after the hatchery became a critical part of the recovery plan. Full production was reached in 2003–2005 after the benefits of the Phase I hatchery upgrades were realized. Annual releases have averaged 22,000 fish per year from 2003–2005. Annual releases typically include fish from multiple brood years because fish from any given brood year require different intervals to reach tagable sizes.

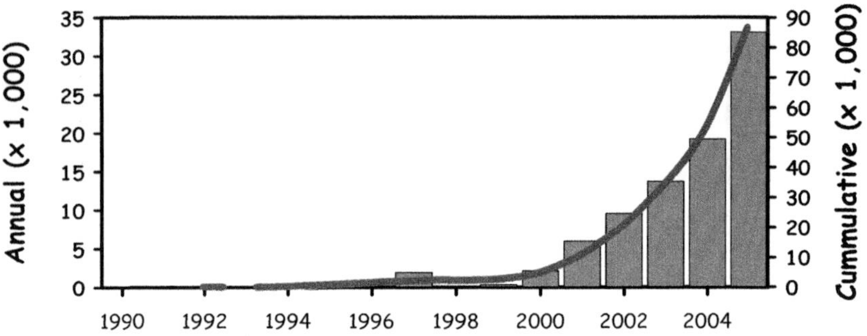

Figure 5. Annual (bars) and cumulative (line) numbers of juvenile white sturgeon released into the Kootenai River and Kootenai Lake

Through 2003, annual release goals were 1,000 to 1,500 sturgeon (age 1+ or 2) per family for up to 12 families. Total numbers and number per family were limited according to the initial captive breeding plan to avoid genetic swamping of natural production with large numbers of hatchery fish contributed by a few parents (Ireland et al. 2002). Target release levels were projected to produce an effective hatchery-produced population in the next generation of 200 adults or about 20% of the estimated 1990 population size (four breeding pairs per family at age 20 based on assumed annual survival rates). As with original broodstock target numbers, initial release goals were based on an assumption of significant natural production which has not been met.

Production limitations were reduced in 2004. Current plans call for maximizing release numbers and family sizes within the constraints of the existing hatchery facilities as a precaution for continued natural recruitment failure. Up to 10,000 fish per family may be released at 10–15 g as age 0+ in fall rather than 30 g at age 1+ or 2. Previous production levels were constrained by the need to raise all fish to sizes suitable for Passive Integrated Transponder (PIT) tag placement and to rear families separately so that family sizes could be equalized upon release. More recent evaluations have concluded that low population size in the next generation is a much more acute demographic and genetic risk than unequal family contributions in the following generation. The benefits of individual marking for evaluation purposes are also exceeded by the need to avoid a population bottleneck. The hatcheries have the capacity to raise greater number of sturgeons.

Survival

Recapture of hatchery-reared fish in the monitoring program confirm excellent survival in the wild (Ireland et al. 2002). Survival is estimated at 60% during the first year as hatchery sturgeons adapt to the wild environment and 90% per year thereafter based on analysis of mark-recapture data. Observed survival rates are substantially greater than initial program expectations of 50% in the first year and 60–70% in subsequent years (Ireland et al. 2002). Effects of size and season of release on survival are unknown but will be evaluated in the future to guide hatchery release practices.

Five year plan 2005–2010

During the 5 years covered by this Plan (2005–2010), a multi-faceted adaptive management approach will be applied to Kootenai River sturgeon habitat and recovery needs (Walters et al. 2005). This approach involves simultaneous implementation of multiple remedial actions to achieve the desired outcome in the shortest amount of time possible. In this case the desired outcome is reestablishment of natural recruitment, or on smaller experimental scales, in-river production of viable larvae. If desired outcomes occur, individual measures may be terminated to determine their specific effects on in-river larval production and recruitment. This approach differs from a common approach of sequentially implementing and evaluating single recovery treatments and adding new treatments if initial treatments fail to provide in-river larval production or restore recruitment.

This implementation plan provides detailed measures, tasks, and activities planned from 2005 through 2010 based on strategies identified in the Kootenai Subbasin Plan, the White Sturgeon Recovery Plan, and new information that has been developed since 1999 when the USFWS White Sturgeon Recovery Plan was adopted. The new Plan has four main components corresponding to the four primary strategies in the 1999 Recovery Plan: (1) Recruitment restoration, (2) Conservation aquaculture, (3) Monitoring and evaluation, and (4) Recovery Plan adaptation and revision (Figure 6).

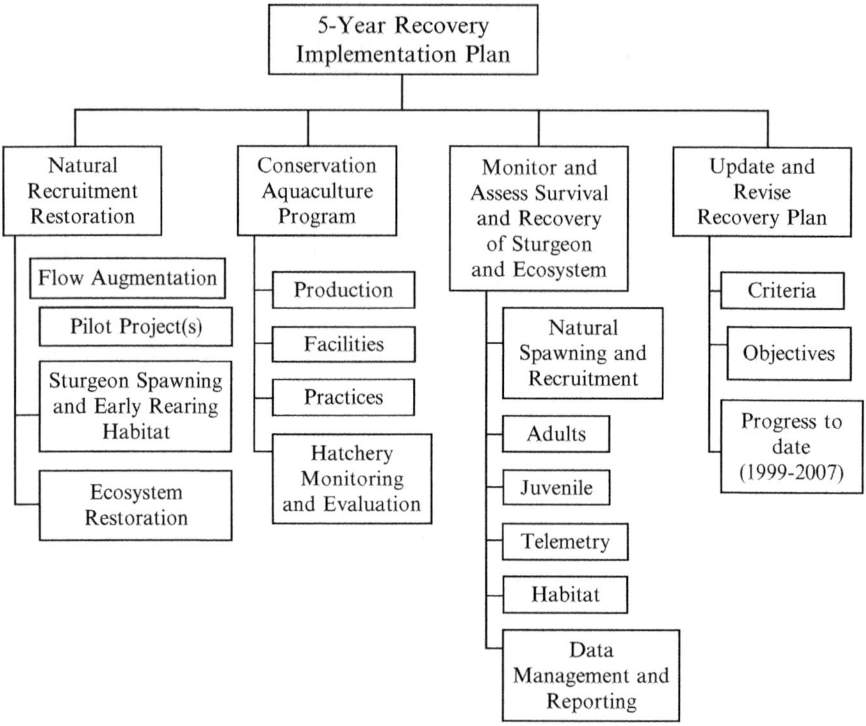

Figure 6. Five year recovery implementation plan 2005–2010

Sturgeons in the other states of USA

The Alabama sturgeon is a slender, golden-yellow, freshwater fish that was historically widespread in the Mobile River Basin of Alabama and Mississippi. It grows to about 30 inches in length and weighs 2–3 lbs. It was once so abundant it was caught and sold commercially. Biologists attribute the decline of the species to over-fishing, loss and fragmentation of its habitat due to navigation-related development, and decline in water quality. Scientific evidence supports the Alabama sturgeon as a distinct species. Both national scientific organizations, the American Society of Herpetologists and Ichthyologists and the American Fisheries Society, recognize the Alabama sturgeon as a separate species (USFWS 2005).

The Gulf sturgeon, also known as the Gulf of Mexico sturgeon, is a subspecies of the Atlantic sturgeon. It is an anadromous fish with a sub-cylindrical body imbedded with bony plates or scutes. The snout is greatly extended and bladelike with four fleshy chin barbels in front of the mouth which is protractile on the lower surface of the head. The upper lobe of the tail is longer than the lower lobe. Body color is light brown to dark brown and pale underneath. The species grows to a maximum length of about 8 feet and is over 200 lbs in weight (Anonymous 2002).

The USFWS stocked pallid sturgeon in the Missouri River below Fort Randall Dam near Running Water, South Dakota, this fish, averaging 12–13 inches in length, were spawned and reared at hatchery near Yankton, South Dakota. Gavins Point National Fish Hatchery is unique in that it is the only facility in the nation that maintains all of the future broodstock for the continuation of the recovery of the pallid sturgeon. The endangered pallid sturgeon is an ancient fish that can grow up to 6 feet long and weigh 85 lbs, with a lifespan of up to 100 years. Currently, it is found only in the Missouri River, the Mississippi River downstream of the Missouri River, the lower Yellowstone River, and Atchafalaya River. Current range-wide populations are estimated at 6,000–10,000 (CITES 1997).

USFWS announces the listing of all sturgeon and their products in the Convention on International Trade in Endangered Species of Wild Fauna and Flora (CITES). Now all sturgeon species and fishery product is covered by the Service's regulations regarding import or export of wildlife and also importation of caviar and other sturgeon products into the United States [Author: Dr. Rosemarie Gnam, Office of Management Authority, USFWS].

Lake sturgeon is the only sturgeon species endemic to the Great Lakes basin and is the largest freshwater fish indigenous to that system. Lake sturgeon can be considered a nearshore, warmwater species with water temperature and depth preferences of low 50s–mid-60° F and 15–30 feet, respectively. Lake sturgeon is benthivores, feeding on small invertebrates such as insect larvae, crayfish, snails, clams, and leeches.

Short nosed sturgeon (A. brevirostrum), Baltic sturgeon (A. sturio), Atlantic sturgeon (A. oxyrhynchus) and American paddlefish (Polydon spathula) were listed in CITES. Five of the newly listed species: Beluga, Russian, stellate, Siberian (A. baerii), and ship or spiny (A. nudiventris) sturgeons; were listed because of their population status and trade levels. All other species of sturgeons were listed because of the similarity of appearance of their caviar to that of the Caspian Sea species such as the white sturgeon (A. transmontanus) from North America. The end result is that all sturgeon species worldwide, are now covered under the provisions of CITES. The listing of sturgeon will provide a regulatory mechanism for the import and export of sturgeon and their products, thereby curtailing the illegal caviar trade and detriment to the wild populations, notably those of the Caspian Sea. It will ensure sustainable use and management of wild sturgeon stocks (Ireland et al. 2004).

Conclusion

Overfishing, poaching, and loss of virtually all spawning habitat to destruction by dams, pollution, and habitat modification have left a grim legacy for sturgeon populations around the world. Suspending trade in beluga caviar through an endangered species listing is just one positive step along the road to rebuilding beluga sturgeon. The conduct of regular and scientifically sound stock assessments, strengthening management and enforcement programs, and maintaining the operations of sturgeon hatcheries is very essential. In order to rescue the sturgeon, a critically endangered species, the joint efforts of international community are essential.

Acknowledgement

I am thankful to the advisory committee to award partial travel grant for my travel and stay.

References

Artyukhin, E.N., and Romanov, A.G., 2000. Old Problem of a New Fish Farm. *Fish Culture and Fishing* 1:14–15.

Artyukhin, E.N., Barannika, I.A., and Romanov, A.G., 1999. Russian Strategies for Conservation and Reproduction of Rare and Common Caspian Sturgeons. *Journal of Applied Ichthyology*, Blackwell Synergy, Berlin, Volume 15, Issue 4–5, pp. 191–192.

Anonymous, 2002. Caspian Ecology Teeters on the Brink. *Science* 295:430–434.

CITES, 1997. Listing of all Sturgeon and Paddlefish Species and Their Products in the Appendices to the Convention on International Trade in Endangered Species of Wild Fauna and Flora (CITES) and its Implementation by the United States, Federal Register, 62/234, Proposed Rules, 64347–64348.

Conte, F.C., Doroshov, S.I., Lutes, P.B., and Strange, E.M., 1988. Hatchery manual for the white sturgeon *Acipenser transmontanus* Richardson, with application to other North American Acipenseridae, Cooperative Extension, University of California, Division of Agriculture and Natural Resources, Publication 3322.

DFG, 2006. DFG Sturgeon Meeting report (October 16, 2006).

Ireland, S.C., Siple, J.T., Beamesderfer, R.C.P., Paragamian, V.L., and Wakkinen, V.D., 2002a. Success of hatchery-reared juvenile white sturgeon (Acipenser transmontanus) following release in the Kootenai River, Idaho. *Journal of Applied Ichthyology* 18:642–650.

Ireland, S.C., Anders, P.J., and Siple, J.T., 2002b. Conservation aquaculture: an adaptive approach to prevent extinction of an endangered white sturgeon population (*Acipenser transmontanus*), in: W. VanWinkle, P. Anders, D. Dixon, and D. Secor, eds., *Biology, Management and Protection of North American Sturgeons*, American Fisheries Society Symposium 28, pp. 211–222.

Ireland, S.C., Anders, P.J., and Beamesderfer, R.C.P., eds., 2004. *An Adaptive Multidisciplinary Conservation Aquaculture Plan for Endangered Kootenai River White Sturgeon.* Hatchery Management Plan prepared by the Kootenai Tribe of Idaho with assistance from S.P. Cramer and Associates, KTOI (Kootenai Tribe of Idaho), 56 pp.

Kincaid, H., 1993. Breeding Plan to Preserve the Genetic Variability of the Kootenai River White Sturgeon, U.S. Fish and Wildlife Service Report (Project 93-27) to Bonneville Power Administration, Portland, OR.

PWI, 2000. Kootenai River Tributaries, Water Quality Summary, 1998–2000, Power Water Inc., Boise, ID.

KTOI (Kootenai Tribe of Idaho), 2005. Kootenai River White Sturgeon Conservation Aquaculture Program, 1990–2005, Bonners Ferry, Idaho, Report prepared by S.P. Cramer and Associates, R. Beamesderfer and P. Anders, 75 pp.

Paragamian, V.I., Kruse, G., and Wakkinen, V., 1997. Kootenai River White Sturgeon Investigations. Annual Progress Report FY, Prepared for U.S. Department of Energy, Bonneville Power Administration, Contract No. DE-AI79-88BP93497; Project No. 88-65, Portland, OR.

Paragamian, V.I., Kruse, G., and Wakkinen, V., 2000. Kootenai River White Sturgeon Investigations, Annual Progress Report FY 1996, Prepared for U.S. Department of Energy, Bonneville Power Administration, Contract No. DE-AI79-88BP93497; Project No. 88-65, Portland, OR.

Paragamian, V.L., and Wakkinen, V.D., 2002. Temporal Distribution of Kootenai River White Sturgeon Spawning Events-Effect of Flow and Temperature. *Journal of Applied Ichthyology,* 18(4–6), 542.

Paragamian, V.L., Beamesderfer, R.C.P., and Ireland, S.C., 2005. Status, Population Dynamics, and Future Prospects of the Endangered Kootenai River White Sturgeon Population with and Without Hatchery Intervention. *Transactions of the American Fisheries Society* 134:518–532.

Smith, T.I.J., 1986. Culture of North America Sturgeons for Fishery Enhancement, 15th Joint meeting on Aquaculture, Koyto, Japan.

USFWS, 1999. US Fish and Wildlife Service, White Sturgeons: Kootenai river population Acipenser transmontanus, Recovery Plan, US Fish and Wild Life Services, Region 1, Portland, OR.

USFWS, 2005. US Fish and Wildlife Service release (September 2, 2005).

Walters, C., Korman, J., Anders, P., Holderman, C., and Ireland, S., 2005. Draft Kootenai River Adaptive Management Plan, Report prepared for the Kootenai Tribe of Idaho, 11 pp.

RIVER REHABILITATION: A NEW APPROACH TO THE DESIGN OF FISH PASSES THROUGH DAMS

VLADIMIR LAGUTOV*

DonEco Research and Consulting, Russia

VIKTOR LAGUTOV

Department of Environmental Sciences, Policy and Management
Central European University, Hungary

Abstract Ecosystems of most European rivers were significantly altered by human activity throughout the 20th century with the aim of facilitating economic development and settlement protection. The ecosystem needs and environmental flow requirements were neglected, which resulted in river ecosystem degradation and biodiversity loss. As a rule, river rehabilitation programs are costly and complicated processes. While aiming at overall river ecosystem rehabilitation, recently initiated restoration programs should take into account regional peculiarities and specificities. One of the relatively cheap and easy technologies to deal with some aspects of river rehabilitation and securing environmental flow needs is suggested. It is based on restoration of the migration routes for the most sensitive and vulnerable river basin species: migratory fish, such as sturgeons. The existing fish passing facilities in South Russia are inefficient and require urgent replacement. The proposed technologies for fish passage through the dams and barrier complexes are based on the "non-forced" principle, unlike the predominant 20th century approach to fish transfer.

Keywords: Sturgeon, beluga, fish pass, dam, Azov, Don, Kuban', fishlock, regulators of variable perforation

Rivers degradation

Rivers and river valleys have always been the cradles of human civilizations. Rivers provided not only food and freshwater, but also shelter and means of transportation. Societies traditionally learned to adapt to changing river hydrological regimes; however, the growing needs of human societies accompanied by growing abilities have caused significant river alterations

* Vladimir Lagutov, DonEco Research and Consulting, Dachnaya 1, 2, Novocherkassk, Russia; Lagutov@sturgeons.ru

and ecosystem changes. Up until the end of the 20th century environmental needs were sacrificed for the sake of economic development, resulting in river contamination, biodiversity losses and general riverine ecosystem degradation. Disappearance of the Aral Sea is a perfect example of inadequate water resource management strategies in inflowing river basins.

Large scale river impoundments occurred in the second half of the 20th century worldwide and had, probably, the highest negative impact on the river ecosystems. The cascades of large dams and weirs were erected on every large European river (Figure 1) resulting in substantial modifications of the environmental conditions. The impact of dams on the river ecosystem is complex and versatile (Craig 2000; Larinier 2000; Marmulla et al. 2001; McAllister et al. 2000; Nislow et al. 2002). It affects various aspects of functioning of the ecosystem: biodiversity loss, habitat fragmentation, change in hydrological regime, decrease in self-purification service, siltation, eutrophication, pollution accumulation and many others.

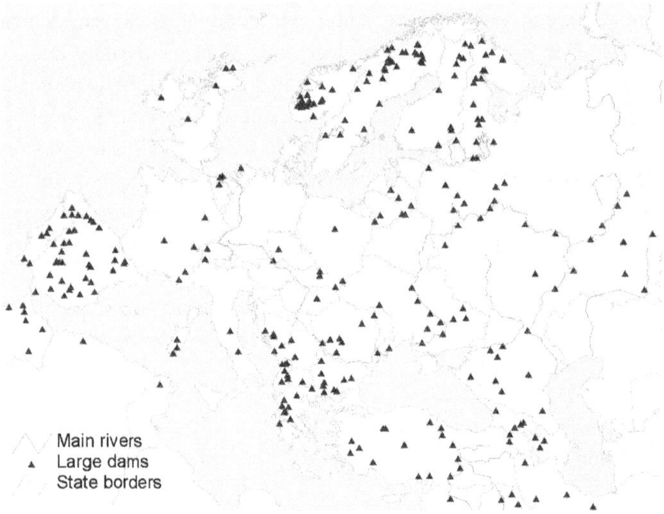

Figure 1. The large dams on the European rivers. Most of them were constructed in the second half of the 20th century

Thus, to help with river rehabilitation programs, an integrated comprehensive environmental indicator is needed to evaluate the status of the condition of river ecosystem health. The anadromous sturgeon species migrating through the whole river basin can play such an indicator role[1] (Russian State Duma 1995; Lagutov 1995, 1996). Historically, sturgeon

[1] See the discussion on sturgeon as an indicator in this volume.

species inhabited almost all rivers of the Northern hemisphere and, if river rehabilitation programs are successful, should ideally return to all these habitats.

At the same time dams and wires (accompanied by commercial overfishing) are the main reason for sturgeon extinction in the European rivers. Constructed in the lower river streams, the dams break the fish migration routes, cut off the spawning grounds or destroy them by submerging. For instance, 100% of spawning grounds for the Beluga (*Huso Huso*), the most valuable sturgeon species, were lost in the Volga and Don rivers by construction of the high pressure Volgograd and Tsimliansk dams. Deprived of access to the spawning grounds and thus their ability to reproduce, the long-lived sturgeon stock was further depleted by constantly increasing fishing efforts over the course of several decades.

Existing approaches

Induced rather by high economic value of sturgeon fishery than by environmental concerns, numerous programs on sturgeon stock replenishment have been launched throughout the region. Though the main efforts were focused on stock replenishment through hatchery-reared sturgeon fingerlings, some attention was paid to providing the migrating species with the means of overcoming the barriers to get access to their spawning grounds.

Generally speaking, the best approach for the rehabilitation of damaged river ecosystems in general and restoration of migration routes, in particular, is to remove constructed structures (Cowx and Welcomme 1998). This would not only eliminate habitat fragmentation and restore migration routes, but also might resume the functioning of spawning grounds submerged by impoundment. Unfortunately, removal of these structures is often not a practical option in environmental management. It is often impossible partly due to the fact that these structures are essential parts of the regional economy (navigation, agriculture, electricity generation, etc.). Another factor preventing removal of dams is the high cost involved and the large scale territory recovery works (silt/sludge removal, etc.) needed as a consequence of such an action.

The problem of migratory fish transfer through dams is a relatively well-studied and discussed subject. However, in many cases the primary target for these efforts are salmonids. This migratory species inhabiting small northern rivers is a very strong swimmer and can easily overcome various fish pass constructions. The sturgeon species, possessing much lower swimming abilities and historically inhabiting bigger rivers, are practically extinct in the European rivers. Not being a subject for commercial fishery, sturgeon was not considered as a species to take care of during construction of the dams outside of the former Soviet Union. The problem of restoration of sturgeon habitat connectivity (i.e. restoration of migration routes) was not actively discussed in the literature and no practical projects were carried out. Only recently have some attempts been made to conduct feasibility studies on sturgeon reintroduction in the big European rivers (WWF 2002). The Danube River can be mentioned as an example of such a situation. The Iron Gates I and Iron Gates II – high pressure dams, constructed on the Danube in the 1970s – were not equipped with any kind of fish passages, which resulted in cutting off the sturgeon spawning grounds and elimination of even the chance for restoration of the (by that time already overharvested) Danube sturgeon population.

A similar trend for river ecosystem degradation can be observed on the territory of the former Soviet Union. The Sea of Azov and Caspian Sea contained almost the whole world sturgeon stock, with 90% located in the Caspian Sea (CEP 2002). According to some estimations the Sea of Azov was the most productive sea in the world (AzovBas 2002; Russian State Duma 1995). The drastic decline in sturgeon harvest occurred in a decade (one sturgeon generation) after the establishment of the barrier complexes at the main Azov sea tributaries, the rivers Don and Kuban' (Figure 2).

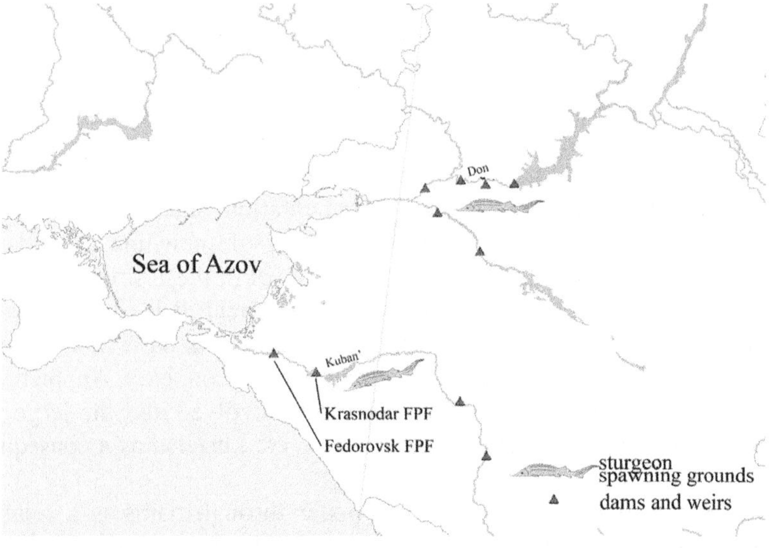

Figure 2. The tributaries of the Sea of Azov with main dams and sturgeon spawning places

Though environmental concerns did not have any priority in water management strategies, fish passages of various designs were incorporated into the initial hydroscheme design of many dams in the former Soviet

Union. In particular, even the Volgograd dam, one of the biggest in Europe, had migratory species transfer incorporated into the dam design. Significant capital investments were directed to the research, design and implementation of fish protection (in particular, passing) facilities.

Unfortunately, the efficiency of most of these devices was extremely low (Lagutov 1995), while the Soviet Ministry of Fishery was interested in concealing the inefficiency of the created complexes and the investments involved.

For instance, at the time of the Soviet Union's collapse spawning in the river Kuban' upstream dam complex had not been observed at least for a decade due to lack of producers, while the fish pass constructions on these dams were officially reported as working effectively and transferring hundreds of thousands of sturgeon migrants to the spawning grounds (Russian State Duma 1995; Lagutov 1995). The situation became known to the public only in the 1990s after several notorious lawsuits on fish mafia involving high-level fishery managers and bureaucrats.

It should also be mentioned that during all these years the sturgeon hatcheries in the Azov sea operated in the region to replenish the stock. As a result, the sturgeon in the basin of Sea of Azov should be considered as an extinct species except for the freshwater *sterlet*, which stays in the Tsimliansk reservoir its whole life cycle.

The same situation can be observed in other rivers of the former Soviet Union. The only large river free of dams and weir barriers from the point of view of sturgeon migration is the Ural river.

If sturgeon species are to be returned to the European rivers, the dams on all other rivers require installing (or re-equipping with effective) fish pass constructions to secure natural sturgeon reproduction, and to enable it to be used as an indicator of river ecosystem health.

Design of fish passing facilities

There are currently two main recognized approaches to the design of fish pass facilities (FPF). The first technology is based on independent ("unforced") natural fish passage through the barriers, while the second one suggests using various compulsive ("forced") techniques to transport fish through the dams.

In this way all possible varieties of FPFs securing fish transfers during upstream migration can be divided into two broad categories: natural *fishways* and force-based *fishlifts/elevators/locks*.

There are different types of fishways (pool, weir, Denil, ladders, bypass channels, etc.). They are usually applicable for the strong swimmers, i.e. salmonids. In the case of South Russia the number of migratory and semi-migratory species, requiring transfer through the barriers, was high, and each species possessed different swimming abilities. Because of this and other reasons (e.g. abundance of high pressure dams), natural fishways in Russia and the former Soviet Union are not widely spread on the rivers that are not inhabited by salmons. Instead, fish lifts/elevators and fish locks are mostly used.

In relation to the sturgeon species transfer in former Soviet countries, there is not a single effectively working device based on the second "forced" approach known to the authors in Russia. The fish passage facility installed at the Nizhne-Tulomsk Hydroscheme is considered to be the most effective FPF built in the former Soviet Union.

The designers of the fish pass facilities have to solve three main problems:

- To help fish find the entrance to the FPF. The number of fish entering a FPF is proportional to the ratio between the amount of water released through the FPF and transit volume through the whole hydroscheme.
- To provide hydrological conditions for fish transfer through fishways or to drag it with a fishlift.
- To secure fish exit to the head (upper) pool and prevent them from washing back down to the lower pool.

Each of the existing fish pass designs has its own shortcomings with regards to the indicated tasks. For instance, there are a number of well-recognized problems with fish-ladders and fish passages, i.e.:

- Inability of migrants to find the entrance to the fish passage because of insufficient attracting water discharges through fishways. Usually the water discharge through a fishway is a few percent only.
- Difficulties with fish transfer through the fishway itself because of inadequate hydraulic conditions along it (i.e. high stream velocity).
- Silting up of the fishpassage.
- Fish disorientation in turbulent currents.
- Fish inability to enter the upper water from FPF for the whole possible range of water level drop at the head pool and inappropriate stream velocities.
- Migrants washing down from the head pool with main water discharge.

Cyclic fish lock and *fish lifts, typical* "forced"-based technologies, which have most of the mentioned shortcomings, are traditionally widely used in South Russian rivers. Though having higher efficiency, the later introduced *flat lock* did not substantially increase the efficiency of sturgeon transfer since the hydraulic conditions required by sturgeon cannot be created with this FPF design. The fish locks have interrupted water discharges, which causes interruptions in attracting flow discharge and a lower percentage of fish approaching the FPF. The main idea behind this type of FPF is to capture the fish with a metallic close-meshed net/screen and mechanically deliver it to the head pool by means of some elevators/transporters. Depending on operation regime and FPF design, the operation cycle can take several hours. Mechanically dragged over hundreds of meters the fish experiences significant stress, i.e. it might be compressed, exposed to the open air, or injured by moving metallic and concrete surfaces. As a result, when the fish is released to the head pool it is unable to swim actively and is often washed back down to the lower pool. This type of FPF is an extensive construction, consuming a lot of energy and working in a cyclic manner. FPF functioning also requires an operator.

In addition to that, most FPF designs suggested for sturgeon transfer are characterized by extensive size as well as high construction and maintenance costs. For instance, a cyclic FPF consists of a number of elements (upper lock, lower lock, incentive screen dragging fish upstream, etc.) mounted within a long (150–200 m) flume made of armored concrete (Figure 3).

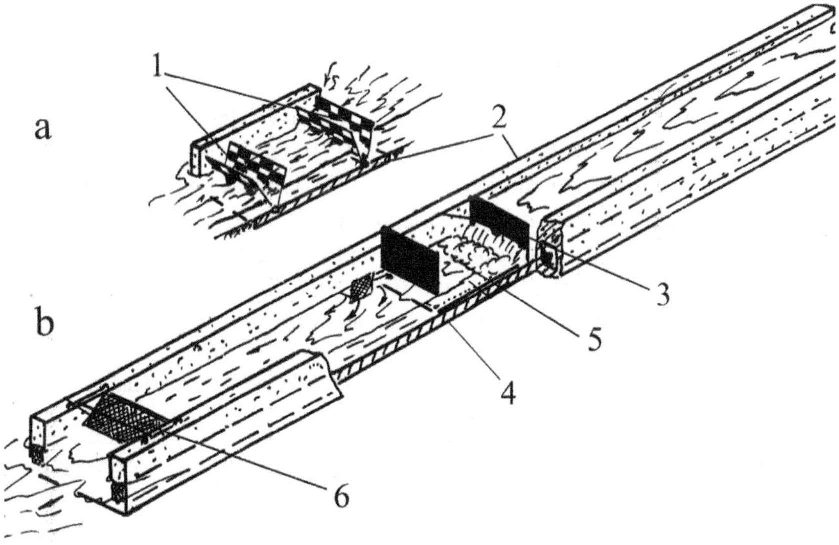

Figure 3. The design of FPF based on the radial regulators of variable perforation (RRVP) (a) and traditional cyclic fishlock (b) 1 – RRVPs; 2 – canal made of armored concrete; 3 – upper lock; 4 – incentive screen; 5 – "ichthyological platform" for fish counting; 6 – screen for fish transportation to "ichthyological platform"

Another typical shortcoming of fish transfer technologies arises from the process of counting of migrants. To provide such a possibility fishlocks are usually equipped with so-called "ichthyologic platform", which drags the captured fish to the open air where operators can count it manually. If it is not collected by operators, the stressed fish is disposed to the head pool from where it can again be easily washed down to the lower water.

Taking into account all these considerations, calculation of FPF efficiency is a very challenging task. According to the prevailing approach in the USSR and the one currently applied in Russia, the efficiency is calculated as the number of fish transported to the head pool. The washing down of the fish and its recurring attempts to go through the FPF after recovery downstream the dam were never taken into account. Thus, counting the same fish over and over again overestimates not only the efficiency of the FPF, but also the size of the stock (in particular sturgeon), resulting in higher fishing quotas and faster stock depletion.

Currently there are some modern techniques of counting of migrants. However, these have still not been introduced in Russian FPFs even now, let alone the 1970s–1980s. In this situation, probably the only way to evaluate an FPF's efficiency is to compare the number of sturgeon migrants approaching upstream FPF with the number of transported migrants by downstream FPF. A convenient situation occurs at the rivers Don and Kuban', emptying into the Sea of Azov. A cascade of closely located dams equipped with fishlocks can be found on these rivers, and the fishlock on each dam has an "ichthyological platform" for precise fish counting. Using sturgeon transfer statistics the efficiency of FPFs was calculated. The highest efficiency for the fishlocks at the Don river dams was about 3.4% and 3.8% at the river Kuban' (Lagutov 2005). Table 1 shows an excerpt from the statistics used for the Fedorovsk FPF (river Kuban'). It should be noted that the distance between the selected FPFs is short; both of them are located on the main river stream and no spawning grounds can be found within the distance between these two FPFs. In addition, the selected years were characterized by a high level of state control in fishery and public

Table 1. The number of migrants transported by the fishlocks at the river Kuban' and estimation of fishpassing efficiency

Years	Number of sturgeon migrants transported at Fedorovsk FPF	Number of sturgeon migrants transported at Krasnodar FPF	Efficiency (%)
1983	798	24	3.0
1984	1,015	61	6.1
1985	605	20	3.5
1986	1,092	43	4.0
1987	2,139	47	2.3

life, so the effect of poaching can be ignored, as well. Thus, due to these and other considerations, the sturgeon losses can be mainly assigned to the efficiency of the operating FPFs.

Taking into account the fact that in order to reach the spawning grounds sturgeon has to pass several FPFs, the drastic depletion of sturgeon stocks in the Azov Sea can be easily explained even without the impact from commercial overfishing. In fact, by the late 1980s no sturgeon was spotted at the upper FPFs during the years with low and middle water availability when all the dams were fully operational during the sturgeon spawning migration[2] (Lagutov 1995). As a result, the total sturgeon catch in the Azov fishing zone in 1995 was only 1.5 t in comparison with thousands of tons a decade ago. Since 2000 no official sturgeon catch was registered. Starting from 2006 the fishery statistics were completely deleted from the Rostov region Annual Statistical Report.[3]

It should be mentioned that all of the FPFs considered ineffective are located on the low-pressure dams where water level drops up to 3 m only. The Don and Kuban' rivers run through the steppes of South Russia, where the creation of high-pressure dams is a complicated task. Moreover, the fish passage to the upper reservoir on the high pressure dams (more than 10 m of water level drop) is usually a pointless task since migrants heading upstream to the spawning grounds, according to stream currents, would not be able to find their way in the standing water of the reservoir.

The only high-pressure dam in the Azov basin, the Tsimliansk dam, is not even considered as a subject for fish passage.

In this way, conventional FPF designs installed at the dams in the former Soviet Union (South Russia rivers in particular) cannot be considered as effective tools for the restoration of sturgeon migration routes and river rehabilitation programs. On the contrary, by utilizing enormous amounts of money that could be spent otherwise on environmental protection, installation of these FPFs can be considered as an environmentally damaging action.

Proposed technologies

The improvement of existing fish pass facilities is an urgent task. Following this demand, an attempt to design a new FPF for low-pressure dams was carried out by the author. The main idea behind the research was to

---

[2] It should be noticed that during the high water years with large-scale flood typical for the steppe rivers in the South Russia the main gates at the dams were open and successful sturgeon spawning was recorded.

[3] Rostov region is the Russian Oblast having the biggest share of the Azov sea shore and fishery facilities.

create such hydrological conditions in the stream that migrants (in particular, sturgeon species) could independently find the entrance to the FPF and pass it without any external incentives or enforcement.

As a result, the new approach to fish transfer through the barriers having these features was elaborated. It is based on the usage of the surfaces with variable perforation to the FPF design. Using coarse-perforated surfaces a new class of hydraulic locks, named regulators of variable perforation, was designed. The new technology, based on multijet stream energy suppression using course perforated surfaces, lays a foundation for the development of a new class of hydrological constructions. Ten inventor certificates were patented in fish protection using the developed technology.

One of them, a fishway based on *radial regulators of variable perforation* (RRVP) was created and thoroughly tested in hydrological and ichthyological laboratories, as well as field experiments, showing good working characteristics. The results obtained confirmed the device's ability to maintain the given water level drop between lower and upper water pools and, at the same time, secure the hydraulic conditions needed for fish passage.

This technology provides secure, inexpensive and highly effective fish passage through low pressure dams with a water level drop of a few meters.

There are muplitple environmental and economic benefits of using these devices, including:

- Application of multijet approach results in creation of natural hydrological characteristics in water streams, which is very important for the migratory fish species.
- Increase of open cross-sectional area from 3–5% on existing to 40–50% on suggested fish passages allows the transit flow rate through the device to be increased tenfold. As a result, not only attracting water flow can be increased tenfold, but also fish species of different types, various sizes and swimming abilities obtain equal chances to pass a dam simultaneously in both directions.
- Fish can overpass a dam independently in a non-forced continuous way without being stressed with consecutive washing downstream. The passage time is limited to 10–15 min compared to the usual several hours.
- The supported water drop between head and lower pools in case of using multijet RRVP-based FPF can be three times higher than based upon a traditional single hole.
- Usage of new technology can drastically cut down the investments needed for installation, maintenance and reconstruction of hydraulic facilities. The size and building time of hydraulic constructions can also be significantly decreased. In particular, in comparison to other fishways, the length of the fishpassage canal can be decreased as much as ten times.

- Migrating fish can be easily counted, identified and measured using intact methods.
- An RRVP-based fish passage can be easily mounted on old dams in regulated water streams to assist in restoration of affected river ecosystems, to avoid fragmentation of aquatic biodiversity and to secure reproduction of migratory species.

Apart from the extensive laboratory experiments and tests, several feasibility studies for the installation/re-equipping of the existing FPFs with new regulators were conducted for the dams on the rivers Don, Kuban' and Terek. Detailed comparison with existing FPFs was carried out which revealed that the proposed technology has much better environmental and economic characteristics than all other existing FPF types. Figure 3 presents two FPF designs based on regulators of variable perforation (a) and cyclic fishlock (b).

The crucial feature for every fish passage is the fish's ability to overcome the stream velocity while moving upstream through the device. In case of RRVP the range of stream velocities is quite wide depending on the path location (close to the canal bed or to the water surface) which secures safe passage for many different migratory and semi-migratory fish types. With regards to sturgeon species the field experiments in the rivers Don and Kuban' proved that though the conventional recommended speed for designing sturgeon passes is 1 m/s, the sturgeon can develop and hold for short periods (up to 2 h) a speed up to 2 m/s. (Umanetc 1977). In fact, this research simply confirms the basic ichthyological concept of fish's ability to achieve high acceleration in a short period of time. The time needed for various fish types to pass RRVP is a split second, which was confirmed by laboratory experiments. Thus, even weak swimmers can easily overpass the construction. The design of fish protection constructions using a multijet approach gives even bigger freedom and opportunities for strong swimmers, like salmonids.

Figure 4 shows the passage of the RRVP by fish during the laboratory experiments. Using video materials shot during these experiments the speed developed by some fish types was estimated as up to 4 m/s.

As noted earlier, the installation of RRVP-based fish passes does not require either significant capital investments, or a long construction period. For instance, a new fish passage through Kargalinsk Hydroscheme on the river Terek (inflowing the Caspian Sea) could be constructed by means of a pontoon dock in only one working season. Moreover, re-equipping of the existing dams with newly designed FPFs does not require changes to be made in the dam design and could be done using pre-created canals and fish passages.

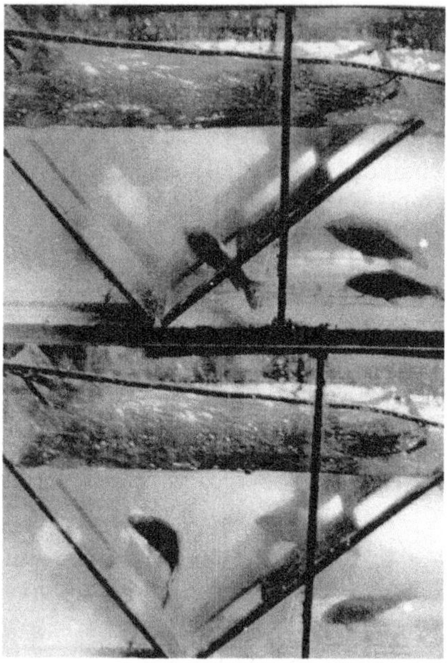

Figure 4. The fish passing RRVP during the laboratory experiments

Unfortunately, the design and feasibility studies of the project on re-equipping the South Russian rivers inhabited by sturgeons with this kind of FPFs was finalized shortly before the collapse of the Soviet Union. This and the following periods in the former Soviet Union and Russia, in particular, are characterized by economic collapse and low priority of environmental concerns in water management strategies. As a result, the programs on FPFs improvement were delayed and are yet to recommence.

Conclusion

The existing fish passing facilities on South Russian rivers have extremely low efficiency in terms of sturgeon species to say the least. It might be stated that these constructions have contributed significantly towards sturgeon stock depletion in the Azov and Caspian seas. The proposed fish pass technology secures cheap and effective fish transfer through low-pressure dams and weirs with water level drops of a few meters. In particular, it allows for safe sturgeon passage and guarantees the restoration of its migratory routes. On this basis a holistic program of integrated water resource management in a watershed and river rehabilitation strategies can be gradually built up. Urgent measures on improving existing FPFs by utilizing regulators of variable perforation can not only restore natural sturgeon reproduction and save the sturgeon gene pool, but also maintain the natural hydrological conditions and ecosystem services in deteriorating river streams. Until the time when natural sturgeon reproduction on the rivers of Azov and Caspian Sea basins is assured, the Ural river, the only non-regulated free-flowing river with a viable self-sustaining sturgeon population, should be internationally protected to preserve the sturgeon gene pool.

References

AzovBas. 2002. Proceedings. Azov Sea Basin Workshop. Novocherkassk, Russia, Novocherkassk: Green Don.

CEP. 2002. *Transboundary Diagnostic Analysis for the Caspian Sea.* Caspian Environment Programme, Baku, Azherbaijan.

Cowx, G.I. and Welcomme, R.L. (eds). 1998. *Rehabilitation of Rivers for Fish.* A study undertaken. by the European Inland Fisheries Advisory Commission of FAO. Oxford: Blackwell Science.

Craig, J. 2000. Large Dams and Freshwater Fish Biodiversity. [on-line] Dams, ecosystem functions and environmental restoration. World Dam Commission. http://www.dams.org/docs/kbase/contrib/env246.pdf [cited 17 April 2004].

Russian State Duma. 1995. Committee on Ecology, Russian State Duma. Hearing on "Improvement of fishery policies in the Azov Sea Basin". Report by Dr. V. Lagutov.

Lagutov, V. 1995. *Mekhanizm unichtozheniya riubniuh zapasov Yuga Rossii i puti ih spaseniya* [The mechanism of extermination of fish stocks in South Russia and possibilities of their restoration]. Novocherkassk: Donskaia Rech.

Lagutov, V. 1996. Imitacionnoe modelirovanie zhiznennogo tsikla osetroviuh v basseine reki Don. [Simulation of Sturgeon Life Cycle in the Don River Basin]. MSc thesis, Department of Applied Mathematics, South Russia State Technical University, Novocherkassk.

Lagutov, V. 2005. Dams and Fish. Novocherkassk: Uralbas. [in Russian].

Larinier, M. 2000. *Dams and Fish Migration.* Toulouse, Institut de Mecanique des Fluides.

Marmulla, G., Larinier, M., et al. 2001. *Dams, fish and fisheries: opportunities, challenges and conflict resolution.* FAO Fisheries Technical Paper 419, Fishery Resources Division. FAO Fisheries Department.

McAllister, D., Craig, J., et al. 2000. Biodiversity Impacts of Large Dams. [on-line] World Dam Commission. http://www.dams.org/docs/kbase/contrib/env244.pdf [cited 17 April 2004].

Nislow, K.H., Magilligan, F.J., et al. 2002. Effects of Dam Impoundment on the Flood Regime of Natural Floodplain Communities in the Upper Connecticut River. *Journal of the American Water Resources Association* 38(6): 1533–1548.

Umanetc, I.K. 1977. *Opiut ekspluatacii riubopropusknogo shliuza Kochetovskogo girdouzla na reke Don.* Novocherkassk: Novocherkassk Amelioration Academy.

WWF. 2002. *Sturgeon 2020. Initiative for Providing Habitats for Sturgeons.* Program despcription and schedule, Budapest, WWF.

PART II
THE URAL RIVER BASIN

2.1 WATER RESOURCES

THE URAL RIVER BASIN: HYDROLOGY, CHARACTERISTICS AND WATER USE

VIKTOR LAGUTOV *

Department of Environmental Sciences, Policy and Management
Central European University, Hungary

Abstract The Ural river is a unique ecosystem with a mostly undisturbed hydrological regime and is crucial for the preservation of the Caspian sturgeon species. This is the third longest river in Europe and the last large river in Europe unaffected by river regulation or damming. Nevertheless, this fact is not well known to the broader scientific and environmental communities (UNEP 2002). The present paper gives an introduction to the Ural river's hydrology, climate, land use types and other basin characteristics. An analysis of the river flow statistics, human activities' influence and regional climate change is undertaken. The administrative watershed division and the problems of transboundary water management are also discussed.

Keywords: Ural, sturgeon, beluga, river basin, watershed, Caspian Sea, Orenburg, Ilek, Sakmara, Iriklinskoe

The Ural river and its role in the Caspian Sea

The Ural River, the third longest river in Europe, forms the traditional boundary between Europe and Asia. According to different sources the total river length varies from 2,428 to 2,534 km. It rises in the South-eastern slopes of the Ural Mountains at 640 m above sea level and runs through Russia and Kazakhstan into the Caspian Sea. The length of the Kazakhstan portion of the main Ural stream is 1,084 km. The total catchment area is about 220,000 km^2.

The Ural river plays a special role in the Caspian basin (AzovBas 2002; Lagutov 1995), being the only free-flowing river in the region with a non-regulated hydrological regime in its lower and middle flow.

* Viktor Lagutov: Department of Environmental Sciences and Policy, Central European University, Nador ut.9, H1051, Budapest, Hungary; lagutov@gmail.com

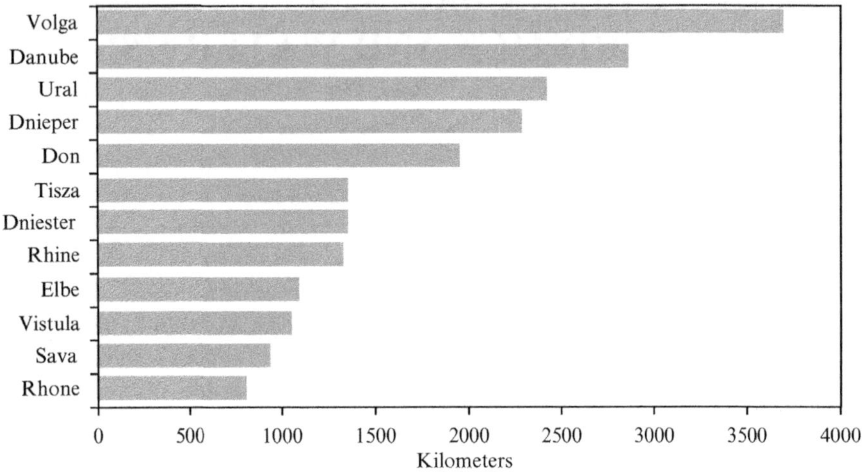

Figure 1. Comparative length of European rivers

While it is the second longest river in the basin after the Volga river (Figure 1), total freshwater influx delivered by the Ural is only 3% against 80% by the Volga and a total of 8.8% by the rivers Kura (6.3%) and Terek (2.5%). In absolute values the mean total flow in the Ural is 9–10 km^3, while the Volga has 260 km^3. In other words, the Ural's flow is 25–30 times smaller than the Volga's. However, the riverine system productivity in terms of fishery is as high as that of the Volga: 11,000 t.[1] Moreover, sometimes the yield from the Ural has been even higher (up to 15,000 t) (KaspNIRH 1999) (Figure 2).

The sturgeon spawning grounds in the Ural River are much more efficient and the Ural sturgeon population was much more productive prior to its current depletion. This is a very interesting phenomenon which has still not been paid due attention in the literature and in Caspian environmental programs.

The greatest attention has been paid to the Volga River, as the biggest Caspian tributary and the habitat for the greatest number of species in the region. For instance, currently according to the research conducted by Caspian Fishery Research Institute the contribution of the Volga ecosystem to the sturgeon stock in the Caspian Sea is 69.8%, the Ural's 29.7%, while

[1] These evaluations are based on the maximum registered catch occurred in 1970s of last century in the Ural river. After this level of overfishing sturgeon population has never recovered.

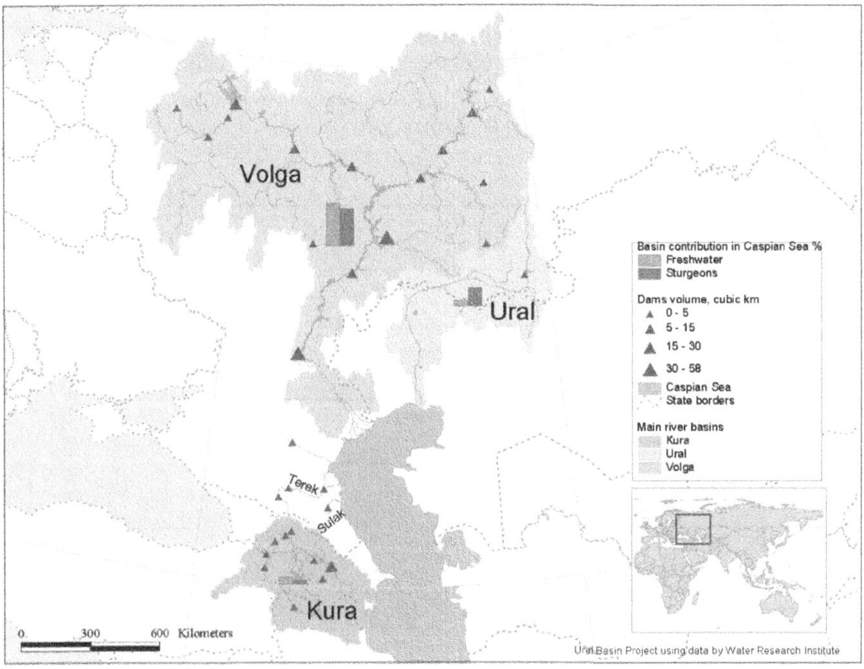

Figure 2. The main basins of the Caspian basin Sea and their contribution to the sea freshwater influx and sturgeon population

Kura and Terek together contribute only 1.4% (KaspNIRH 1999). Nevertheless, these estimates should be treated with due reservations and care. They are based on hypothetical numbers of survived fingerlings released by the hatcheries in the Volga delta in the framework of the hatchery-based restocking program, but the efficiency of this program as well as the importance of these fingerlings to the fishery is challenged by many researchers. Thus it is possible that the contemporary role of the Volga river in sustaining the Caspian sturgeon population has been overestimated.[2] Historically, the proportion of the Volga sturgeon catch was much higher than the Ural's. However, after the Basin Rivers' impoundment the historical spawning grounds in the Volga and other Caspian rivers were cut off. The Volga river lost 100% of the spawning grounds for the Beluga, while the decrease in spawning grounds for all sturgeon species was 85%.[3] The spawning in the Volga river occurs in the areas under the Volgograd Dam characterized by low efficiency (Lagutov 1995, 1999). Moreover, in a 40 year period the lower Volga regulation spawning grounds were flooded

[2] See paper on the Ural sturgeons in this volume.

[3] See paper on the Ural sturgeon in this volume.

only in 13 years (Dubinina and Kozlitina 2000). Figure 3 shows the areas of available sturgeon spawning grounds according to the Caspian Environment Program (CEP 2002). However, the situation in other basin rivers is even worse – no sturgeon spawning was observed in the Terek or Sulak since the mid-1980s (KaspNIRH 1999).

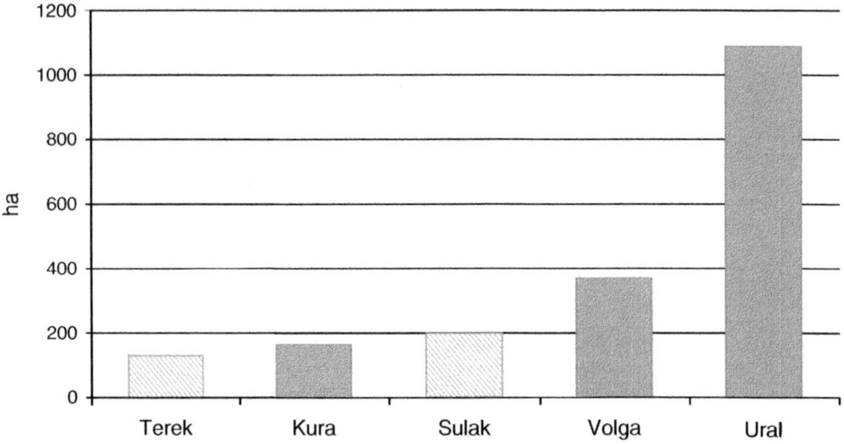

Figure 3. Sturgeon spawning grounds available in the rivers of the Caspian basin (UNEP 2006)

At the same time the Ural river is the only river with natural hydrological flow where all sturgeon habitats are intact. Successful sturgeon spawning was observed in historic highly productive spawning grounds in the Ural river even in 2007 (KamUralRybVod 2007). In this way, the observations suggest that the predominant role in natural sturgeon reproduction belongs to the Ural river. This makes the Ural river a unique ecosystem of high importance.

Despite its importance, it is not only poorly represented in the National and International Caspian Action Plans, but also in the scientific literature. For example, out of 6.75 million articles available at ScienceDirect, one of the biggest online collection of scientific publications, the "Ural river" is mentioned in only three papers (Elsevier 2008).

The river's morphology

The Ural river is a typical steppe river. The specific feature of the Ural river is the highly uneven distribution of the river flow through the year with 80% of the annual flow occurring during the spring flood. Another feature is the substantial fluctuations in total annual flow.

Three distinctive parts of the river can be distinguished: upper, middle and lower stream. The characteristics of the river hydrology, alimentation sources, river bed and other factors within these intervals are more or less homogeneous.

The *upper stream*, which runs along the eastern slope of the Ural Mountains for 750 km, has characteristics of a typical mountain river with a turbulent current. The average depth is 0.5–0.6 m with a maximum of 3 m. The elevation drop in the upper stream is 450 m for 750 km, 1.3 m for 1 km on average.

The biggest water reservoir Iriklinskoe is situated in the lower part of this river section (1,800 km from the delta) and does not have a substantial influence on the river's hydrological regime.

The *middle stream* of the Ural lasts for 850 km from east to west. The average elevation drop for 1 km drops to 10–20 cm, which results in a significant decrease of flow velocity. The river runs through a wide valley, containing a lot of meanders, lakes and wetlands. The width of the valley varies from several hundred meters to tens of kilometers. The hydrological characteristics in this river course are of a typical plain river. There are 65 rifts in this river stream. The maximum depth is 4–5 m.

The main tributaries of the Ural, Sakmara and Ilek, are located in this river section (Figure 4).

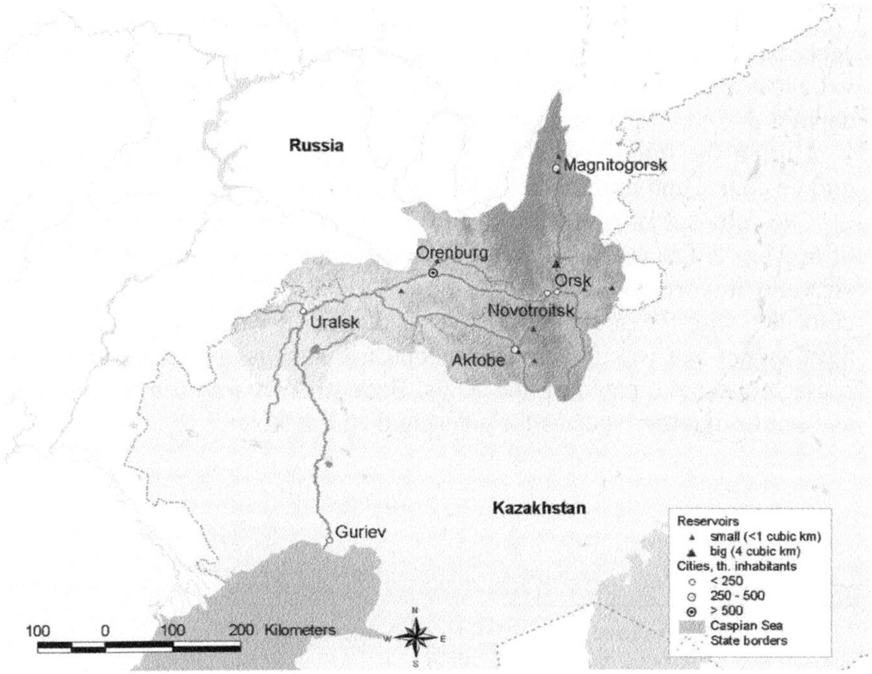

Figure 4. The Ural river basin

The length and watershed area of the Sakmara river are 760 km and 29.1 km^2 correspondingly. The Sakmara watershed, fully located in Russian territory in the northern part of the Ural basin, is covered with forests and characterized by high precipitation. A lot of snow is accumulated in this area.

The Sakmara river has a strong influence on the hydrology of the Ural. Being the tributary of the Ural and having a smaller watershed, the Sakmara river has a higher average water flow than the Ural. At the rivers' junction near Orenburg the annual water discharge in the Sakmara is 4.4 km^3, while in the Ural it is only 3.3 km^3 (Dmitriev and Vasilenko 2007).[4] The average water flow increases from 33 m^3/s near the city of Orsk to 110 m^3/s at Orenburg.

The Ilek tributary is shared by Russia and Kazakhstan. The watershed area is 41,300 km^2 and the river length is 700 km. Despite the bigger catchment area the quantity of water delivered by the Ilek is much smaller than that delivered by the Sakmara. The average discharge in this river segment is 40 m^3/s.

The lower part of the Ural river runs through the steppe and deserted steppe areas for about 500 km. No single regular tributary joins the Ural till the Ural mouth. Moreover, some rivers (e.g. Kushum) are taking some water away from the main stream to be lost in deserted areas. The average river discharge at the gauge station Kushum reaches 380 m^3/s.

In the lower Ural stream two river segments can be distinguished according to the river bed conditions. There are 98 rifts at this river interval, most of which are located in the upper part of the stream. The average depth is 3–5 m with a maximum of 7–20 m.

During the flood season the Ural can reach a width of 10 km in the middle course and more then 20 km in the delta.

The delta starts from the town of Guriev (Atyrau) and occupies an area of 600 km^2.[5] There are two main delta distributaries of more or less equal size and importance: the Zolotoj and Yaickij. The Zolotoj distributary coincides with the ancient Ural river bed, shaped when the water level in the Caspian Sea was lower. The second tributary is connected to the fish passage channel created in the 1960s. Both streams are used by migratory and semi-migratory species for entering the Ural River.

---

[4] Other sources indicate as annual water discharges for the Ural and Sakmara river 4.23 and 4.47 km^3 correspondingly (Uralbas 2007).

[5] According to other sources 700 km^2 (Fashchevsky 2003).

Climate conditions and river alimentation

The Ural River basin has an arid, sharply-continental climate. Most of its territory is covered with steppe, either cultivated and converted to crop-lands or wild. The precipitation level in this area is 350–500 mm/year, while the average evapotranspiration over water surfaces is up to 650–690 mm/year.

Figure 5 presents the Bailey ecoregions (after Gerasimov, 1964) on the territory of the Ural basin. A substantial part of the basin is covered by various types of steppe. The lower course of the river crosses semi-deserts and deserts, while the Northern part of the basin is covered with forests.

Most of the water flow is formed in the northern (Russian) part of the watershed in the upper and middle stream (Figure 6). Snowmelt is the main source for the total river flow, which depends on snow availability and the water content of snow cover in winter period. Though the basin shares of Kazakhstan and Russia are almost equal, 72% of the Ural's total runoff is formed in the Russian part of the basin (KamUralRybVod 2007).

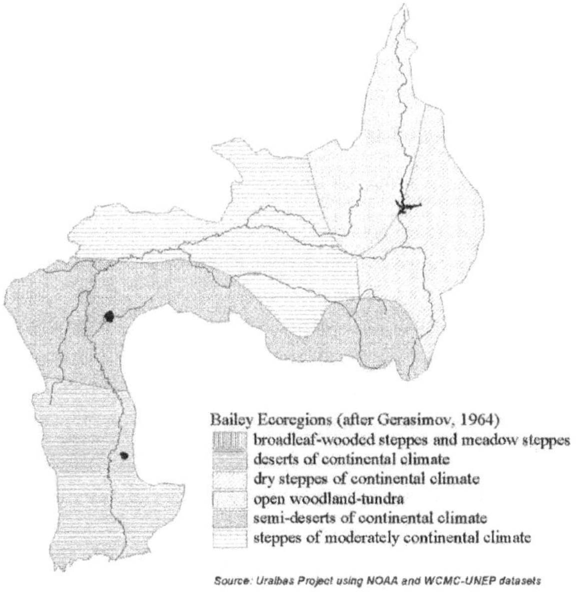

Figure 5. The ecoregions in the Ural River Basin according to Gerasimov-Bailey classification. Source: Ural Basin project using NOAA and WCMC-UNEP datasets

The Ural's freezing up starts in November–December and lasts for 80–160 days. After the ice drift the water temperature in the river increases rapidly. At the end of May, the water temperature can reach 20°C, which creates favorable conditions for sturgeon spawning development and larvae growth. The highest water temperature occurs in July, matched by the lowest water level.

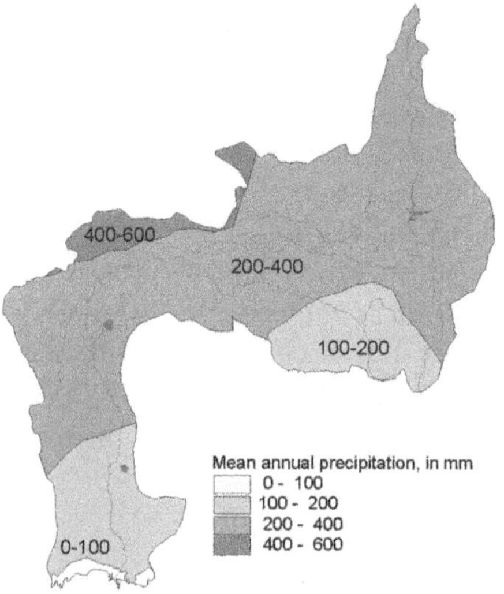

Figure 6. Mean annual precipitation for the period 1970–2002 (UNEP 2008)

Hydrological regime

As indicated above, the specific features of the Ural river are extreme fluctuations in the total annual flow and annual flow distribution.

A more than tenfold difference in total annual water flow was observed during the period of river hydrology monitoring from 1915 to 1985s. The total flow in 1946 was 26 km^3, while in 1976 it comprised only 1.9 km^3. The total river flow measured at the gauge station of Kushum since 1915 is shown in Figure 7.

The annual water distribution is extremely uneven due to snowmelt-induced river alimentation.

The water level regime in the river is different in the South and North parts of the basin due to the variance in regional temperature regimes and the river's greater extent. The water level rise downstream typically occurs in March–April, while flooding in the upper branches occurs in April–May.

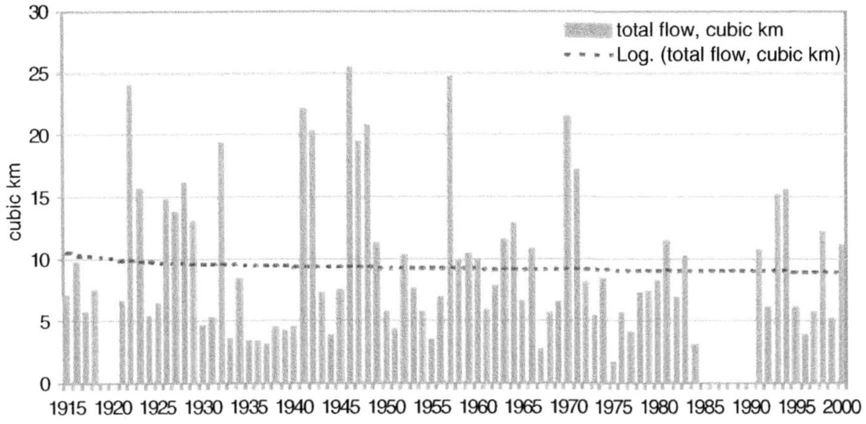

Figure 7. The annual Ural River total flow for the period of observation since 1915 at gauge station Kushum (Vörösmarty et al. 1998; CEP 2002)

Since most of the precipitation occurs during the winter period, the spring flood takes up to 80% of the annual river discharge (Dmitriev and Vasilenko 2007; Tlenbekov 1967). In particular, the mean flood discharge in April–May through 1936–1965 was 78% of annual discharge (Tlenbekov 1967). Summer and fall are characterized by low-water. Summer water effluent comprises about 10% of annual flow, while winter contributes only 1–3%.

The timing and duration of flooding varies from year to year. Average flooding in the Ural delta starts on the 10th of April, reaches its peak on May 12th and finishes on July 6th (Tlenbekov 1967). The mean flood duration is 89 days (Nesterenko 2006).

The monthly discharge for the average year based on the monitoring records for 1915–1984 is presented in Figure 9 (Vörösmarty et al. 1998). The mean seasonal discharges for the upper and middle Ural can also be found in Figure 10.

The Ural River is a unique river not only in the Caspian Sea Basin, but also in Northern Eurasia as a whole. The biggest water reservoir, Iriklinskoe, which is situated far upstream (1,810 km from the river mouth), and a number of other smaller reservoirs in the upper branches (on the rivers Kumak, Chernaja and Ilek) do not have a major influence on the hydro-ecological regime of the river (Cowx et al. 2004; Uralbas 2007), though some changes in flood duration and annual water flow redistribution can be observed (Dmitriev and Vasilenko 2007).

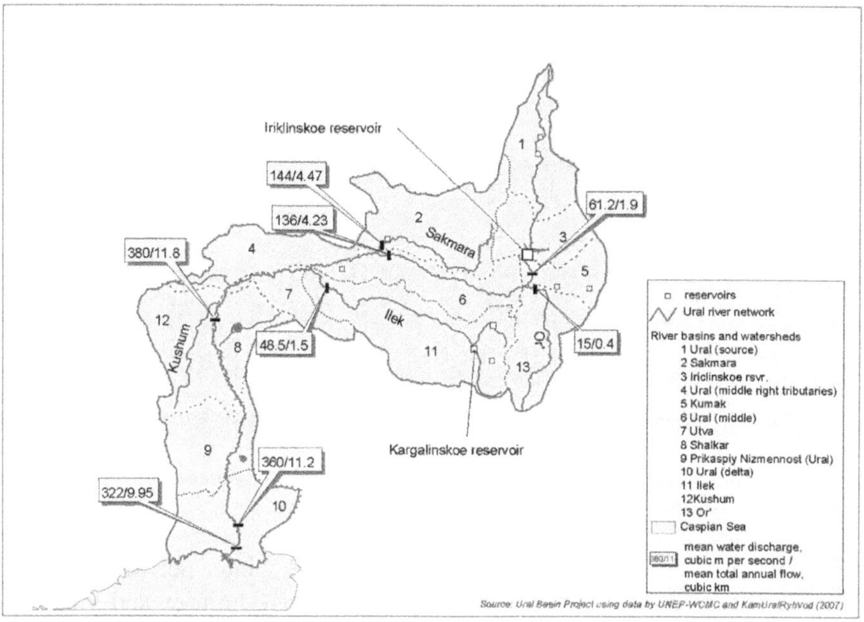

Figure 8. Hydrological network of the Ural river with mean water discharge and mean total flow in different subbasins (UNEP-WCMC and KamUralRybVod 2007)

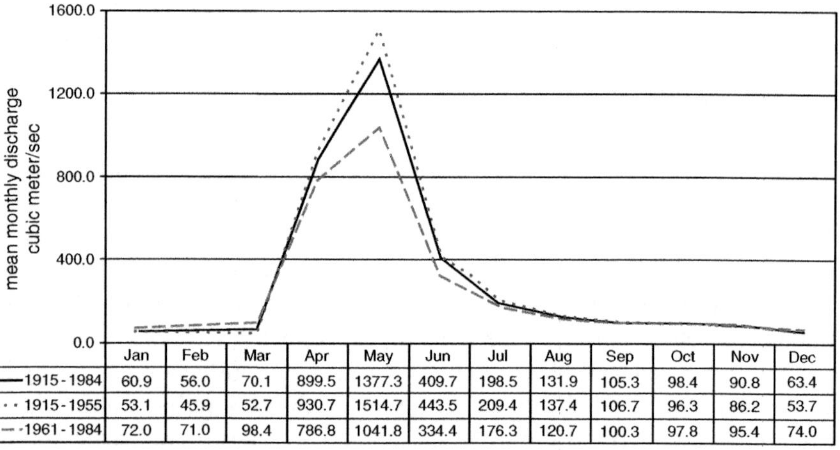

	Jan	Feb	Mar	Apr	May	Jun	Jul	Aug	Sep	Oct	Nov	Dec
—— 1915 - 1984	60.9	56.0	70.1	899.5	1377.3	409.7	198.5	131.9	105.3	98.4	90.8	63.4
···· 1915 - 1955	53.1	45.9	52.7	930.7	1514.7	443.5	209.4	137.4	106.7	96.3	86.2	53.7
– – - 1961 - 1984	72.0	71.0	98.4	786.8	1041.8	334.4	176.3	120.7	100.3	97.8	95.4	74.0

Figure 9. Changes in the Ural river hydrological regime after completion of the Iriklinskoe reservoir

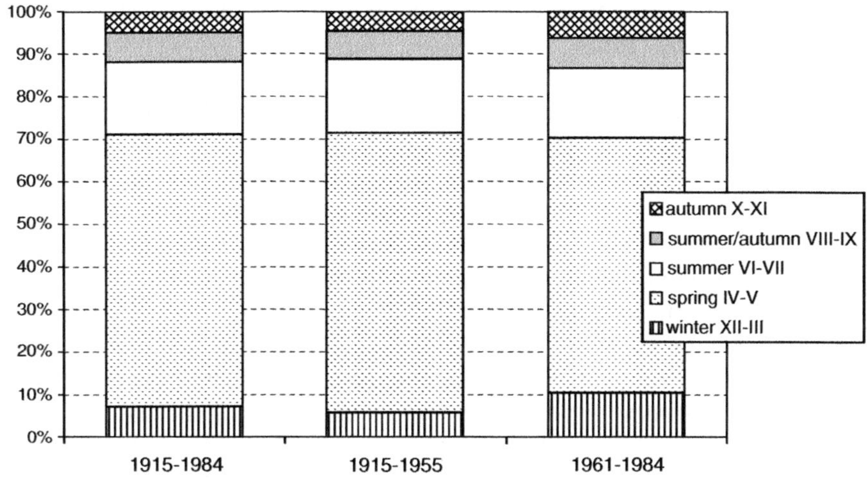

Figure 10. Proportion of seasonal discharges in the Ural

The river hydrological regime, the natural pattern of annual water distribution and the essential floodplain ecosystems (i.e. meadows) are still preserved. Generally speaking, this feature is unique not only for the Caspian basin, but also for most of the major water streams in the Northern hemisphere, most of which have undergone severe anthropogenic alterations (i.e. damming, channeling, etc.).

Thanks to its natural environmental flow regime the Ural's aquatic biodiversity has not deteriorated as much as that of other big rivers. The Ural River contains the only available spawning and wintering habitats of worldwide famous sturgeon species which are protected under numerous international conventions.

Iriklinskoe reservoir

The Ural river basin contains 91 reservoirs with a total volume of 4.9 km^3 (Fashchevsky 2003). The biggest reservoir on the Ural River and the only one having considerable influence on water regime and ecosystem both downstream and upstream is the Iriklinskoe reservoir, fully located at the eastern part of the Orenburg oblast (Russia) in the dry steppe area. The water drop at the dam is 30 m; the length of the reservoir is 73 km; and the maximum width is 10 km. The average depth is 12.5 m, while the deepest part next to the dam is 38 m. The minimal discharge from the Iriklinskoe reservoir is 15 m^3. The filling of the reservoir, which had a much greater degree of impact on the river ecosystem downstream, took place in 1956–1960 (Malik et al. 2000).

The main water users are the industries and households in the cities of Orsk, Novotroitsk and other significant uses are for irrigation, energy production and navigation. The planned energy production is $7*10^7$ kW/h.

Though the Iriklinskoe reservoir has some influence on the Ural River hydrology the negative impact of this influence is decreased by the free-flowing Sakmara River downstream of their junction. Hydrological records from 1915 to 1984 for the gauge station Kushum (Vörösmarty et al. 1998) located 450 km from the river mouth were reviewed to study the influence of the reservoir on the river's hydrology. Preliminary analysis reveals changes in the Ural River hydrological characteristics since the dam's construction (Appendix I). However, similar analysis for the gauge station Aktiubinsk at the river Ilek, one of the two biggest Ural tributaries unaffected by the Iriklinskoe reservoir, also shows a decrease in total annual flow as well as in monthly discharges, suggesting there may be other factors involved.[6]

Figure 10 represents the proportion of seasonal discharges in the Ural River at the Kushum gauge before and after the 1956–1960 filling of the Iriklinskoe reservoir. The first column shows the mean proportion in seasonal discharges for the entire period of observations from 1915–1984, while the other two columns show the mean seasonal discharges before the Iriklinskoe's establishment (1915–1955) and after it (1961–1984). The changes in seasonal discharges are noticeable (total flow in April–May decreased from 65% to 60%), though they cannot be considered significant alterations.

It should be stressed that, as is the case for any steppe snowmelt-fed river, the average annual flow in the Ural River fluctuates greatly over the years. Figure 11 shows mean total flows for the decades in the 20th century. Though mean total flow for the 1990s (9.05 km^3) is slightly less than mean total flow for the entire period of observations, it is much higher than in most other decades. From Figure 11 it also follows that in the 1970s–1980s the total annual flow and water availability was much lower then in the previous decades or in the 1990s. At the other end of the scale, the 1920s and 1940s saw extremely high water availability, which is reflected in the statistics on seasonal water distribution in the river from 1915 to 1955.

---

[6] The analysis is complicated by the presence of Kargalinskoe reservoir at the river Ilek. Though having much smaller size and impact on hydrological regime downstream and riverine biodiversity its influence should be taken into account.

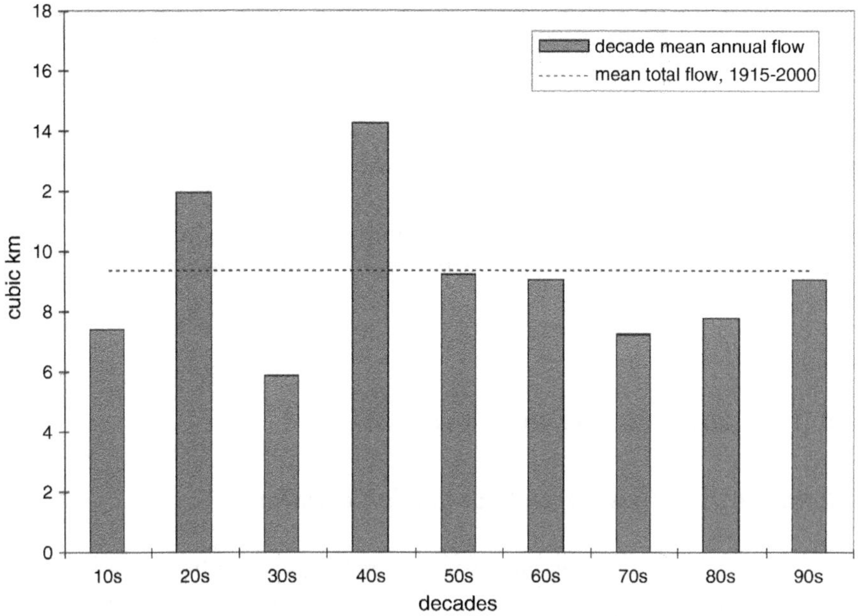

Figure 11. Mean annual flow for the Ural river (Kushum) by decades

In addition, changes in precipitation patterns over the entire basin, which definitely influence the total river runoff, can be observed through the last decades of the 20th century[7] as well.

Another contributing factor is the natural course of climate change in the region and its shift towards an arid climate zone. Moreover, according to some authors the Ural river basin is likely to experience aridization due to climate change[8] (IPCC 2001).

Consequently, the changes in the total river flow and seasonal water distribution can be caused by natural water availability cycles or changing climate conditions rather than impact from the Iriklinskoe reservoir.

In addition, the growing needs of the regional economy and related irreversible water intakes should also result in changes in total water flow patterns over a long period of time. The second half of the 20th century in the region was characterized by rapid agricultural and industrial development, accompanied by population growth. The steppes in the Ural watershed were converted to arable lands in the 1950s and 1960s (*tselina*). Nowadays, croplands comprise more then half of the basin's territory,

[7] See chapter on regional climate conditions in this paper.

[8] IPCC provides completely opposite forecasts for this region from significant decrease in water flow to significant increase. See corresponding chapter below.

while in the first half of the 20th century arable lands were not widespread in the region. The increase of ploughed land in the watershed decreases surface runoff and water inflow to the river streams.

Though regional practitioners (Uralbas 2007) and statistical data show gradual changes in the Ural river's hydrological regime and a decrease in total flow by the end of the 20th century, the analysis of the historical records suggests these changes have largely been caused by regular cycles in water availability and economic activities in the region. It can be concluded that the impact from the Iriklinskoe reservoir on the river ecosystem has been small-scale.

Land use and water usage in the basin

Administrative division and population dynamics

The Ural watershed is spread through three Russian administrative units (*oblasts*), namely Orenburg, Cheliabinsk and Bashkortostan, and three in Kazakhstan: Western-Kazakhstan (Uralsk), Atyrau (Guriev) and Aktiubinsk Oblasts. The administrative basin division is shown in Figure 14, while the basin territory share by administrative units can be found at Figure 12. The biggest basin share and major Ural River streams belong to the Orenburg oblast.

In the river basin there are several cities with a population above 100,000, the biggest being Orenburg, Magnitogorsk, Orsk, Novotroitsk, Uralsk, and Guriev (Atyrau). The average population density in the basin is 17 people/km^2.

There are no official national statistics on population according to watershed territorial division. Statistics in both Kazakhstan and Russia are organized by administrative division, which does not match the river basin borders (Figure 13).

The calculation of the Ural basin population and population density was carried out by the authors using statistical data from FAO (UNEP 2008). The resulting population estimates show a good correlation with available official statistics for the administrative units.

The analysis shows that in 2005 the total population of the basin was 3.76 million inhabitants. The Russian and Kazakhstan basin population shares were 2.6 million and 1.16 million inhabitants respectively.[9]

---

[9] By author calculations.

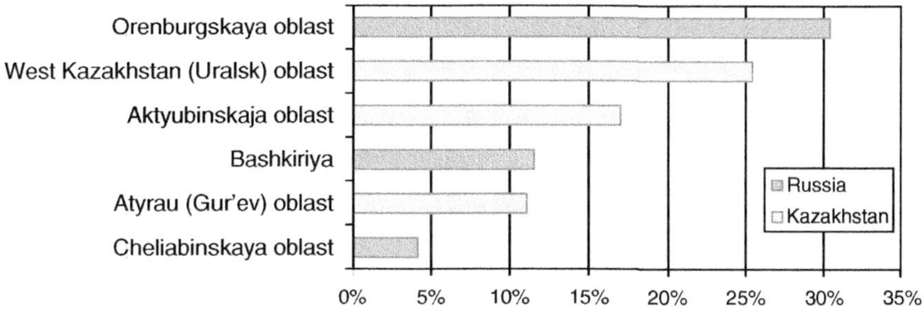

Figure 12. Ural river basin shares by administrative units

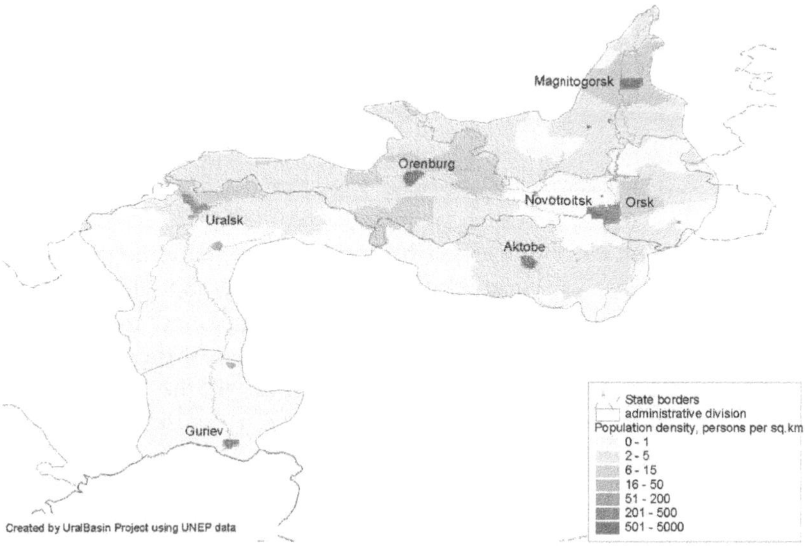

Figure 13. Population density in the Ural basin (Ural Basin Project using FAO and UNEP data)

Based on FAO global population forecasts we estimate that the basin population drop to 3.713 million inhabitants by 2015. This decline will come about solely due to the decrease of population in the Russian part of the basin, while the Kazakh population is predicted to grow. Specifically, it is anticipated that the population in the Russian watershed will decrease to 2.548 million, while in Kazakhstan it will increase by 10,000 inhabitants (up to 1.166 million).

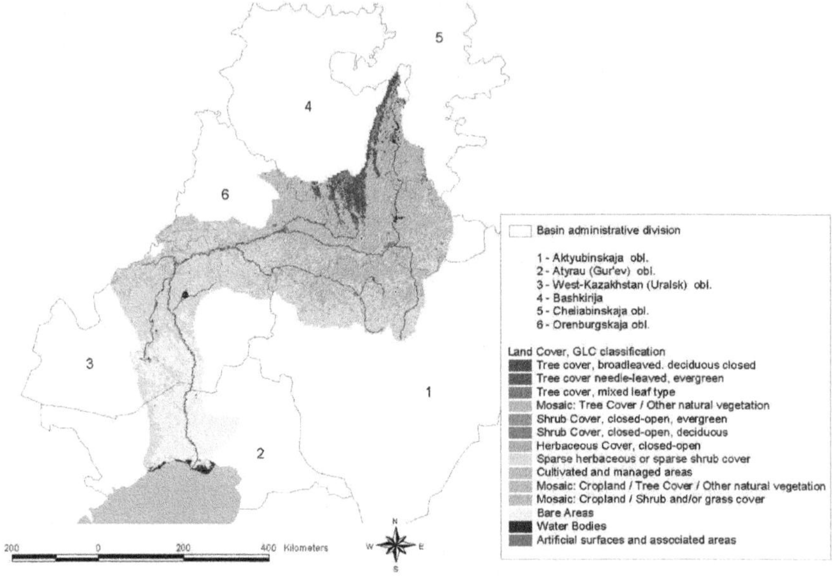

Figure 14. Administrative division and land use categories in the Ural river basin

This statement can be justified by the fact that the demographic situation in Russia clearly indicates a trend towards de-population of remote country areas (like the Orenburg oblast) accompanied by the general population aging and continuing demographic crisis. At the same time, the forecast increase of the Kazakhstan population in the Caspian region would be in line with the growing regional economy caused by oil exploration in the region.

Land cover

Apart from direct water management options such as water intakes, damming or sewage release, many other types of human activities in the watershed have an impact on the water streams (Lagutov 2003). In particular, patterns of land use in the watershed have a substantial effect on river flow formation, water consumption and quality.

As indicated above, the difference between environmental conditions in the South and North of the basin are significant. The semi-desert and arid desert ecological regions close to the Caspian Sea are not favorable for agriculture development. At the same time agriculture is very well developed in the Northern basin areas. According to research conducted by the

Water Resources Institute (WRI 2003) more then half of the basin territory is cropland (59%), 0.9% of which is irrigated. Other specified land use types are grassland and shrubland (33.4%), urban and industrial areas (4.2%), forest cover (2.3%) and wetlands (0.2%) (Figure 15).

An independent attempt to analyze land use patterns in the basin was conducted by the authors using the Global Land Cover 2000 dataset, derived from the regional satellite images (UNEP 2008). The resulting land use map is presented in Figure 14.

Based on this map the land cover statistics were derived and the main land cover categories are represented in Figure 16. This analysis slightly differs from the values originating from WRI, but it confirms the general picture of different types of croplands in the watershed.

Similar analysis conducted using an alternative land cover dataset – UMD (UNEP 2008) – confirmed the general pattern of prevailing cultivated areas and croplands in land use categories.

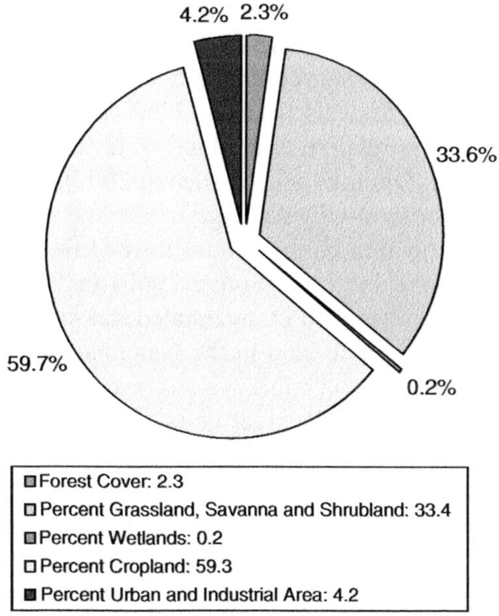

Figure 15. Land cover in the Ural basin (WRI 2003)

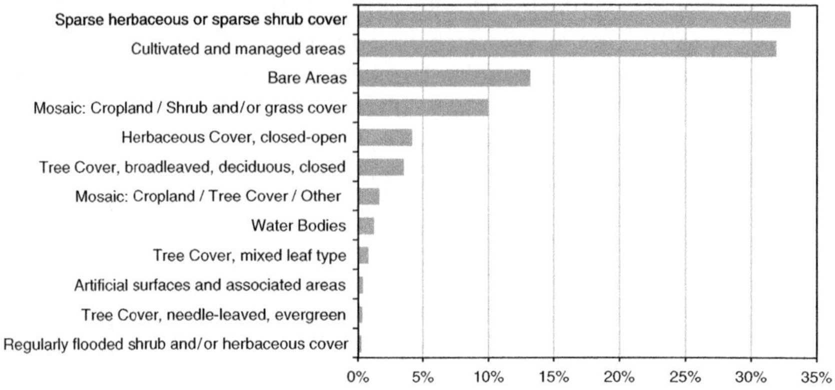

Figure 16. Main land cover categories in the Ural river basin (GLC/UNEP 2008)

Water use

The average water supply per person in the basin according to the research conducted by the Water Resource Institute was 2003 m^3/year (WRI 2003). The ratio of water withdrawals from the Ural river for industrial, agricultural and municipal use relative to the total water flow is the highest in the whole Caspian Basin (Dmitriev and Vasilenko 2007). Moreover, the volume of water intakes is constantly increasing.

According to some authors the annual flow of the Ural river is reduced due to irrigation intake by 23% (Sonne 2000) or 30–40% (Fashchevsky 2003). These estimates seem to be overstated, taking into account the low percentage of irrigated arable land in the watershed and the lack of industries with high levels of irreversible water consumption. The highest level of anthropogenic impact on the river ecosystem, including water intakes, occurs in the Orenburg Oblast. However, the intensity of water intakes in the oblast is still on a much smaller scale than indicated levels.

At the same time, the low river stream is a zone of water losses due to natural factors. From the Kushum gauge to the Ural delta up to 20% of water flow can evaporate or be diverted by distributaries. This fact is well reflected by the difference in average river discharges between the river delta and the Kushum gauge shown in Figure 8.

Role of the Orenburg oblast (Russia) in the Ural watershed's management

Though the Ural River basin spreads across several administrative units in Russia and Kazakhstan (Figure 14) the biggest share in Russian territory belongs to the Orenburg Oblast – 65.7%, which comprises about 31% of the total basin area.

Most of the basin population lives in the Orenburg oblast, and most of the industries and other human activities affecting the river ecosystem are located here as well.

Moreover, as follows from the presented maps and statistics, the formation of the river flow occurs in this territory. In addition, the biggest reservoir Iriklinskoe is fully located within the territory of the Orenburg oblast.

The most productive sturgeon spawning grounds are found in the Ural tributaries (Ilek and Sakmara) and the Ural stream in this area (KamUralRybVod 2007; Uralbas 2007).

Thus, the oblast has the biggest impact on river ecosystem health and plays an important role in integrated watershed management and preservation of migratory species.

Though the Orenburg Oblast itself is spread across several watersheds (the Ural, Volga, Ob' basins and drainless area), the Ural river basin comprises the biggest share. The share of the Ural River basin in the total oblast area is 63%, while the Volga constitutes around 31%.

Most of human activities in the oblast are conducted within the Ural River basin, as is reflected in the statistics of surface water intake over the Orenburg oblast (Table 1):

Table 1. Water consumption in the Orenburg Oblast by basins (Sobolin et al. 2007)

River basin	Water intake (million m^3)
Ural river basin	1783,340
Volga river basin	42,450
Ob river basin	0.48
Drainless regions	0.11

As this table shows, about 97% of the total water intake in the oblast, which was 1,826 million cubic meters in 2006, occurs in the Ural basin. The composition of the water use intensity by user is shown in Table 2.

Table 2. Water intakes by users in the Orenburg Oblast (Sobolin et al. 2007)

Water users	Intake (million m^3)
Industry	1643.29
Household	141.1
Other purposes	26.72
Irrigation	11.3
Other agricultural water supply	4.72

These tables show that the water supply for all types of human activities in the Orenburg oblast is based mostly on the Ural river basin.

The main water sources for household and municipal purposes are underground aquifers – in 2005 about 90% of municipal water intakes were covered by underground water sources. There are 138 underground water aquifers in the oblast, and the total withdrawal capacity of these sources is about 53,000 m^3/day.

The percentage of irrigated arable lands in the oblast is 0.8%, while the share of the total agricultural yield collected from irrigated arable lands is about 5%. Taking into account the low land fertility and the arid continental climate some regional academics have suggested increasing water withdrawal for irrigation purposes in order to increase the regional agricultural yield (Uralbas 2007).

This claim seems to be problematic considering the limited available water resources.

Water quality

According to the statements by the Orenburg Branch of the Federal Agency of Agricultural Inspections (FAAI), specialized systematic hydrochemical monitoring in water bodies used in fishery in the Orenburg oblast has not conducted in the last few decades (Uralbas 2007).

At the same time there are alternative means to evaluate the water quality. One of them is HydroMeterological Centers in basin countries carrying out water quality monitoring. The following data was presented by FAAI during the First Ural Basin workshop (Uralbas). These centers classify the water pollution level according to the following scale: unpolluted – slightly polluted – moderately polluted – highly polluted.

According to the HydroMeteorological Centre of Russia the water at the monitoring point located at the border with Cheliabinsk Oblast (station Berezovskij) is highly polluted. In 2006 the concentration of ferrous iron and oil products was 1.6 times above the maximum allowable concentration (MAC). The biggest polluters upstream are the Sibajsk Copper-Sulphur Factory, the Sibajsk Waterworks, the Baimaksk Engineering Plant and the Buribaevsk Ore Plant.

The water quality in the Ural River at the monitoring point next to the city of Orsk is categorized as "moderately polluted". The sewage from nine main pollutants in Orsk contributes to water pollution of the Ural river. The biggest pollutants in Orsk are the "OrskNefteSintez" Oil Refinery, Orsk Railways, the Orsk-Khalilovsk Metallurgical Plant, and the Gaj Ore Mining and Processing Industry.

Orenburg contributes to the pollution of the river Ural through sewage from several industries. Among the pollutants are ammonia nitrogen, nitrite nitrogen, copper, ferrum, and oil products. As a result at the monitoring station downstream from Orenburg the water is categorized as highly polluted.

The water quality in Iriklinskoe Reservoir is moderately polluted by sewage from the Iriklinskoe Hydropower Plant, runoff from neighboring agricultural areas and pollutants brought by the inflowing "highly polluted" rivers.

The river Sakmara is moderately polluted by the Sakmara Power Plant and the Orenburg Thermal Network Company.

The river Ilek is a tributary of the Ural which forms the border between Russia and Kazakhstan. The water in this river is categorized as "highly polluted". The average annual concentration of hexavalent chrome in 2006 was 2.3 times higher than the MAC, while the maximum registered concentration was 4.3 higher. A trend towards a decrease of the annual maximum concentration can be observed here: in 2004 this value was 8.3 times and in 2005 7.5 times above the MAC. The main polluter on this tributary is the Chrome Factory in Aktiubinsk (Kazakhstan). Other pollutants in the Ilek river in 2006 included copper compounds (2 times above MAC), ferrum (3.3 times MAC), nitrite nitrogen (1.8 times MAC), sulphates (2.4 times MAC).

As a result the water at the trailing river monitoring station at the border with Kazakhstan is categorized as "slightly polluted".

At the same time Kazakhstan State Reports on the environmental situation in the Caspian region characterize the water in the Ural river as "clean". The only pollutant mentioned in the reports as exceeding the MAC is phenol (KazGidroMet 2007).

Concentrations of microelements and organic matter do not exceed the maximum admissible concentrations (Fashchevsky 2003).

Ural flow and salinity of the Northern Caspian

Though average sea salinity has remained stable (~12‰) for a long time there is a big difference between north and south as well as between west and east. Historically, salinity in the North Caspian varies from 0.1‰ at the mouth of the Volga and Ural rivers to 10–12‰ in the middle Caspian. Such a difference is the result of the inflow of freshwater from the northern rivers, mainly the Volga River.

The Northern Caspian Sea is a very shallow region (Figure 17). The average depth is only 5 m with a maximum up to 20 m (UNEP 2006). Due to the huge amount of freshwater delivered to the sea the Volga desalinates a significant area of the Northern Caspian, while the contribution from the Ural is enough to desalinate only the areas close to the river mouth.

Nevertheless, it should be noted that during the years with high water availability the Ural's freshwater influxes can decrease the salinity in the shallow waters of the northern-eastern Caspian region. During such years the salinity of the big areas in the eastern-northern Caspian drops to 0–4‰. The influence from the freshwater Ural's influx to the Caspian Sea extends for 400–450 km along the Ural's depression towards the Middle Caspian (Dmitriev and Vasilenko 2007).

These hydrological conditions have resulted in high productivity in the Northern Caspian. Prior to the installation of the Volgograd dam the entire area close to the Volga and Ural deltas between the Kizlyar bay and the

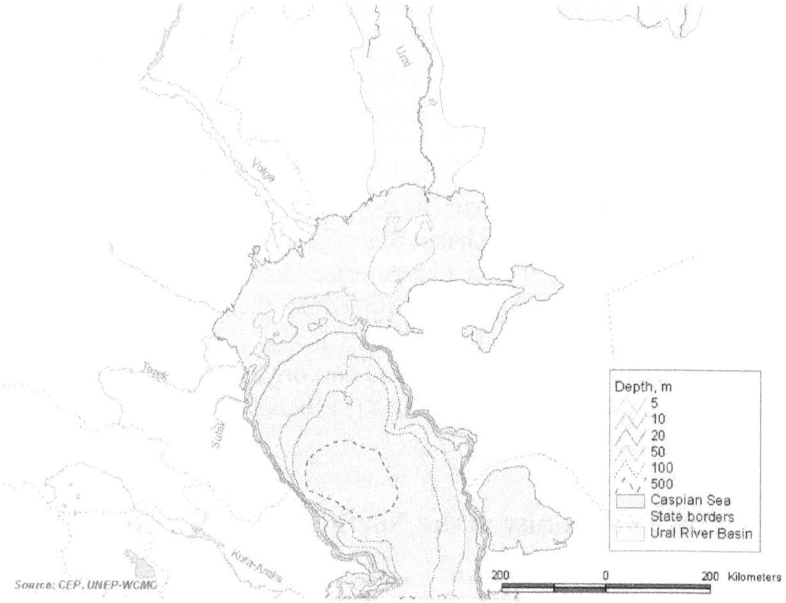

Figure 17. Bathymetry of the Northern Caspian

river Emba was one continuous highly productive system with more or less homogenous conditions and salinity (0–5‰) (Dmitriev and Vasilenko 2007).

The regulation of the northern Caspian rivers in the 20th century has resulted in significant changes in the salinity of the Northern Caspian. Since the beginning of the 1930s a number of high pressure dams were constructed on the sea tributaries of the Volga, Terek, Sulak, Samur, and Kura. Among many other negative environmental impacts this development has significantly altered the volume and timing of freshwater inflow and consequently the salinity regime of the shallow sea waters (CEP 2002; Dubinina and Kozlitina 2000; ZIN 2006).

The completion of the Volgograd Hydropower station in 1958 has resulted not only in a decrease of freshwater inflow, but also in shoaling of the Volga delta and redistribution of remaining water flows. In particular, the water flow through the eastern delta branches has decreased to 43% (Uralbas 2007). The water exchange between the eastern and western parts of the Northern Caspian has been significantly impaired and the salinity of the western part has started to grow at much faster rates.

The sea salinity in areas next to the Ural river delta has increased to 12‰. The decline in freshwater influx from the eastern Volga distributary resulted in conversion of the sea areas to the east of the Ural mouth into the type of ecosystems typical for the middle Caspian Basin. The benthos biomass drastically decreased, depriving fish populations of their feeding grounds and breaking the food chains in the ecosystem.

The salinity of the estuaries in general and the shallow Northern Caspian in particular is a good indicator of the sustainability of the water management in the watershed. Excessive water intakes in the river basin upstream without taking into account the needs of downstream ecosystems/ countries is one of the important reasons for salinity changes in estuaries.[10]

Water management

One of the typical factors limiting the transboundary water cooperation in the river basins between upstream and downstream countries is a lack of the incentive for such cooperation on the part of upstream countries. The rivers are traditionally considered to be a unidirectional flow. There is no effective feedback mechanism to link the damages and losses of downstream countries so that the upstream countries treat them as their own.

[10] Such an analysis in the Caspian is complicated by sea level fluctuations. Increase in sea level may also result in an increase in salinity of the Northern Caspian as well.

Though usually it is hard to identify feedback from the delta and coastal area to the ecosystems upstream, in the case of the Caspian, and the Ural river in particular, sturgeon species can play the role of such a feedback mechanism. The sturgeon species, migrating through the whole river network, link together the situation in the river basin and sea and introduce bi-directional causal relationships among the basin countries.

The upstream country, Russia, can be made interested to actively cooperate in integrated watershed management and other related areas, to rehabilitate the vanishing sturgeon, not least because they are a source of significant commercial income.

If transboundary understanding is achieved and practical joint water management plans are to be implemented then two activities influencing water balance in the river can be distinguished: water intakes and annual flow redistribution by means of reservoirs upstream. As indicated above irreversible water intakes from the Ural river are not significant. Though the only water-consuming regional activity characterized by irreversible water withdrawal, irrigation, is not large-scale, there are some initiatives to increase it (Sobolin et al. 2007). Correspondingly, this issue is a subject for close cooperation and monitoring.

The second management option, annual flow redistribution, is implemented by means of small reservoirs on Ural tributaries, including the biggest, the Iriklinskoe Reservoir. The water usage in the Iriklinskoe reservoir is regulated in accordance with the "Water Usage Regulations for Iriklinskoe Reservoir" adopted in 1962. This document defines the priorities of the Iriklinskoe reservoir as water supply for industrial facilities and population in the cities of Orsk, Gai and Novotroitsk as well as the "creation of a favorable river flow regime for the spawning of sturgeon species and successful agriculture in the Kazakhstan Soviet Socialist Republic". For the needs of the users mentioned above the water discharge is redistributed through and over the years.[11]

A new version of this document was developed and adopted in 1972 answering the increased water demand for industry and municipalities. Since that time the regulations were not changed.

During the Soviet period the Ural River Basin was managed by centralized water management authorities. Despite some shortcomings, the centralized water management and planning system allowed the river basin

---

[11] As a matter of fact the favorable conditions for sturgeon reproduction are not only hard to achieve by alterations of natural hydrological regime, but also sturgeon migrations and spawning are significantly undermined by this influence. The closer the river hydrological regime to the natural one the higher the chance of initiation of sturgeon upstream migrations, spawning and juvenile survival during downstream drift (Lagutov 1996; Sulak and Randall 2002). Often supposedly environmentally-friendly regulations cause damage to migrating sturgeon when they are implemented in practice.

to be managed effectively in accordance with the development priorities established by the State economy. Nowadays these priorities in water management in the river basins are considered to be inadequate and require significant revision. However, using centralized water management system it was possible to implement an integrated approach to the management of water and biological resources.

The collapse of the Soviet Union resulted in the breakdown of transboundary water cooperation. The negative impacts of anthropogenic activities (i.e. biodiversity loss, habitat degradation, etc.) previously arose mostly because top priority was given to the economic, industrial or military sectors. Today in addition to these factors the situation is worsened by the lack of any cooperation between the two basin countries. Industrial and agricultural development projects are often established without consultation with or even notification of the other party.

For instance, the Iriklinskoe reservoir is utilized for inland fishery using introduced species. Correspondingly, higher water discharges from the reservoir, which are beneficial for sturgeon reproduction on the territory of Kazakhstan, result in economic damage for the inland reservoir fishery in Russia. The Russian economic benefit from harvesting low-value fish in the reservoir is obvious, while the environmental needs along the river and conservation of sturgeon species which do not reach the territory of Russia are of low priority.

In this situation it is easy to understand why the chief expert of the Orenburg Oblast on protection and restoration of aquatic biological resources is more interested in creating favorable conditions for the Russian fishery in the Iriklinskoe reservoir, than in the sturgeon habitats further downstream or in river hydrology restoration (Uralbas 2007).

As reported by the managers of the Iriklinskoe reservoir (Uralbas 2007), in accordance with these needs water discharge during May–November is held equal to 15 m^3/s, which is the official minimal discharge following the "Water Regulations". For the rest of the year the average discharge is 25 m^3/s. In this way during the summer the reservoir is filling up to the maximum possible water level, while during the winter period the water level drops to the levels much lower then specified in the regulations.

Moreover, following this logic the requests of Kazakhstan authorities for additional winter discharges from the Iriklinskoe reservoir to prevent fish kill in the lower branches of the Ural were not granted in the last few years.

The Russian side managing the Iriklinskoe reservoir should be made interested in transboundary water cooperation. One and probably the most effective way of doing this is to guarantee the mutual benefit from the restoration of the Ural sturgeon population by securing the natural hydrological regime and favorable conditions in sturgeon habitats.

Climate change

The scale and importance of climate change and its impact on river ecosystems is actively discussed nowadays. In fact, changing climate can significantly alter regional precipitation and temperature regimes, inducing changes in the river's hydrological regime and sturgeon habitats. However, the specific effects of global climate change trends on a particular river system are often not considered or estimates differ drastically depending on the research group.

Preliminary analysis of the Ural river flow statistics was undertaken by authors using the results of water flow monitoring at the Kushum station over almost 70 years[12] (Vörösmarty et al. 1998), and a trend towards decline in total annual flow was identified. This observation agrees to some extent with research conducted by other authors stating that water level and total flow in the Ural river has dropped significantly over the last few decades (Uralbas 2007).

However, as discussed earlier, the changes in river flow observations are caused by a combination of factors, such as water intakes for different purposes, river stream regulation, human activities in the watershed, land use patterns, etc. For instance, the territory of the basin in the second half of the 20th century has undergone significant anthropogenic influence. At the beginning of the century the area was mainly inhabited by Ural Cossacks and Kazakhs, who did not practice agriculture, while at the moment more then half of the territory is covered with croplands (UNEP 2008; WRI 2003). Correspondingly, the water balance of the territory has completely changed, since ploughed areas have different infiltration characteristics (Lagutov 1997; Nesterenko 2006).

A holistic integrated analysis of all the factors and their interconnections is needed to evaluate changes in river flow. Nevertheless, preliminary analysis can be carried out using historic meteorological data, such as precipitation, evapotranspiration, temperature, etc. Such an analysis was conducted using annual precipitation and temperature statistical data for the period 1970–2002, which was obtained from the United Nations Environmental Program (UNEP 2008). The analysis suggests that climate conditions which can influence the river flow and sturgeon population have

[12] See discussion of impact on the Ural river hydrology by Iriklinskoe reservoir.

changed over the last decades. In particular, changes in annual precipitation and temperature patterns have been identified.

The changes in precipitation over the Ural basin are depicted in Figure 18 as the differences between mean annual precipitation for the period 1970–1986 and 1987–2002.

According to these calculations the annual precipitation increased over the entire territory of the basin. In some parts of the basin the increase was as high as 20% relative to the average precipitation for 1970–1986 or in absolute terms up to 43 mm/year.

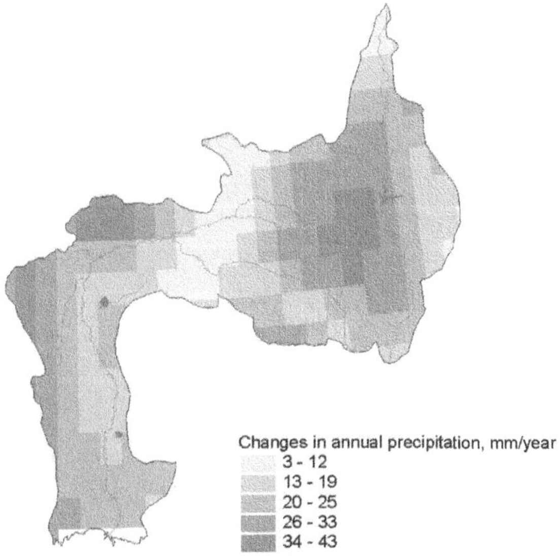

Figure 18. The change in mean annual precipitation for the periods 1970–1986 and 1987–2002

On the one hand, an increase of precipitation should result in an increase of water stream flows. From this point of view the last observation seems to contradict the previous conclusion on the decrease in the Ural river flow. On the other hand, the connections between precipitation and river flows are not always clear for large watersheds under rapidly changing anthropogenic pressure. The period of the observations in the Ural watershed is characterized by drastic growth in agriculture and ploughed areas, increased water intakes for new industries and households, temperature increase and corresponding flow redistribution, etc.

Figure 19 presents the changes in annual temperature for the same observation period. The mean annual temperature increased through the entire basin. The absolute increase values depend upon the geographic position in the basin and range from 0.21°C to 0.61°C.

Both calculations reveal higher rates of change in environmental conditions at the North of the basin. Taking into account that the Ural river flow is formed mainly in this area, the changes in the river flow, e.g. smaller snow accumulation during winter period and annual flow redistribution, should be of a higher order of magnitude. A deep analysis of the interlinkages between climate conditions in the region and river flow is needed.

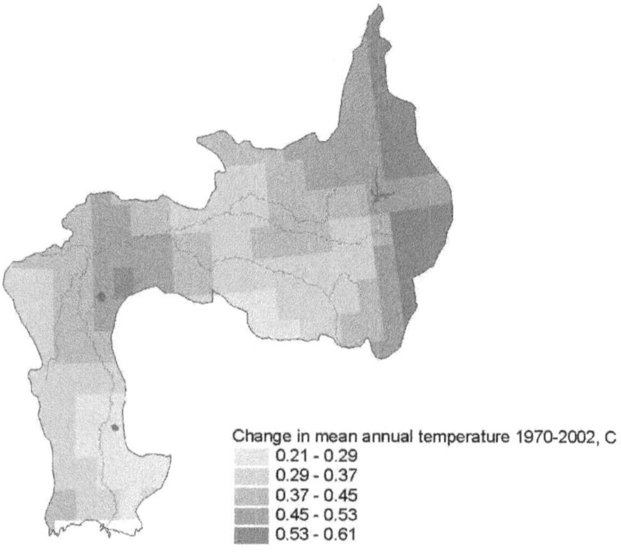

Figure 19. The change in mean annual temperature over the Ural Basin for the period 1970–2002

Numerous attempts to predict the regional consequences of global climate change have been carried out. However, there is no standard methodology and no commonly shared vision on both rates of global climate change as well as its local impacts.

Often the analysis made even by the same agency can produce contradictory results and serve as justification for different management strategies. For example, experts from the International Panel on Climate Change (IPCC) utilize different scenarios and modeling tools for the predictions of climate change. The simulations of the change in the average annual runoff in the Ural basin under HadCM2 and HadCM3 give completely opposite results.

Both experiments have similar climate conditions but differ in evaluation of the Ural River runoff. The first one predicts 50–150 mm/year increase, while the second one shows a possible decrease in precipitation of nearly the same amount (IPCC 2001).

Such a high uncertainty in environmental conditions serve as justification of any management strategy based on decision-makers' bias or today's interests. For instance, national water management strategies in the Ural Basin in Russia and Kazakhstan are grounded on different assumptions. The Kazakhstan Ministry of Environmental Protection counts on an increase of water discharges and availability (RK 2003), while Russian water authorities believe otherwise (Uralbas 2007).

Preliminary analysis suggests that the environmental conditions in the Ural basin are changing at a significant rate. Global models are not capable of considering various regional specifics (e.g. land use patterns or changes in cropland areas). Detailed analysis utilizing local expertise is required for the evaluation of climate change impacts on the regional water streams and affiliated ecosystems.

Conclusion and discussion

The Ural river is a unique ecosystem of global importance. All other rivers of the Caspian basin have been severely altered by various hydrological projects and their ecosystems have been significantly impaired. According to EU WFD, a lot of effort should be invested in order to try to return European rivers to a state of "slight variations from the ecosystem with minimum anthropogenic impact".

Unlike other large European rivers the Ural river is still a mostly undisturbed ecosystem. The lack of dams and other waterworks on the Ural river preserved the natural hydrological flow and the only natural sturgeon habitats in the Caspian basin. These habitats are crucial for the preservation of the worldwide renowned sturgeon species. The current environmental state of the ecosystem should be preserved and maintained.

However, following the disintegration of the Soviet Union, the situation relating to water resources and associated river biodiversity is rapidly worsening. The region is also a subject for climate change impact. Coupled with growing anthropogenic impacts, this endangers the riverine ecosystem.

Successful water resources management depends upon many factors and brings together numerous aspects of basin wide human activities and development strategies. The river network with its affiliated sturgeon population is an indicator of the sustainability of these activities. A holistic integrated approach to watershed management should be implemented to secure the Ural's ecosystems' and habitats' preservation.

A functional monitoring system and regular information exchange is badly needed in the basin. Current management strategies are based on research conducted several decades ago.

Unfortunately, the current management system needs major improvement or even re-establishment. It can be stated that integrated watershed management of the Ural river water and biological resources does not currently exist.

In contrast to other transboundary basins (e.g. the Danube) the Ural river basin is shared by only two countries, Russia and Kazakhstan. This gives an excellent opportunity for the development of a model case of transboundary integrated watershed management. The situation can be improved only by close transboundary cooperation. Practical river-protection steps are urgently needed to improve the situation and conserve the unique sturgeon habitats.

The existing water usage regulations (in particular the management strategy for the Iriklinskoe reservoir) are extremely outdated and require significant and urgent adaptations to today's conditions. Both basin countries should participate in the development of an integrated water usage scheme which takes into account the interests of all water users.

However, this issue should be approached with care considering the fact that environmental flow needs (biodiversity, self-purification, etc.) are often neglected by regional practitioners and decision-makers. To secure proper consideration of environmental needs the involvement of international organizations is desirable.

Appendix I

Hydrological characteristics of the river Ural and its tributaries on the territory of the Orenburg Oblast (Sobolin et al. 2007)

River/number of tributaries	Destination	Length (km)		Catchment area (km^2)	Discharges						Summer minimal discharge, m^3/sec	Year of certification	Administrative units
		Total	In oblast		P = 95% ($km^3/year$)	(m^3/s)	P = 75% ($km^3/year$)	(m^3/s)	P = 50% ($km^3/year$)	(m^3/s)			
Ural/10	Caspian Sea	2,428		231,000									Cheliabinsk oblast Orenburg oblast, Kazakhstan
Suunduk/27	Ural	174	174	6,430	0,008	0.25	0,047	1.49	0,115	3.66	0.18	1979	Orenburg oblast
Bolshoj Kumak/55	Ural	212	212	7,900	0,012	0.59	0,070	2.21	0,174	5.53	0.28	1979	Orenburg oblast
Guberlja/66	Ural	111	111	2,410	0,039	1.23	0.10	3.15	0.17	5.36	0.11	1979	Orenburg oblast
Sakmara/55	Ural	798	435	30,200	1.45	46.0	2.84	90.1	4.27	135	12.4	1978	Bashkortostan, Orenburg oblast
Salmish/179	Sakmara	193	193	7,340	0,228	7.21	0,477	15.1	0,745	23.6	3.70	1978	Orenburg oblast
Ilek	Ural	623		41,300									Orenburg oblast, Aktiubinsk oblast (Kazakhstan)

References

AzovBas. 2002. Proceedings. Azov Sea Basin Workshop Novocherkassk, Russia, Novocherkassk: "Green Don Publishing".

CEP. 2002. *Transboundary Diagnostic Analysis for the Caspian Sea*. Caspian Environment Programme, Baku, Azerbaijan.

Cowx, I.G., Almeida, O., et al. 2004. Value of river fisheries. Proceedings of the Second International Symposium on the Management of Large Rivers for Fisheries Sustaining Livelihoods and Biodiversity in the New Millennium, Bangkok, Thailand, FAO Regional Office for Asia and the Pacific.

Dmitriev, A.N. and Vasilenko, A.V. 2007. The influence of the hydrological regime on sturgeon restoration in the Ural river. Ural Basin Workshop Orenburg.

Dubinina, V.G. and Kozlitina, S.V. 2000. Water resources management of the southern rivers of Russia with reference to fisheries requirements. *Fish. Manage.* 7. (1): 157–165.

Elsevier. 2008. Science Direct Online Journal Collection. Elsevier.

Fashchevsky, B. 2003. Human impact on rivers and fish in the Ponto-Caspian Basin. Proceedings of the Second International Symposium on the Management of Large Rivers for Fisheries, Phnom Penh, Kingdom of Cambodia, FAO.

IPCC. 2001. *IPCC Third Assessment Report*. Intergovernmental Panel on Climate Change. Geneva, Switzerland.

KamUralRybVod. 2007. *Vozdeistvie godovogo stoka na vosstanovlenie osetrovih v reke Ural*. Orenburg, Orenburg Department of Federal State Agency "KamUralRybVod".

KaspNIRH. 1999. *Osetrovie-sostoianie bioresursov, tendencii, prichiny istoschenija, predlojenija po vosstanovleniu*. Astrakhan: Caspian Fisheries Research Institute.

KazGidroMet. 2007. *State Bulletin on the Ecological Situation in the Caspian Region*. Ministry of Environmental Protection of Kazakhstan.

Lagutov, V. 1995. *Mekhanizm unichtozheniya riubniuh zapasov Yuga Rossii i puti ih spaseniya* [The mechanism of extermination of fish stocks in South Russia and possibilities of their restoration]. Novocherkassk: Donskaia Rech.

Lagutov, V. 1996. Imitacionnoe modelirovanie zhiznennogo tsikla osetroviuh v basseine reki Don. [Simulation of Sturgeon Life Cycle in the Don River Basin]. MSc Thesis, Department of Applied Mathematics, South Russia State Technical University, Novocherkassk.

Lagutov, V. 1997. Simulation Model of Sturgeon Life Cycle as a Tool in Sustainable Water Management. MSc Thesis, Department of Environmental Sciences, Central European University, Budapest.

Lagutov, V. 1999. New environmental policy. Sustainable development: from myth to reality. 13th Global Biodiversity Forum, San Jose, Costa Rica.

Lagutov, V. 2003. *Modelling of Radionuclides Washout with Surface Water Runoffs from Radioactive Waste Storage Site*. Luxemburg, International Institute for Applied Systems Analysis.

Malik, L.K., Koronkevich, N.I., et al. 2000. *Development of Dams in the Russian Federation and other NIS Countries*. Cape Town, Secretariat of the World Commission on Dams.

Nesterenko, Y.M. 2006. *Vodnaja Komponenta Aridnnih Zon. Ekologicheskoe i Khoziaystvennoe Znachenie*. Ekaterinburg: UrO RAN.

RK. 2003. *National Report on the State of the Environment in Kazakhstan 2003*. Ministry of Environmental Protection of the Republic of Kazakhstan.

Sobolin, G.V., Gulyayev, A.I., et al. 2007. Water Management in the Orenburg Oblast. The First Ural Basin Workshop, Orenburg, Russia.

Sonne, I.S. 2000. Three centuries at the Caspian. In *The Synchronism of Major Historical Events of XVIII-20 cc.* ed., Moscow.

Sulak, K.J. and Randall, M. 2002. Understanding sturgeon life history: enigmas, myths, and insights from scientific studies. *J. Appl. Ichthyol.* 18. (4–6): 519–528.

Tlenbekov, O.K. 1967. Raspredelenie stoka vody po vodotokam i buduschee delty Urala. *Trudy GOIN.* 89. 125–139.

UNEP. 2002. United Nations Environment Programme: Vital Water Graphics. [on-line] http://www.unep.org/dewa/assessments/ecosystems/water/vitalwater/23.htm [cited 15 February 2008].

UNEP. 2006. *Caspian Sea, GIWA Regional Assessment 23.* Kalmar, Sweden: University of Kalmar.

UNEP. 2008. GEO Data Portal. UNEP/DEWA/GRID-Geneva.

Uralbas. 2007. Proceedings. First Ural River International Workshop Orenburg, Russia.

Vörösmarty, C.J., Fekete, B., et al. 1998. River Discharge Database, Version 1.1 (RivDIS v1.0 supplement). Available through the Institute for the Study of Earth, Oceans, and Space/University of New Hampshire, Durham, NH.

WRI. 2003. Water Resource Institute. Watersheds of the World. [on-line] IUCN. http://www.iucn.org/themes/wani/eatlas/html/about.html [cited 2007].

ZIN. 2006. *Report on Caspian Sea Biodiversity Project.* St. Petersburg, Zoological Institute of the Russian Academy of Sciences.

RIVER FLOW FORMATION IN THE RUSSIAN SOUTH URALS

YURY NESTERENKO* AND MAXIM NESTERENKO

Geoecology Department of the Orenburg Scientific Center
Russian Academy of Sciences (Ural branch), Moscow

Abstract Features of small rivers' flow formation in the Orenburg region of Russia are analyzed. The flood flow dynamics of the small rivers of the South Urals are presented, in relation to changes in the system of agricultural land tenure in the period 1936–1995. Consideration is given to how the balance of water from melted snow on the slope is influenced by relief shapes and holding types, and an estimate of the impact of winter tillage on the melted snow flow factor is provided.

Keywords: River flow, anthropogenic influence, snowmelt, water-scarce zones, arid zones, surface flow, underground water

The Orenburg region is a typical water-scarce region of the South Ural area with a varying environment, and has a high level of industrial and agricultural development. This means a double anthropogenic pressure on its natural environment – the influence of the industrial zones around cities, occupying about 2% of the region's territory, and of the activities on agricultural lands, occupying about 90% of its territory.

The intensity of anthropogenic influence on a unit of area used in agriculture is less than on an equivalent area used in industry. But the average agriculture loading of the total area considerably exceeds that of the industrial area. Agricultural land tenure has resulted in changes of the region's hydrophysical properties and soil quality. As a consequence, the water balance and quality of the active water exchange zone has changed. Changes in natural waters have caused changes in the processes going on in the top part of the earth's crust, in vegetation, soil, ground evolution, in ecology and in nature management.

In the second half of the 20th century in the South Urals the influence of human activities on the basic natural components of the steppe zone considerably increased (including influence on soil, aboveground and

* Yury Nesterenko, Geoecology Department, Orenburg Scientific Center, Russia, 460014 Orenburg, p.o. 59; geoecol-ONC@mail.ru

underground water flow formation factors, and their balance and quality). In water-scarce zones water controls the processes taking place in these zones. Changes in the water component of steppe cause greater changes both to the steppe itself and to the mode of river flow than any other factors.

The arrival of the water balance in the Orenburg region averages 44 km^3 per year. This includes the inflow of river waters from outside the region of 3.3 km^3 per year and atmospheric precipitation (AP) of about 41 km^3. The basic part of water expenditure in the region's balance is evaporation, which accounts for about 80% of the annual AP. The share of AP spent on evaporation varies from 0.4 in the south to 0.65 in the north, while the river flow consumption of falling AP is 20%. The supply of river flow in the Orenburg region comprises 20–25% from underground waters and the rest (80–75%) comes from the surface flow of melted snow and storm waters. Within the latter category it should be noted that the share of the storm flow is low, only 1–2 mm. (2–3% of the flow).

One of the major factors of surface and underground river flow formation is the ratio of the total annual atmospheric precipitation and evaporability. If AP exceeds evaporation, areas and zones of superfluous watering are formed, the distinctive features of which are close deposition of underground waters (UW) and high soil humidity in aeration zones. In zones and territories where evaporation is much higher than total annual atmospheric precipitation, a significant downturn in aeration zones' soil humidity is observed and UW sink to greater depths. Deep deposition of UW causes different conditions of water's entry into the ground and of surface flow formation than those corresponding to close deposition. A scarcity of atmospheric precipitation also changes the problems of agricultural tillage fundamentally. The main task of farming becomes accumulation of moisture in soil by reducing the surface flow of melted snow waters. To this end, fields are ploughed in autumn. In most of the agricultural lands in the steppe rivers' basins all changes in runoff factors influence the flow formation in the whole basin and the conditions of the flow's passage in water streams. These features of managing river basins in water-scarce territories cause considerable changes in flood and low-water flow formation.

Research into the South Ural rivers' spring flow have revealed its strong dependence on the share of autumn plowing of the basin lands, since this increases infiltration capability of the top 20–30 cm of soil cover. Besides, considerable changes in economic activities both within the watersheds and corresponding infiltration properties of the soil occur every 5–10 years (Table 1).

Table 1. Surface flow originated from melted snow in small watersheds the South Urals (the Orenburg region) as a function of the intensity of agricultural land use in the watershed use

Years and the period of farming	Winter precipitation (A) (mm)	Melted snow flow (Y) (mm)	Flow factor (Kd = A/Y)	Tillage share on the basins (Ka)	Winter tillage share (K3)
(1936–1941 I)	91	47	0.52	0.40	0.12
(1942–1945 II)	76	86	0.74	–	0.05
(1946–1954 II)	129	68	0.53	0.48	0.12
(1955–1965 III)	154	58	0.38	0.63	0.20
(1965–1975 IV)	133	35	0.25	0.67	0.52
(1976–1985 IV)	130	34	0.26	0.64	0.56
(1986–1990 IV)	161	49	0.30	0.63	0.53
(1991–1995 V)	165	78	0.48	0.63	0.36

In 1936–1941 the winter tillage share (K3) was equal to 12%, while the flow factor (Kd) comprises 0.52. The period of Great Patriotic War (1942–1945) is characterized by low K3 (5%), but the highest Kd (0.74). During the following period 1946–1954 when K3 was restored to the prewar period (12%), Kd was also restored (0.53).The next period 1955–1965 was the period of *tselina* (virgin lands reclamation) and characterized by increase in winter tillage share K3 (20%) and Kd = 0.38. Starting from this period until the economic depression of 1990s the winter tillage share was increasing: 1966–1990: K3 = 52–56%, Kd = 0.25–0.30. In 1991–1995 K3 reduced to 36% with Kd increased up to 0.48. Similar flow changes were observed in the other sub zones of the South Urals.

The results of research into the dependence of melted snow waters' flow on basin-wide agricultural land tenure systems show that the statistical method of river flow forecasting can be applied only for territories with natural vegetation which has not been changed by anthropogenic activity, or for territories with stable and constant economic activities within a basin. However, intensifying human influence on the conditions of the majority of basins in water-scarce areas and the rather rapid changes in the directions of the basins' agricultural use (ploughing and grassing, changing methods of tillage, creating forest shelter belts, moisture and snow accumulation on the fields etc.) make it incorrect to apply statistical methods of hydrological calculations to a long series of river flow observations without introducing appropriate correction factors taking into account the various anthropogenic changes for the analyzed period and expected changes for the forecast period.

In conditions of intensive activities on a basin it is expedient to apply landscape-genetic principles of determining melted snow surface flow parameters based on determining the interdependence between surface flow and both natural and anthropogenic factors. Figure 1 shows the dependence of the melted snow surface flow factor on the share of autumn plowing within a basin (correlation coefficient = 0.84).

The diagram of the interrelationship between the share of winter tillage on a basin and melted snow flow factor makes it possible to forecast changes in the flow factor and to calculate the necessary share of winter tillage on a basin in order to provide the required regulation of flood flow.

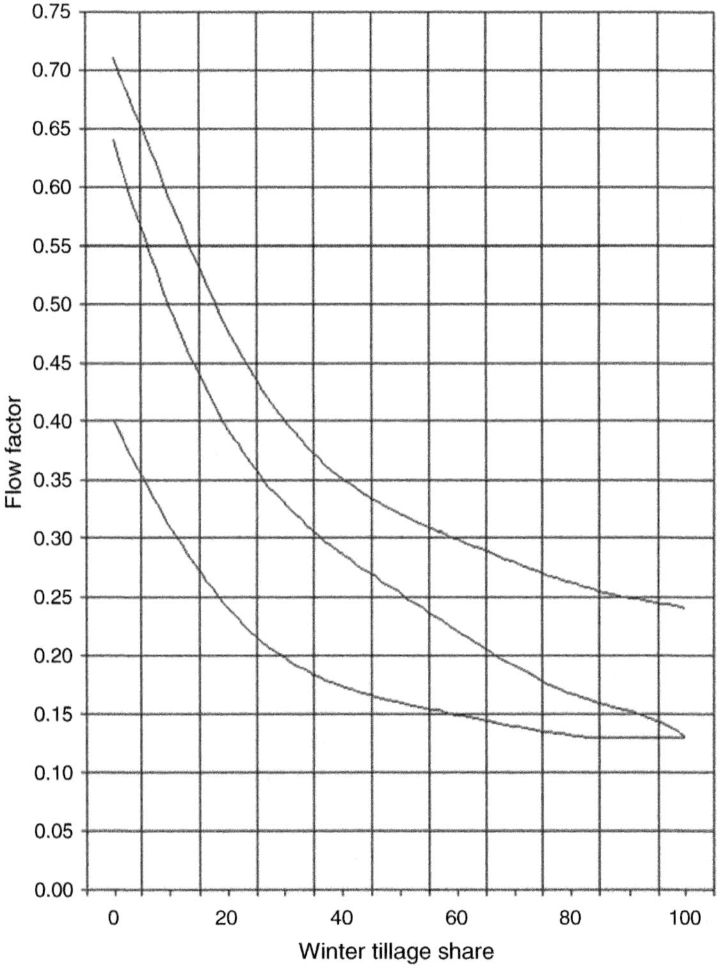

Figure 1. Dependence of surface melted snow flow factor (Kts) on winter tillage share on a basin in the South Preduralie (1) and the South Zauralie (2 – overgrazed virgin soil; 3 – not enough overgrazed virgin soil)

These calculations can be executed with the formulae:

$$Kip = Kcp.p + (Ki - Kcp.f)\ Kcp/Kcp.f \tag{1}$$

$$Уip = Уi\ Kip/Ki = Azi\ Kip \tag{2}$$

Where Kip is the flow design factor of the i-th year at the winter tillage design share on a basin

Ki is the actual flow factor of the i-th year at the winter tillage actual share on a basin

Kcp.f is the long-term average flow factor calculated by the dependence diagram for winter tillage's actual share on a basin

Kcp.p is the long-term average flow factor determined by the dependence diagram for winter tillage's design share on a basin

Уi is the actual flood flow of the i-th year

Уip is the design flood flow of the i-th year

Azi – the sum total of winter precipitations and snow melting period precipitations of the i-th year

With the help of the dependence diagram Kts ↔ K3 and formulae (1, 2) for the investigated Samara river basin located in the Southern Preduralie, design flow layers Уip and design flow factors Kip for all years of the analyzed 60 year period are calculated at a constant winter tillage share on a basin equal to 0.1; 0.2; 0.3; 0.4; 0.5 and 0.6 from the basin area graphically shown as a family of probability curves in Figure 2.

The annual water balance analysis of the agricultural croplands on water-scarce territories shows that surface flow formation and infiltration within the aeration zone limited by the rooting layer occurs on these lands only during spring melting. In summer and in autumn atmospheric precipitation usually does not form surface flow and is spent totally in evaporation and in partially replenishing stocks of moisture in the rooting soil layer. Therefore special attention should be given to the period of spring melting when analyzing water balance on water-scarce territories.

The second distinctive feature of agricultural lands on steppe zone basins is the high variability of surface soil humidity within years before melting – from withering humidity (WH) up to least moisture capacity (LMC) at low humidity and close to WH deeper than 10–20 cm.

An important feature of the Southern Urals is the low spring humidification of soil with winter atmospheric precipitation on open watersheds and slopes. Stocks of water in snow at full imbibitions into soil can humidify it up to LMC only 50–80 cm deep, and are completely spent for total evaporation by the beginning of summer. Hence, losses of melted snow for infiltration outside the layer of its possible consumption by plants can occur only in places of snow accumulation formed by winter winds or in

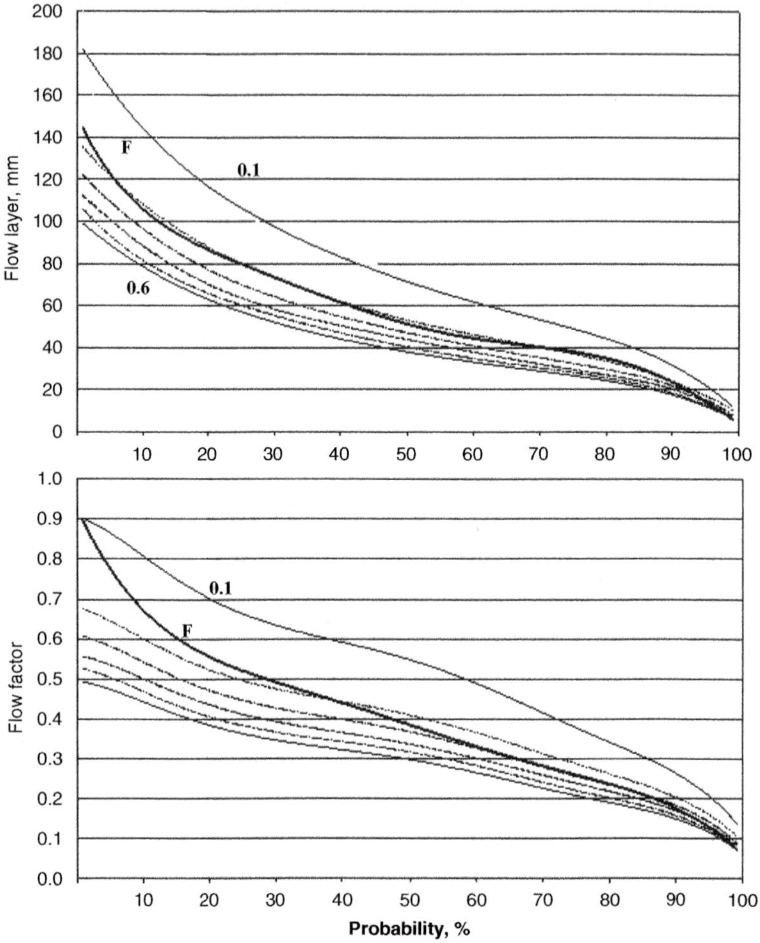

Figure 2. Actual (F) and design probability curves of melted snow surface flow and of its factor for Samara river at various winter tillage shares on the basin

F ▬▬; 0.1 ——; 0.2 - - - - - - -; 0.3 – · – · –; 0.4 — — —; 0.5 — · · · –; 0.6 ——

the presence of surface flow and underground flow (plough soil) of melted snow from adjoining eminences and slopes into closed and open-ended landscape depressions.

The above-mentioned features of the agricultural areas' water balance indicate the importance of studying it, taking into account the small landscape features and emphasizing the melted snow balance on the closed landscape depressions and its balance on adjoining eminences and slopes.

The water balance for relief lowering and eminences within its features can be calculated with the formulae:

$$Az + A_B = Mac2 - Mac1 + E_B + Lips - Yp + Fac \qquad (3)$$

Where Az is water stocks in snow; A_B is quantity of precipitation during melting period; Mac1 is moisture stocks in the active layer of the aeration zone before melting; Mac2 is moisture stocks in the active layer of the aeration zone after melting; E_B is evaporation during melting; Lips is melted snow water flow; Yp is inflow of melted snow from eminences to lowering; Fac is underground water supply or infiltration outside the active layer of the aeration zone (the layer from which consumption of moisture by plants and evaporation are possible).

The amount of water in snow before melting in the watershed of the central Orenburg region was determined on eminences and depressions. Evaporation, AP, moisture stock changes in soil and in subsoil considered during the melting period. The general flow from the catchment area was considered with the help of Thomson's triangular spillways. Measurements have shown that the effective area of the closed depression ($S_п$) on tilled areas averages 10%, while on virgin sites it takes from 15% to 20% of the total basin area. 'The effective area of the closed depression' is the area within the bounds of which the filtration of water outside the active layer of soil is observed (fed into underground waters).

Table 2 shows basic elements of the melted snow balance on eminences and depressions at various holdings of the laboratory site Pokrovsky near Orenburg. The surface flow of melted snow waters (Lips) on the tillage on eminences was 28 mm. The lowering of the relief occupying 11% of the ploughed slope area stopped the surface flow from eminences almost completely. As a result the total flow from the tillage into hydrographic network was only about 1 mm. On the not overgrazed virgin soil the melted snow water flow from eminences was almost three times as much (78 mm), producing 41 mm of flow into the hydrographic network. On the virgin soil prevailing on the landscape before its intensive development in the South Ural an average 8 mm of melted snow flow into the hydrographic network was found (Nesterenko 2006).

Feeding underground waters by melted snow on steppe holdings occurs only on relief lowering. Melted snow was absorbed under eminences as well, but did not leave the limits of the layer of possible moisture consumption by vegetation and partly flowed along the plough soil down into the closed depression. Over a 4 year period the feed of underground waters on the tillage depressions averaged 268 mm, on overgrazed virgin soil 96 mm and on non-overgrazed virgin soil 361 mm.

Table 2. Major elements of the melted snow balance on agricultural holdings in the South Ural (mm)

Holding	Observation years	Eminences				Depressions						Average on a slope						Kfac
		Az	Ust	Mac2-Mac1	S (%)	Az	Yp	Ust	Mac2-Mac1	Fac	S (%)	Az	Eв-Aв	Lips	Mac2-Mac1	Fac	Kст	
1	2	3	4	5	6	7	8	9	10	11	12	13	14	15	16	17	18	19
Tillage	1997	113	35	68	90	181	315	0	151	335	10	120	10	0	76	34	0	0.28
	1998	141	27	99	90	211	243	0	153	286	10	148	15	0	104	29	0	0.20
	1999	111	13	78	90	170	117	0	146	121	10	117	20	0	85	12	0	0.10
	2000	116	39	45	85	245	221	15	99	320	15	135	32	2.3	53	48	0.02	0.36
	Average	120	28	73	89	202	226	4	137	268	11	129	19	1	80	29	0.01	0.22
Over-grazed virgin soil	1997	123	85	28	85	164	481	327	234	74	15	129	10	49	59	11	0.38	0.09
	1998	143	89	39	77	215	297	163	237	97	23	160	15	39	84	22	0.24	0.14
	1999	112	72	20	85	171	408	214	222	96	15	121	20	37	50	14	0.30	0.12
	2000	102	65	42	80	243	260	195	194	19	20	130	-5	39	72	24	0.30	0.18
	Average	120	78	32	82	198	355	225	222	96	18	134	10	41	66	17	0.30	0.13
Non-over-grazed virgin soil	1999	192	38	134	90	261	342	50	156	377	10	199	20	5	136	38	0.025	0.19
	2000	152	68	61	81	227	290	58	104	332	19	166	23	11	87	63	0.07	0.38
	Average	172	53	98	86	244	326	57	130	361	14	182	22	8	102	50	0.044	0.27
Forest shelter belt	1997											567	10	0	284	273	0	0.48
	1998											570	15	0	39	216	0	0.38
	1999											370	20	0	314	36	0	0.10
	2000											356	15	0	312	29	0	0.08
	Average											466	15	0	312	138	0	0.30

The average spring feeding of underground waters on tilled areas was 29 mm, on overgrazed virgin soil 17 mm and on the non-overgrazed areas – 50 mm. Underground water feeding caused by large amounts of snow gathered by wind and lack of surface flow occurred along the whole forest shelter belt and averaged 138 mm.

The suggested techniques of studying the melted snow balance allow historic water performance and flows of the small rivers basins of the region to be calculated and enable forecasts to be made of likely future changes arising from economic activities in a basin.

Thus, we can conclude the following: (1) agricultural activity within the South Ural river basins has caused significant changes in surface flow and melted snow infiltration; (2) the role of mesorelief in the water stock formation in a steppe zone has increased in importance; (3) the developed research techniques and calculation of surface flow and infiltration allow changes to be studied and forecast, taking into account mesorelief and anthropogenic changes within a basin.

References

Nesterenko, Y.M. 2006. Vodnaja Komponenta Aridnnih Zon. Ekologicheskoe i Khoziaystvennoe Znachenie. Ekaterinburg: UrO RAN. [In Russian]

[illegible]

[illegible]*

Department of Geography, Moscow State University, Russia

[illegible]

Department of Earth System Science and Policy, University of North Dakota, USA

Abstract: While previous studies have focused on impacts of average climate change on food production and water resources, this study takes into account the impact of changing frequency and spatial heterogeneity of extreme climate events, first of all, droughts. We analyze impacts of the IPCC A2 and B2 climate scenarios with the use of the GLASS model (containing the GAEZ crop production model and the WaterGAP water resources model). We evaluate future risk of extreme climatic events for food production and water availability for two important regions of Russia – North Caucasus and Urals. Under climate normal conditions it is estimated that "food production shortfalls" (a year in which potential production of the most important crops in a region is below 50% of its average climate normal production, taking into account production in food exporting regions) occur roughly one to three years in each decade. This frequency will double in the both regions in the 2020s, and triple in the 2070s. The assessment of climate impacts on water resources indicates an increase in average water availability in Russia, but also a significantly increased frequency of high runoff events in much of central Russia, and more frequent low runoff events in the South. Unlike the food production, the situation with water resources looks very different for North Caucasus (Kuban and Don river basins) and Urals (Ural river basin) regions. The results suggest the increasing threat to the water resource of the North Caucasus and more stable water flow in the Urals in the new climate.

Keywords: Average climate change, extreme climate events, food shortfalls, high runoff events, water availability

* Nikolai Dronin, School of Geography, Moscow State University, Moscow 119992 GSP-2, Russia; ndronin@gmail.com

Introduction

Russia belongs among those countries that are the most vulnerable to climate variability due to unfavorable natural conditions. Russian farming is characterized by its extreme northerly location. The center of Russia lies at roughly the same latitude as Hudson Bay, and St. Petersburg is actually at the same latitude as southern Alaska. The thermal condition of Russia is unfavorable from the point of view potential for agriculture. At the same time large parts of Russia are unfavorably placed in regard to moisture. The continental position of Russia means a low rainfall generally and a more restricted water supply. About four-fifths of cropland lay in a zone of risky agriculture. The steppe zone of European Russia is characterized by frequent droughts which could bring crop failure and shortage of water for irrigation and other needs.

A general tendency found in many publications on climate change impact on agriculture is that under a warmer climate the agriculture output would increase in the higher latitudes and decrease in lower latitudes (IPCC 2001). In Russia, there is a widespread public belief that climate change is generally favorable for agriculture, which is also supported by global model-based studies for which regional results are available (Fischer et al. 2000). On the base of regional models several authors have concluded that climate change will increase average agricultural production over most of the territory of Russia because of increasing CO_2 and/or more favorable temperature and precipitation conditions for crop growth (Pegov et al. 2000). Decline of harvests in drier districts would be compensated by increasing yields in regions located to the north. Likewise, assessments of climate impacts on water resources have shown an increase rather than decrease in water availability over most of Russia (see, e.g., Alcamo et al. 2000b; Vörösmarty et al. 2000).

Yet these crop and water assessments were based on an evaluation of average changes in climate. However, there is no sense to operate with average figures only for such large country as Russia if several southern regions of the country with population of 12 million people would suffer from increasing aridity of climate, decline of potential of agricultural production and diminishing runoff of local rivers. Our paper focuses on such problematic regions. We define problematic regions in relation to climate change as those expected to become experiencing worse climate in terms of both food production/supply and water availability. Our models show that at least North Caucasus and South Urals are founded among them. In these regions more frequent crop failure because invasion of droughts are projected. However, major river basins of two regions – Kuban, Don (North Caucasus) and Urals (South Urals) – are different in terms of future water availability as their water regimes are depended on winter precipitation in up stream flow areas which are located in very different geographical zones (wooded steppe and taiga).

□ □□□ □n □ □e□□□s

In our research of climate change we used the results of the third version of East Anglia University GCM HADCM3. Two climate integrations we used for our research are based on IPCC socioeconomic scenarios SRES-A2 and SRES-B2, with SRES-B2 corresponding to somewhat lower global temperature increase. To test our findings for consistency, we also made a few additional model runs with the data from another GCM, namely ECHAM4/OPYC3 model, developed jointly by Max Planck Institute and Deutsches Klimarechenzentrum (Roeckner et al. 1996).

For our analysis we use the IPCC “A2” and “B2” scenarios. These two scenarios were selected because they provide a range of estimates of both socio-economic drivers (needed for estimating future population exposed to food production shortfalls, and future water withdrawals and water stress) and future climate change (needed for estimating potential crop production). By analysing these two contrasting scenarios we can cover part of the uncertainty of future estimates. The A2 scenario assumes economic and population trends consistent with a regionalized and economically-oriented world (IPCC 2000). The B2 scenario also assumes economic and population trends consistent with a regionalized world but with a strong environmental focus. In 2075, the population in Russia is 185 million under the A2 scenario and 109 million under the B2 scenario. Income in 2075 is approximately US$25,000 under A2 and US$54,000 under B2 (Table 1). These contrasting driving forces lead to a range of estimates of future water withdrawals (Table 2).

Table 1. Assumptions of scenarios for Russia (GOSCOMSTAT 1998a, b; CIESIN 2002a,b; Lutz and Goujon 2002; Gordon et al. 2000; Pope et al. 2000; Roeckner et al. 1996)

Scenario and year	Description	Population (millions)	GDP per capita per annum (US$/cap-a)	Temperature increase (°C)* HADCM3/ECHAM
1995	–	150.6	3,202	–
A2 2025	Regionalized,	148	7,409	1.8/2.2
2075	economically-oriented	185	24,810	4.8/6.1
B2 2025	Regionalized,	131	8,342	2.1/2.8
2075	environmentally-oriented	109	54,055	3.9/5.3

* Average increase over Russia relative to climate normal period (1961–1990).

Table 2. Water withdrawals in 2025 in Russia (km^3/a)

Sector	1995	Scenario 2025	
		A2	B2
Domestic	14	29	10
Industry	47	31	16
Irrigation	21	22–24	24
Livestock	1	1	1
Total	82	83–86	52

Population and economic data used for these calculations differ slightly from the data in Table 1.

We used the GCM decade mean climate parameters to emulate cumulative climate change by 2020s and 2070s. For 2020s, we estimated climate change as the difference between 2021–2030 and 1961–1990 climates. For 2070s, we used 2071–2080, correspondingly. For temperature, this difference was measured by subtraction of the values, for precipitation – by division. The results were rescaled into 0.5° x 0.5° grid. Finally, we generated two sets of scenarios, describing the climate of 2020s and 2070s, which took into account natural variability. In these scenarios, variability of temperature and precipitation for each month is described by a set of 30 values.

Calculations of impact of expected climate change on crop productivity and water resources were carried out with the GLASS model which was developed for analyzing global change impacts on food and water security, taking into account extreme climate events (Alcamo et al. 2000a). In this paper, GLASS is used to process the climate scenarios from climate models and to analyze climate impacts on agriculture and water availability.

To analyze climate impacts on crop productivity we use the GAEZ model which is nested in the GLASS model. The GAEZ model (□lobal □gro-□cological □ones) was developed at the International Institute of Applied Systems Analysis (IIASA) in Austria in conjunction with the Food and Agriculture Organization for analyzing potential production of different crops on a 0.5° latitude x longitude global grid (Fischer et al. 2002). Each of the grids is characterized by the soil, relief, and six parameters of climate: precipitation, temperature (monthly average and monthly variability), relative air humidity, incoming solar radiation, and wind speed. The climate parameters were synthesized for each month of 1901–1995 period using CRU climate database (New et al. 1999). For the synthesized future climate (2020s and 2070s) we assumed that monthly temperature variation, air humidity, solar radiation and wind speed were characterized by 1961–1990 data. For those years, we also adopted synthesized monthly precipitation and average temperature ensembles, described above. The GAEZ model describes the production process of 154 crop species and varieties for the whole world. We modified GAEZ model for the simulation of Russian agriculture. While the base model simulates agriculture of different countries, we adopted the existing subdivision of Russia into 89 administrative subjects of Federation. We limited the list of crops by three of the most important cereals: wheat (including four spring and four winter varieties), rye (including four spring and four winter varieties), and corn (including six varieties).

To compute climate impacts on water availability we use the WaterGAP model (□ □**er**-□lobal □ssessment and □rognosis) which is also nested within GLASS. The WaterGAP 2 model was developed at the Center for Environmental Systems Research at the University of Kassel in Germany and is a flexible tool that can compute many different indicators of water use and water availability on a 0.5° latitude x longitude global grid or on a river basin level (Alcamo et al. 2003a, b; Döll et al. 2003). The model computes water use in domestic, industrial and agriculture sectors. In the domestic sector, water use is computed by relating changes in national income to changes in the amount of water used per person, and in the industrial sector to the water withdrawn per unit electricity generated. These calculations also take into account the saturation of water demands at high incomes, as well as continuing improvements in water use efficiency due to technological change. Water requirements for irrigated crops are computed by taking into account the location of irrigated areas, local climate, and crop and management variables. Water availability (equivalent to the sum of surface runoff and groundwater recharge in each river basin) is computed from daily water balances of the vegetation canopy and soil. Water balance computations are driven by precipitation, temperature, and other climate data. A water balance is also performed for open waters, and river flow is routed through a global flow routing scheme.

WaterGAP 2 calculations of withdrawals have been calibrated against historical data provided by Shiklomanov (2000). WaterGAP first computes country-scale domestic and industrial water use and then uses the geographic distribution of population and other data to downscale these values to a geographic grid (0.5° latitude x longitude). Because of the uncertainty of the downscaling procedure, Russian-total estimates have a lower uncertainty than grid estimates. With regards to the agricultural water use sector, irrigation water use is computed on the grid basis, and hence there is no uncertainty involved in downscaling country estimates. On the other hand, Alcamo et al. (2003a) report that estimates of irrigation water use in Russia are based on incomplete information about the location of irrigated land in the country. To sum up, estimates of grid-scale water withdrawals in this paper have low to medium certainty, although statements about overall trends of water withdrawals for particular scenario assumptions have much higher reliability.

The runoff calculations of WaterGAP have been calibrated to the long term average of multi-year river discharge measurements representing most of the river basin area of Russia (Alcamo et al. 2003a; Döll et al. 2003). After calibration, the model was "validated" by comparing computed and measured time series of annual average river discharge. Alcamo et al. (2003b) and Döll et al. (2003) showed that the "modeling efficiency" (Nash-Sutcliffe coefficient) of this comparison was 0.5 to 0.7 using Russian runoff data, which is considered satisfactory. (The modeling efficiency is a measure of the goodness-of-fit of a model to the variance of the measurement data.) In general, model estimates of year-to-year variation in water availability (river discharge) are judged to have a medium level of certainty.

Res□□s □f □ □□e□n□ □f □□e c□□□ □e c□□n□e

Climate change scenarios show a significant increase of temperature and a moderate increase of precipitation (Table 3). By 2020s, HADCM3 and ECHAM4 SRES-B2 scenarios demonstrate a fairly higher temperature

Table 3. Climate change scenarios for agricultural zones of Russia

			Annual temperature	Summer temperature	Annual precipitation	Summer precipitation
Current climate			-5.2	12.9	457	187
2020s	A2	HADCM3	-3.4	14.4	486	196
		ECHAM4	-3.	14.1	497	192
	B2	HADCM3	-3.1	14.9	497	198
		ECHAM4	-2.4	14.7	497	186
2070s	A2	HADCM3	-0.4	12.5	546	211
		ECHAM4	0.9	17.4	540	194
	B2	HADCM3	-1.3	16.4	519	206
		ECHAM4	0.1	16.7	527	192

increase than SRES-A2 due to bigger driving force during the first decades of B2 scenario. However, by 2070s the SRES-A2 temperatures are about 1°C higher than SRES-B2 temperatures for both HADCM3 and ECHAM4. Also, temperature increase is lower in HADCM3. For HADCM3, final annual temperature growth is 4.8°C for A2 and 3.9°C for B2 scenario, for ECHAM4, 6.1°C and 5.3°C, correspondingly.

Precipitation follows a similar pattern: both A2 and B2 scenarios show a comparable precipitation increase in 2020s, but in 2070s A2 scenario demonstrates higher precipitation growth. Both HADCM3 and ECHAM4 demonstrate virtually identical increase of precipitation for the same scenarios. Again, this additional precipitation is not distributed evenly. The main part of Russian territory experiences moderate precipitation increase; however, the most intensive agriculture lands in the south of the European part of the country experience precipitation decline (Figure 1). Two regions – North Caucasus and South Urals – will perform differently in terms of change of summer precipitation. The former will experience dramatic decline of summer precipitation while the latter will see no change.

Figure 1. The change of summer precipitation (%) relatively to 1961–1990, scenario A2 for model HadCM3 for 2020s (a) and 2070s (b)

We used the Summer Hydro Thermal Coefficient (HTC) by Seljaninov (1966) to quantify the effect of changing temperature and precipitation on climate aridity. The HTC is computed as a sum of precipitation during the vegetation period, multiplied by 10 and divided by a sum of effective temperatures, with the vegetation period defined as a period with average daily temperatures above 10°C. HTC values usually stay within 0.4 and 2, with lower values corresponding to dryer conditions. HTC values below 0.7 are considered to differentiate droughts, and values from 0.7 to 1 explicate moderately dry summers.

We computed HTC for three time periods: current climate and two time slices (2020s and 2070s) of the climate change scenarios. Since both current and future scenarios of climate supply the monthly averages only, monthly climatology was further transformed into daily values using the stochastic weather generator (Friend 1998). Our computations show that the majority of the principal agriculture regions of Russia will be drier in the future than they are now, especially during the summer months. For the European part of Russia, the temperature will increase by 1–2°C by 2020s and by 3–4°C by 2070s, while precipitation will stay the same or even decrease. For a handful of the regions with intensive agricultural production in the Asian part of the country, temperature increase will be followed by correspondent increase in precipitation; however, more favorable climate here is unlikely to compensate for drier climate in the majority of agricultural regions.

Both the expected frequency of droughts (Figure 2) and drought severity are expected to increase. For example, for Stavropolsky kray (North Caucasus) current drought frequency is 28 dry years in a century, in 2020s – 64, and in 2070s – 89 years (HADCM3; for ECHAM4, 32 and 70 years, correspondingly). For Krasnodarsky kray (North Caucasus), current value is 21 years, in 2020s – 51, and in 2070s – 67 years (HADCM3; for ECHAM4, 32 and 57 years,correspondingly). The severity of droughts also increases: average HTC drops from 1 to 0.65 for Krasnodarsky kray and from 0.9 to 0.5 for Stavropolsky kray (HADCM3). Urals region will also experience more frequent droughts: in Orenburg oblast it will increase from 40% to 60%. However, most dramatic change is expected for North Caucasus in terms of both average climate deterioration and risk of droughts. The SRES-B2a shows a similar development. The northern regions of European Russia and the Far East, on the contrary, do not show any significant change in frequencies of dry summer weather. For example, in both

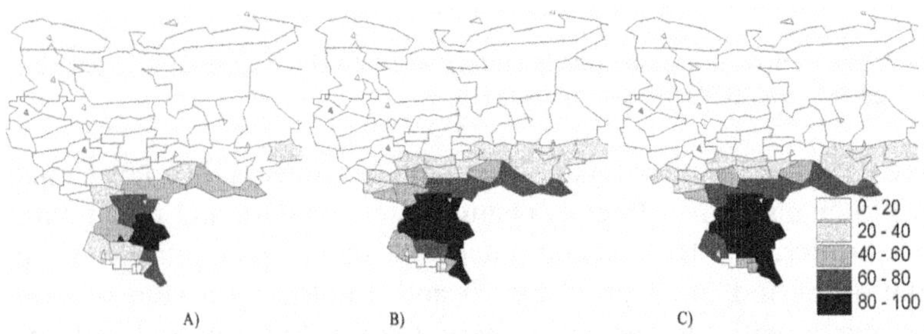

Figure 2. Change in frequency of droughts (%) in the south of European Russia: 1961–1990 (a) and projections for the model HadCM3 for 2020-x (b) and 2070-x (c), scenario A2

Leningradskaya oblast (North-West) and Khabarovsky kray (Far East) the corresponding values have 1–5% probability of dry summer. A lengthening vegetation period and increasing precipitation can be the factors that will benefit agriculture of the regions situated farther north.

Results of modelling of impact of the climate on crop production

Similar to other simulation models of Russian agriculture (e.g., Sirotenko et al. 1997), the model of agroecological zones GAEZ demonstrates a considerable decrease of cereal yields in the currently most production part of Russia. Even though the agricultural production increases in some regions, multiyear average yield decreases considerably due to more frequent droughts in the most production regions, including North Caucasus. At its extreme, in Stavropolsky krai cereal production decreases by 23% in 2020s and by 56% in 2070s (under HADCM3 SRES-A2 scenario). In contrast, yields of cereals in the Central geographic region do not change much, and yields in the northern regions increase significantly. Climate change also benefits grain production in East Siberia, where the climate becomes milder with growing temperature and increasing precipitation (Alcamo et al. 2003c).

However in Russia as a whole, the gains largely balance out the losses. Depending on the scenario, we compute either a 9% loss or a 12% gain in total potential grain production by the 2020s (relative to averages during the climate normal period). By the 2070s, only losses are estimated, ranging from 5% to 12% for net country-wide grain production.

We have arrived at different conclusions after examining the change in frequency of occasional but severe droughts that can lead to temporary but serious shortfalls in food production. We define a "food production shortfall" as an event in which the annual potential (i.e. climate-related) production of the most important crops in an administrative region in a specific year falls below 50% of its climate-normal (1961–1990) average. Under current climate conditions, food production shortfalls (as defined above) typically occur in the main crop growing regions in about one to three years out of every decade, depending on the region Although these are theoretical estimates based on model calculations, they compare well with the actual situation in these regions (Dronin and Bellinger 2005).

Our calculations show that in the 2020s the majority of the main crop growing regions experience more frequent shortfalls because of combined warmer temperatures and declining precipitation (Figures 1 and 2). In some regions the frequency of shortfalls doubles. These results pertain to both the A2 and B2 scenarios, and climate scenarios from both climate models. By the 2070s almost all of the main crop growing regions show large increases in the frequency of shortfalls. Some of these regions will have three times the frequency of shortfalls in the 2070s as compared to current climate conditions. Projected frequency of crop failures for North Caucasus reaches 5–6 times and for South Urals crop failure risk is estimated at 3–4 years for a decade in climate of 2070s. In current climate the former is more stable than the latter. Detailed picture for food situation in Russia in the future climate is presented in recent work by N. Dronin and A. Kirilenko (2008), in which risk of food crisis was estimated for different market scenarios.

Results for changing water availability and withdrawals

Changes in climate that affect crop production will also affect water availability throughout Russia as it is in detail discussed in recent work of J. Alcamo (Alcamo et al. 2007). The WaterGAP submodel nested in GLASS was used to assess the current and future level of water availability, withdrawals and stress under the A2 and B2 scenarios. Based on this modeling assessment the authors estimate that most river basins in the territory of Russia currently have relatively low levels of water stress (Figure 3). This can be explained by the low population density over much of its territory and the resulting modest level of water withdrawals compared to the volume of surface and groundwater available. Of course a low level of water stress does not necessarily mean that there are adequate canals, pumping

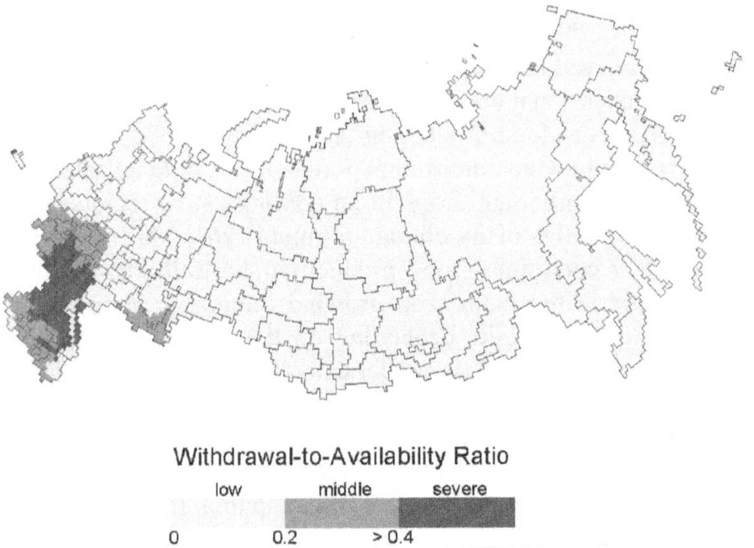

Figure 3. Current water stress in Russia

stations, or other infrastructure to deliver the available water to all users. Also, as explained previously, these calculations are carried out on a 0.5° x 0.5° geographic grid and this spatial averaging overlooks important local water contamination caused by wastewater discharges of municipalities, industries and cropland. Figure 3 shows a particularly high level of water stress over a large area of Southwest Russia caused by a high level of water withdrawals of households, industry and agriculture relative to available water. In these river basins – Kuban, Don and Urals- strong competition is expected between different water users.

How will the water situation change in the future? Figure 4 depicts the change in average water availability in the 2070s using climate scenarios from two climate models. The change in water availability is the result of two opposing tendencies. On one hand, a warmer climate increases evapotranspiration which tends to decrease water availability (river runoff plus groundwater recharge). On the other hand, precipitation is increasing

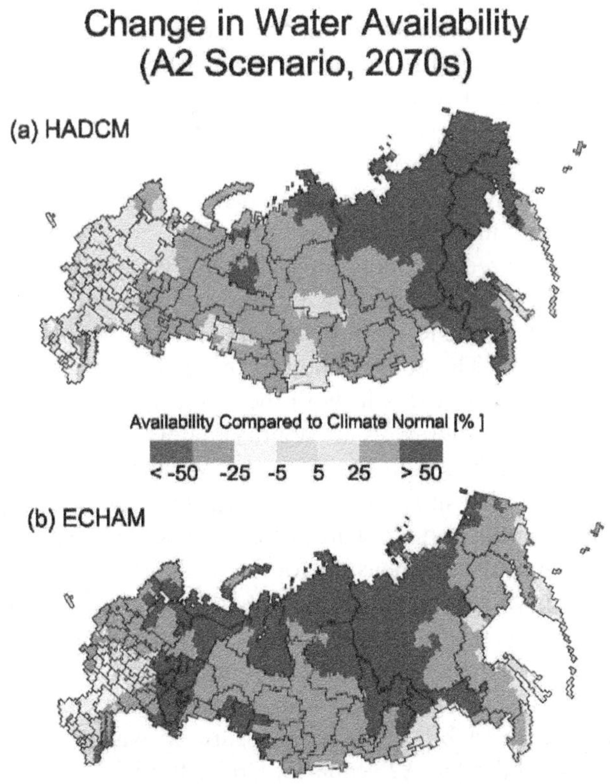

Figure 4. Change in water availability for A2 scenario, 2070s, HadCM (a) and ECHAM (b) models

throughout most of Russia which tends to increase water availability. Figure 4 indicates that the precipitation effect prevails, and water availability increases over more than 90% of the country. An important exception is the North Caucasus where water availability decreases because of a decline in precipitation and increase in temperature. However, Urals river basin will show better water availability due to increase in precipitation (mostly winter) in up stream part of the basin.

But the future water situation in Russia depends not only on the availability of water (including its quality) but also on future water uses. We have found that the assumptions of the A2 and B2 scenarios lead to different trends in water use in the domestic and industry sectors (Table 2). Under the A2 scenario, per capita domestic withdrawals increase because an increase in income leads to the fulfillment of expectations for higher water use. Although income also increases under B2, domestic withdrawals actually decrease between 1995 and 2025 (Table 2). This is because much stronger improvements are assumed in water use efficiency, consistent with the environmental orientation of the B2 scenario. Domestic water withdrawals are also lower under B2 because it has decreasing population (fewer water users) over the scenario period, while population is stable or increasing (more water users) under the A2 scenario (Table 1). Industrial water withdrawals decrease in both scenarios because increasing water use efficiency outweighs the slow increase of industrial production. Since the improvement in water use efficiency is faster under B2, it also has more rapidly decreasing industrial water withdrawals.

Irrigation water requirements only slightly rise because of compensating trends – They tend to expand because of warmer temperatures but shrink because of efficiency improvements and additional precipitation in some irrigated areas. The combination of these two factors leads to a considerable drop in withdrawals in both the domestic and industrial sectors. Trends in irrigation withdrawals are approximately the same as in A2.

How do the combined trends of water withdrawals and availability affect the future water situation in Russia? Under the A2 scenario, withdrawals stabilize in the northern part of European Russia because decreasing industrial withdrawals compensate for increasing domestic withdrawals. Stabilizing water withdrawals combined with increasing water availability lead to decreasing pressure on water resources here under the A2 scenario. Meanwhile, in Siberia and the Far East increasing domestic water use leads to an increase in total water withdrawals which is stronger than the increase in water availability. Hence pressure on water resources in these regions increase under the A2 scenario. Under the B2 scenario, the combination of lower water withdrawals and increasing water availability implies lower pressure on water resources almost everywhere in Russia.

Water resources in the food-growing regions of North Caucasus are, as already noted, in the severe water stress category and remain in this category. Therefore it is questionable whether new sources of water for irrigation can be found here. This should be taken into account when assessing the feasibility of expanding irrigated cropland to enhance food security, as discussed above. Indeed, the irrigated area in Russia has peaked in the late 1980s at 3% of the total agricultural area (Novikova 2006). Half of these irrigated lands was located in the South of European Russia and primarily used for rice production. Half of the rest was located in the Central European Russia. The irrigated land was under considerable degradation and decreased in 1990s due to soil erosion, salinization, and mismanagement. One of the major problems was water availability, low quality of irrigation waters (e.g., industrial pollutions in the surface water and high salinity of the ground waters) and inadequate water management (Novikova 2006). Another complication is associated with an abundance of salinated soils in Southern and Southeastern regions, which present considerable managerial and technological challenges for irrigation projects (Gaponenko 2005). Other problems include water availability, the conflict of interests with electricity production and with water management for sturgeon hatcheries. All of these problems are likely to increase in the future with decreasing water availability in warmer climate and growing water demand from population and the industry (Alcamo and Henrichs 2002).

The occurrence of infrequent droughts plays an important role in food security. In the same way, infrequent events (droughts and floods) also have a significant connection to water security. It was estimated that climate change will cause a major change in the variability of river runoff in many parts of Russia (Figure 5a). One important result is that extremely low runoff events may occur much more frequently in some of the main crop-growing regions in North Caucasus (Figure 5b). The combination of severe pressure on water resources because of large water withdrawals, and more frequent occurrences of low runoff events, may signal a significant threat to the water security of the population living in these regions, as well as limit the expansion of irrigation here. However, no similar problem is expected for South Urals where water availability will be improved due to wetter climate in up stream part of the river basin.

In other areas of Russia, the frequency of extremely high runoff events may increase because of increased precipitation (Figures 1 and 5a). Since some rivers in Central Russia are already subject to regular flooding because of ice flow blockage, the additional risk of high runoff events in this region should be taken into account in future climate impact studies.

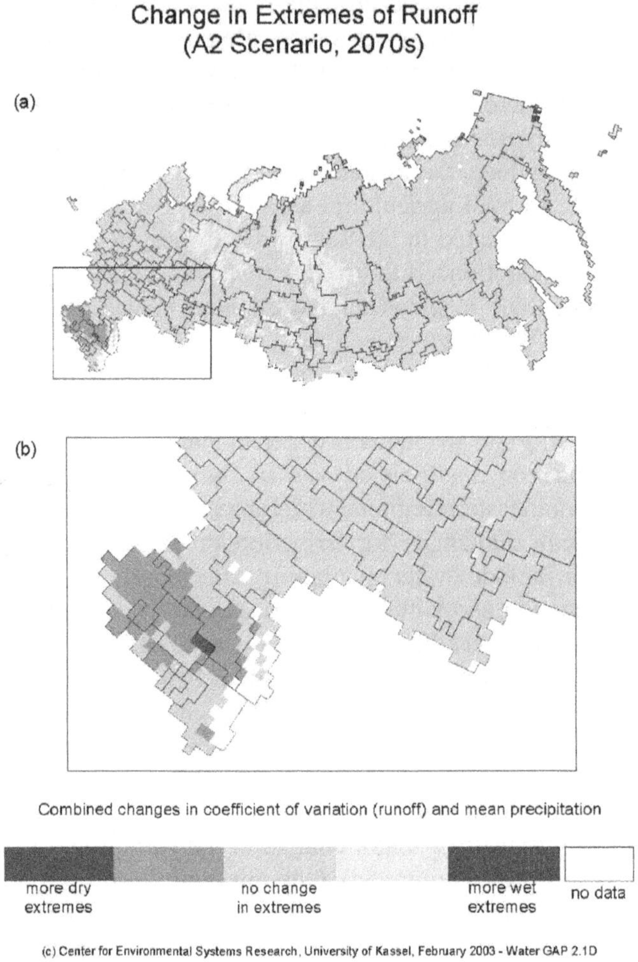

Figure 5. Change in extremes of runoff, A2 scenario, 2070s

Conclusions

In this work we compare performance of two problematic regions, North Caucasus and South Urals, in future climate of 2020s and 2070s. We calculated that climate will become drier in the both regions resulting some decline of potential harvests there. The main problems of the climate change are associated with not average characteristics of climate but with weather anomalies. Our work differed from other modeling exercises by its focus to assessing of risk of droughts for the new climate. Because of changes in the frequency of dry periods, we find the frequency of food production shortfalls could increase in the both regions by the 2020s and, especially, by the 2070s. Such increased risks are likely to require significant adaptive measures in the agricultural system. However, more radical change is expected for North Caucasus then for South Urals.

The second conclusion concerns availability of water resources in the both regions. While the wetter climate over much of Russia is expected to increase water availability on the average, it is also likely to lead to higher extremes in runoff over much of Russia's territory. Conversely, the main crop growing regions in the South could experience more frequent low runoff years. This combined with the already high level of pressure on these water resources, could limit the expansion of irrigated agriculture in this key agricultural region. However, the situation could be very different in South Urals and North Caucasus although the both regions currently have relatively high water stress. In future climate some increase in winter precipitation in upper part of the watershed of the Urals river could bring benefit for water regime in the whole basin while in the Kuban and Don river basin water stress is expected to become acuter then nowadays.

References

Alcamo, J. and Henrichs, T., 2002, Critical regions: a model-based estimation of world water resources sensitive to global changes, *Aquatic Sciences*, 64: 1–11.

Alcamo, J., Endejan, M., Kaspar, F., and Rösch, T., 2000a, The GLASS model: a strategy for quantifying global environmental security, *Environmental Science and Policy*, 4(1): 1–12.

Alcamo, J., Henrichs, T., and Rösch, T., 2000b, *World water in 2025 – global modeling scenarios for the World Commission on Water for the 21st Century*, World Water Series 2, Center for Environmental Systems Research, University of Kassel, Germany.

Alcamo, J., Döll, P., Henrichs, T. Kaspar, F., Lehner, B., Rösch, T., and Siebert, S., 2003a, Development and testing of the WaterGAP 2 global model of water use and availability, *Hydrological Sciences*, 48(3): 317–337.

Alcamo, J., Döll, P., Henrichs, T. Kaspar, F., Lehner, B., Rösch, T., and Siebert, S., 2003b, Global estimation of water withdrawals and availability under current and "business as usual" conditions, *Hydrological Sciences*, 48(3): 339–348.

Alcamo, J., Dronin, N., Endejan, M., Golubev, G., and Kirilenko, A., 2003c, Will climate change affect food and water security in Russia? Report A0302, Center for Environmental Systems Research, University of Kassel, Kurt Wolters Strasse 3, D-34109 Kassel, Germany.

Alcamo, J., Endejan, M., Golubev, G.N., Kirilenko, A.P., and Dronin, N.M., 2007, A new assessment of climate change impact on food production shortfalls and water availability in Russia, in *Global Environmental Change*, 17(3): 429–444.

CIESIN (Center for International Earth Science Information Network), 2002a, Country-level population and downscaled projections for the SRES B2 scenario, 1990–2100, beta version. Columbia University, Palisades, NY; http://sres.ciesin.columbia.edu/tgcia/.

CIESIN (Center for International Earth Science Information Network), 2002b, Country-level GDP and downscaled projections for the SRES A1, A2, B1, and B2 marker scenarios, 1990–2100, beta version. Columbia University, Palisades, NY; http://sres.ciesin.columbia.edu/tgcia/.

Döll, P., Kaspar, F., and Lehner, B., 2003, A global hydrological model for deriving water availability indicators: model tuning and validation, *Journal of Hydrology*, 270: 105–134.

Dronin, N. and Bellinger, Eds., 2005, Climate Dependence and Food Problems in Russia (1900–1990), New York/Budapest: CEU Press, 360 p.

Dronin, N.M. and Kirilenko A.P., 2008, Climate change and food stress in Russia: what if the market transforms as it did during the past century, *Climate Change Journal*, Stanford University 86(1–2): 123–150.

Fischer, G., van Velthuizen, H., and Nachtergaele, F.O., 2000, Global agroecological zones assessment: methodology and results. Internal report IR-00-06, International Institute of Applied Systems Analysis, Luxemburg, Austria.

Fischer, G., Shah, M., and van Velthuizen, H., 2002, Climate change and agricultural vulnerability, Special Report of the International Institute for Applied Systems Analysis as a contribution to the World Summit on Sustainable Development, Johannesburg, 2002.

Friend, A.D., 1998, Parameterization of a global daily weather generator for terrestrial ecosystem modeling, *Ecological Modeling*, 109: 121–140.

Gaponenko, N., 2005, The way towards sustainable water use in food production: Russian vision and framework for actions. Prospective Foresight, June 28 2005. http://www.prospective-foresight.com/IMG/pdf/Paper_WATER_FOR_FOOD_PRODUCTION1.pdf. Last visited: Dec. 9, 2006.

Gordon, C., Cooper, C., Senior, C.A., Banks, H., Gregory, J.M., Johns, T.C., Mitchell, J.F.B., and Wood, R.A., 2000, The simulation of SST, sea ice extents and ocean heat transports in a version of the Hadley Centre coupled model without flux adjustments, *Climate Dynamics*, 16: 147–168.

GOSCOMSTAT, 1998a, Regions of Russia, Statistical Report (in Russian). Vol. 2. Goscomstat, Moscow.

GOSCOMSTAT, 1998b, Population of Russia: 1897–1997, Statistical Abstract (in Russian), Goscomstat, Moscow.

IPCC (Intergovernmental Panel on Climate Change), 2001, Climate Change 2001: Impacts, Adaptation, and Vulnerability, Cambridge: Cambridge University Press, 1032 pp.

Lutz, W. and Goujon, A., 2002, Downscaling of regional population projections to country-level for the SRES A1, B1 and A2 scenarios corresponding to IPCC SRES Report, 1990–2100. International Institute for Applied Systems Analysis (IIASA), Laxenburg, Austria. Available at http://sres.ciesin.columbia.edu/tgcia/ and http://www.iiasa.ac.at/Research/POP/IPCC/index.html

New, M.G., Hulme, M., and Jones, P.D., 2000, Representing twentieth century space-time climate variability, Part II. Development of the 1901–1996 monthly grids of terrestrial surface climate, *Journal of Climate*, 13: 2217–2238.

Novikova, A.F., 2006, Degradation Processes in Irrigated Soils of Russia, the 18th World Congress of Soil Science (July 9–15, 2006), Philadelphia, PA.

Pegov, S.A., Khomyakov, D.M., and Khomyakov, P.M., 2000, Global change impact on socio-economical processes in Russia. In: Kotlyakov, V.M. Global and regional climate change and its environmental and socio-economical impact. Moscow. GEOS: 60–69 (in Russian).

Pope, V., Gallani, M., Rowntree, P., and Stratton, R., 2000, The impact of new physical parameterizations in the Hadley Centre climate model – HadCM3, *Climate Dynamics*, 16: 123–146.

Roeckner, E., Arpe, K., Bengtsson, L., Christoph, M., Claussen, M., Dümenil, L., Esch, M., Giorgetta, M., Schlese, U., and Schulzweida, U., 1996, The atmospheric general circulation model ECHAM4: Model description and simulation of present-day climate. MPI Report No. 218.

Seljaninov, G.P., 1966, Agroklimaticheskaja karta mira. (Agroclimatic map of the world). Leningrad.

Shiklomanov, I., 2000, Appraisal and assessment of world water resources, *Water International* 25(1): 11–32. (supplemented by CD-ROM: Shiklomaonv, I., World freshwater resources, available from: International Hydrological Programme, UNESCO, Paris).

Sirotenko, O., Abashina, E., and Pavlova, V., 1997, Sensitivity of the Russian agriculture to changing climate, CO_2, and tropospheric ozone concentrations and soil fertility, *Climatic Change*, 36: 217–232.

Vörösmarty, C.J., Green, P., Salisbury, J., and Lammers, R.B., 2000, Global water resources: vulnerability from climate change and population growth, *Science*, 289: 284–288.

2.2 FISH RESOURCES

[illegible]

[illegible]*

Central European University, Budapest, Hungary

[illegible]

Ural Basin Project, Russia

Abstract. The Ural river, the third longest river in Europe, has the only remaining spawning habitats in the entire Caspian basin for all sturgeon species. Unlike other large European rivers the river's ecosystem has not been altered and the natural hydrological regime is still intact. The Ural sturgeon yield-to-fishery relative to river discharge was the highest in the Caspian Sea till recently. The environmental conditions to secure natural reproduction are still satisfactory for successful sturgeon reproduction. However, nowadays the catch in all regional sturgeon species is negligible. The Ural sturgeon population dynamics are analyzed along with some anthropogenic and natural factors affecting them. It is argued that legal overfishing (including all legal means of fish removal), based upon (a) faulty estimations of sturgeon stock and catch limits and (b) inappropriate fishery policies are the principal reasons for the stock decline in the Ural. The maintenance of the natural reproduction in the Ural is considered to be the primary strategy for the stock replenishment. If used at all, artificial propagation should be used only as an additional secondary option exclusively at the historical sturgeon habitats upstream the Ural river and not in the river delta, where the hatcheries are located now. Transboundary cooperation of basin countries with active international involvement is essential to prevent further deterioration of the situation.

Keywords: Caspian sea, Ural river, beluga, Huso, Sevryuga, ship, Russian sturgeon, Persian sturgeon, bioindicator, overfishing, impoundment, Cossacks, CITES, hatchery, total allowable catch

---

* Viktor Lagutov, Department of Environmental Sciences and Policy, Central European University, Nador ut.9, H1051, Budapest, Hungary; Lagutov@gmail.com

Introduction

Sturgeons are among the most interesting species in the world. They have successfully survived from the time of the dinosaurs. Extremely high plasticity helped them to adapt to the changing environment through all these millions of years. The historic range of sturgeon species are the main rivers of the Northern hemisphere. Each river basin had a stock of its own with specific features and life cycle characteristics. By now they have vanished from most of them (FAO 2007b; IUCN 2007; WWF 2002a).

The sturgeon is an *anadromous* species, whose reproduction takes place in freshwater river basins with the growing and maturing phases occurring in the sea. After maturation in salted water sturgeons migrate back to freshwater for the purpose of breeding. Particular environmental conditions are required for spawning, depending on species: hard substrate (pebble, gravel, etc.), stream velocity (0.5–2.0 m/s), depth (1–20 m), temperature regime, etc. Spawning habitats are located in the upper branches of rivers. The distance to these grounds can be, depending on species, more then 1,500 km from the river delta. The size of adult specimens varies from 0.5 to 6 m and from 0.5 kg to 2 t. The sturgeon is a long-lived fish standing at the top of food webs.

The extinction of sturgeon species is one of the most tragic and representative examples of the destructive influence of humankind on Nature. Sturgeon, sometimes called the "living fossil" or living "dinosaur" of the fish world, is currently on the verge of extinction solely due to anthropogenic impact.

It is estimated that the number of sturgeons in major basins has declined by 70% over the last century (WWF 2002a). Out of 15 sturgeon species known, most are considered critically endangered or vulnerable to extinction worldwide (WWF 2002a). At the same time some regions are suffering more significant and dangerous trends then others (Pitikch et al. 2005). Sturgeons of the Aral Sea are extinct, while sturgeons of the Sea of Azov are on the verge of extinction (AzovBas 2002; Russian State Duma 1995; Lagutov 1995).

Sturgeons are among the world's most valuable wildlife resources. Gessner et al. (2002) estimate the demand on world export markets for caviar, the delicacy derived from sturgeon roe, at 500 t annually (Gessner et al. 2002). The global caviar trade was a major driving force of the sturgeon fisheries worldwide. The leadership role in international caviar trade shifted from the United States in the 19th century to Russia after the USA's stock's depletion (DeMeulenaer and Raymakers 1996; Pitikch et al. 2005). Russia was the main caviar trader throughout the 20th century due to the active utilization of the enormous Caspian sturgeon stock.

The Caspian Sea is considered to be the world's biggest sturgeon habitat, holding at its peak up to 90% of the world's sturgeon stock (CEP 2002a). Most of the caviar consumed in the world during the 20th century originated in this region. Unfortunately, nowadays these estimations should be treated as outdated. Nowadays despite the active fishing efforts, both legal and illegal, by littoral countries, the catch is miserable. The Caspian sturgeon stock has decreased drastically and some authors claim it to be on the verge of extinction (Chivers 2006; Dulvy et al. 2003; Itoh et al. 2004; Pourkazemi 2007; Uralbas 2007b).

During its history the Caspian Sea went through a series of dramatic changes. Sturgeons could adapt successfully to all challenges: geological transformations, sea level fluctuations, salinity, temperature regime changes, etc. But human activities in the region are about to put an end to the long history of this species.

The drastic decrease in the sturgeon population of the Caspian Basin is caused by various factors (sea level fluctuations, pollution, etc.), but the main ones are believed to be blockage of the spawning places and migration routes by dams and overfishing on the main basin rivers (Uralbas 2007a). The historical worldwide overfishing of sturgeon species throughout Europe and Northern America since Roman times (Keysler 1762) as a reason for stock decline cannot be equally applied to the Caspian basin due to the peculiarities of regional environmental and human history. This region was mostly populated by nomads, not practicing fishing, and the initial number of fish was abundant.

Sturgeon catch as an indicator of the size of the sturgeon population strongly depends upon natural river flows. The variations in catch reflect changes in the numbers able to pass up the rivers to spawn (CEP 2002a). In case of complete blockage or severe reduction in the spawning places the sturgeon population is doomed to extinction even without any fishing efforts. The sturgeon is a marker of both ecosystem health and the sustainability of human activities in the region.

Numerous programs have been launched worldwide aimed at sturgeon restoration. Sturgeon population rehabilitation is a long and complicated process. Success in this challenging task depends upon a wide range of environmental and anthropogenic factors. Thus, only an integrated holistic ecosystem approach to both river basin and related human activities can secure sturgeon rehabilitation.

While some of these programs show some degree of success (namely US-based ones), most fail not only to restore degraded habitats (Buijse et al. 2002; Williot et al. 2002a, b), but even to find a couple of productive breeders to start a restocking program in hatcheries as in European basins (Williot et al. 2000). Furthermore, the effect of artificial propagation as a popular measure to sustain wild sturgeon population is dubious and challenged by many researchers.

From this perspective the Ural River, the third longest river in Europe and second in the Caspian basin, is unique since it contains the only self-sustaining, viable sturgeon population capable of natural reproduction. Though more then 100 rivers empty to the Caspian Sea (Pitikch et al. 2005) sturgeons can reproduce only in major rivers. Every significant Caspian river was impounded in the 1930–1970s, cutting off the sturgeon spawning grounds. The Iranian rivers (Sefidroud, Gorganrud and Tajan), minor in comparison to the f.USSR rivers' contribution to freshwater influx and sturgeon reproduction, have also been dammed recently (Abdolhay 2004).

Moreover, the remaining sturgeon habitats, historical or believed to be appropriate for spawning, often do not have proper hydrological conditions. For instance, during the first 40 years since the lower Volga's regulation the environmental flow conditions at the spawning grounds downstream the Volgograd dams were flooded only 13 times (Dubinina and Kozlitina 2000).

Figure 1 depicts the biggest dams on the main Caspian river basins. The Ural is the only river with non-regulated low and middle water course for more than 1,000 km upstream the delta, which is an historic range of sturgeon spawning and nursing habitats.

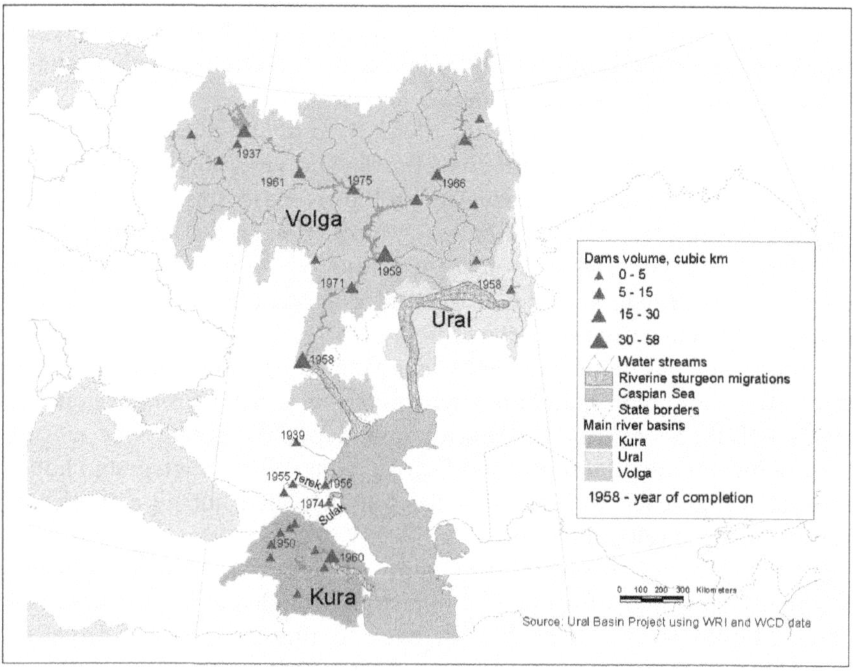

Figure 1. The main river basins of the Caspian Sea with biggest dams and waterworks. The current sturgeon areal in the rivers indicated with dotted areas

The future of the whole Caspian sturgeon stock and worldwide restoration programs depends on the Ural River's spawning and nursing habitats. Till recent times the Ural was able to support abundant sturgeon populations.

However, during the last few decades the catch in the basin has dropped by a factor of 100. Urgent measures are needed to conserve this flagship species and unique ecosystem. Out of six different sturgeon species inhabiting the Ural river basin, five are indicated in the IUCN Red Book as endangered or critically endangered (IUCN 2007). Many authors consider even these conclusions and actions as too optimistic and believe that the "point of no return" towards extinction for most sturgeon populations has been reached (Dulvy et al. 2003; Jonsson et al. 1999; Lagutov 1995; Smith et al. 1993; Stephan and Wissel 1999).

□□□r□e□n s□ec□es □n □r□□

Sturgeons, like other anadromous species, recognize their native river catchments and return there for spawning. Little is known about this phenomenon, called "*homing*". Some theories suggest that homing depends principally on olfactory recognition of streams.[1] As a result, each river basin in the historical range had its own sturgeon stock.

Sturgeon populations in the Caspian's main tributaries possess unique characteristics and life cycle peculiarities. Historically, the specimens originating from different river basins were easily recognized by specialists. Moreover, despite growing and extensively migrating in the sea for 10–20 years upon maturation the sturgeon could identify their own river basins to start spawning.

All sturgeon species living in the Caspian Sea had their distinct populations in the Ural: the Beluga (*Huso huso Linnaeus, 1758*), the Russian Sturgeon (*Acipenser gueldenstaedtii Brandt, 1833*), the Sevryuga (Stellate sturgeon, *Acipenser stellatus Pallas, 1771*), the Ship (*Acipenser nudiventris Lovetsky, 1828*), the Sterliad (Sterlet, *Acipenser ruthenus Linnaeus, 1758*) and the Persian Sturgeon (*Acipenser persicus Borodin, 1897*).

The Beluga Sturgeon or the Great sturgeon is considered to be the most valuable sturgeon species worldwide. The World Wildlife Federation named beluga as the fourth most endangered species on Earth in 2002 (CEP 2002a), but it is still legally harvested and exported from the region.

---

[1] Though being challenged by some researchers homing fidelity is the only explanation to the existence of the river-based sturgeon populations and inherent genetic variations among them after 10–20 years of maturation and active migrations in the sea with numerous incoming river streams.

The Ural River plays a special role in Caspian sturgeon reproduction. Today the river contains the only available spawning grounds in the whole basin. Since 1979 the numbers of beluga spawners entering the Ural have exceeded the number of fish trying to spawn in the Volga (Khodorevskaya et al. 1997). The Beluga catch in the Ural in the 1990s was up to 70% of the total f.USSR beluga catch despite the fact that sturgeon hatcheries in the Ural river did not exist (unlike several in the Volga region releasing hundreds of millions fingerlings annually) (KaspNIRH 1999).

However, the Ural River is a unique ecosystem not only due to its current exclusive position. Even before the regulation of the Caspian rivers the productivity of the Ural ecosystem was as high as that of the Volga, even though total water flow is 25–30 times smaller (!). In particular, the mean total flow in the Ural is 9–10 km^3, while the Volga has 260 km^3. At the same time, average total sturgeon annual yield from the fisheries in the Ural and Volga rivers was roughly equal – 11,000 t[2] from the latter, while the yield from the Ural could reach up to 15,000 t (KaspNIRH 1999).

The sturgeon spawning grounds in the Ural River were much more efficient and the sturgeon population was much more productive. This is a very interesting phenomenon which is still not paid due attention in the literature and environmental programs. Most attention is paid to the Volga River, as the biggest Caspian tributary and the habitat for the biggest number of species in the region. For instance, currently, according to research conducted by the Caspian Fishery Research Institute, the contribution of the Volga ecosystem to the sturgeon stock in the Caspian Sea is 69.8%, the Ural's is 29.7%, while the Kura and Terek together only contribute 1.4%.[3] At the same time, total freshwater influx delivered by the Ural is only 3% against 80% by the Volga and a total of 8.8% by the rivers Kura (6.3%) and Terek (2.5%). Based on these official estimates the river productivity ratio (sturgeon catch/water influx) for the Ural river is 9.9, while for the Volga and Kura/Terek it is 0.87 and 0.126 respectively (Figure 2).

In general all sturgeon species share the same life cycle stages and characteristics with variations in terms of maturation, growth, fecundity, etc. (Detlaf et al. 1981; FAO 2007a).

[2] These estimations correspond to the highest catch in the history of the sturgeon fishery in the region. As a result sturgeon stocks were overexploited and have never restored afterwards.

[3] These estimations by KaspNIRH are based on number of released fingerlings from hatcheries in Volga, Kura and Terek and results of natural reproduction in Ural.

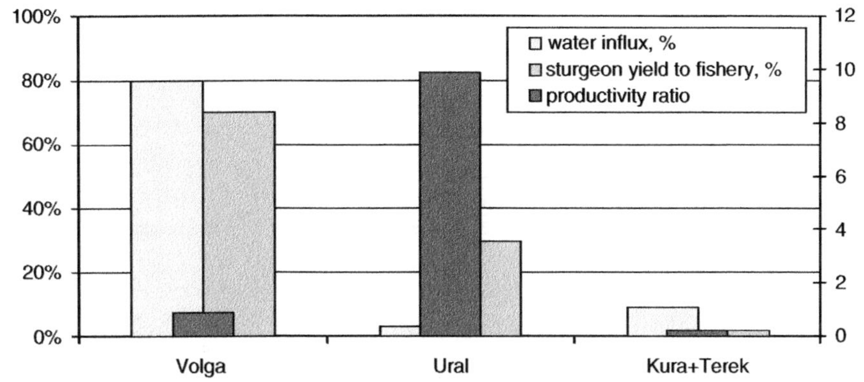

Figure 2. The contribution of main river basins to the Caspian Sea in freshwater influx and sturgeon yield to fishery

Sturgeons have an age-structured population. The development of specimens in every species goes through the same stages:

- Embryonic development
- Prelarvae – until transition to active feeding
- Larvae – able to feed actively (12–14 days after hatching)
- Fry (20–30 days after hatching)
- Juveniles – until maturation
- Adults

Sturgeon species can spawn only in freshwater in the spawning grounds located in upper river branches. Eggs are deposited on hard substrate (stones, gravel, pebble, coarse sand, etc.).The distance from the river delta to spawning grounds depends upon the species. This is an important consideration in the sturgeon life cycle, since larvae and fingerlings migrating downstream, or rather washed down with the water flow, need to reach a certain age to survive in the brackish water of estuaries. Some young sturgeons winter in the river and migrate into the sea in the following year. With regards to mature specimens, sturgeons are mostly euryhaline and eurythermic species.

Sturgeons enter the Ural for spawning in different periods, but in most Ural sturgeon species the vernal (spring) races prevail. Spawners from these races go into the river for spawning in the spring during the flood period. For instance, 80% of the beluga spawning population consists of vernal migrants (Peseridi 1971). It should be noted that spawning itself occurs only once a year. The vernal migrants travel long distances to

Table 1. Some characteristics of Ural sturgeon species (Compiled using materials from CEP 2002b; EPA 2004; FAO 2007a; KaspNIRH 1999; Khodorevskaya et al. 1997; Peseridi 1971)

Species	Maturity		Spawning migrations in Ural (km)	Fecundity (thousand eggs)	Average weight (kg)	Reproductive age	Maximum age	Spawning periods
	Male	Female						
Beluga	12–15	15–20	1,200	680–800	350	11–55	100 and more	Males 3–5 years
Huso huso (Linnaeus, 1758)					Up to 1,500			Females 5–7 years
Russian sturgeon	7	8	1,000	320	50	7–40	50	Males 4–5 years
Acipenser gueldenstaedtii Brandt,					Up to 600			Females 5–6 years
1833 Sevryuga	5–6	7–9		236–253	5–10	5–30	30	3 times in life
Acipenser stellatus Pallas, 1771								
Ship	9–13	13–16	350–650	386–561	20	9–40	32	2–3 years
Acipenser nudiventris Lovetsky,					Up to 80			
1828 Persian sturgeon	7–8	9–10	1,000	320	30	15–38	50	4–5 years
Acipenser persicus Borodin, 1897								
Sterlet	4–5	7–8		4–140	0,5–2	3–17	30	
Acipenser ruthenus Linnaeus, 1758					Up to 15			

spawning habitats from the sea in one attempt. The migrants from the autumn (winter) race enter the Ural in advance, winter in some river bed depressions ("wintering holes") halfway to the spawning grounds and then join the vernal race for the spawning migration in the spring.

Sturgeons are late maturing species. Females mature at the age of 7–20, males from 4–15 years old depending on the species. Fecundity in adult sturgeon females increases with age. Generally they produce a greater number of eggs during each subsequent spawning run (EPA 2004). Besides that, the frequency of spawning runs also increases with age (Dmitriev and Vasilenko 2007).

The Ural river is the most important spawning habitat for the ship, Sevryuga and beluga (CEP 2002a). The first two species are spawned mostly in the Ural, while beluga spawners can be found in the Ural in higher numbers then in the Volga (Khodorevskaya et al. 1997).

The freshwater subspecies of ship and sterlet traditionally inhabited the Ural River as well. This fish did not go to the sea for maturing, but instead stayed in the river during all stages of their life cycle. However, with the river-based fishing strategy which has prevailed in the Caspian fishery since the 1960s, these species were exposed to much higher fishing pressure and have virtually vanished from the Ural. They are very rare species in the Ural nowadays.

The Ural sturgeons are subject to the same general trends as other Caspian sturgeon populations: decrease in numbers, reduced fertility, reduced body weight, decrease in average age.

It should be emphasized that while they originate in different river basins the sturgeon all spend most of their life cycle in the shallow coastal areas of the Caspian sea, actively migrating along the entire Caspian shore. For instance, during winter when the northern Caspian is covered with ice, most of the sturgeon specimens migrate to the south (KaspNIRH 1999). This peculiarity makes sturgeon a common resource for all littoral countries whatever their river of origin.

The same sturgeon species subpopulations often cannot be distinguished by appearance. For instance, the south sevryuga historically inhabiting the Kura differs from the Ural population only by later maturation age and lower fecundity (KaspNIRH 1999). Moreover, sometimes molecular analysis of the species designated upon morphology suggests no difference in them. So, the Persian sturgeon, despite having a different appearance and reproductive behavior, is sometimes not recognized as a species different from the Russian Sturgeon. This is an important consideration for the catch analysis presented later.

Spawning grounds in the Ural river

Six species, including the Persian Sturgeon (Peseridi 1986), historically had their spawning grounds in the Ural river (CEP 2002a).

According to estimates by the Russian Federal State Department on Fishery and Water Resources, the Ural river contains about 1,000 ha of sturgeon spawning grounds (KamUralRybVod 2007; KaspNIRH 1999). The grounds are equally shared between Russia and Kazakhstan. Most of these grounds, especially in the lower parts, are temporarily flooded, meaning that they are available for spawning only during the high water season (KamUralRybVod 2007). The water depth required for sturgeon spawning here is 2–5 m. Such water volumes are not available every year.[4] In low water seasons these grounds are gravel, sandy or limestone fields along the river stream. The natural hydrological regime with its high level spring flood is required for the normal functioning of these spawning grounds.

The spawning grounds of beluga and other sturgeon species in the Ural basin were historically located through most of the Ural River network starting from approximately 500 km from the delta. According to the Orenburg branch of the Federal State Fishery Department (KamUral-RybVod 2007) on the territory of Russia spawning grounds could be found:

- In the Ural and Sakmara rivers up to Kuvandiuk *raion* (small administrative territorial unit in former USSR countries) of the Orenburg Oblast
- In the lower stream of the Salmiush river, a tributary of the Sakmara
- In the mouth of the river Irtek, a tributary of the Ural
- In the river Ilek, and its tributary river Mazanka

In the mid 1980s the Guriev Branch of the State Fishery Institute[5] carried out an assessment of spawning grounds along the Ural River. The results showed much higher viability and survival rates for the juveniles originating from the spawning grounds in the upper Ural branches on the territory of the Orenburg oblast (Figure 3 shows spawning grounds with high productivity). There are different explanations for this phenomenon such as (a) only strong and well-fed specimens can reach grounds located far upstream, (b) larvae and juveniles from lower spawning sites might reach the brackish sea water too early, prior to the development of salinity resistance (Lagutov 1996; Peseridi et al. 1979).

---

[4] See the article on the Ural river hydrology in this volume.

[5] Now Kazakhstan Fishery Research Institute in Atyrau.

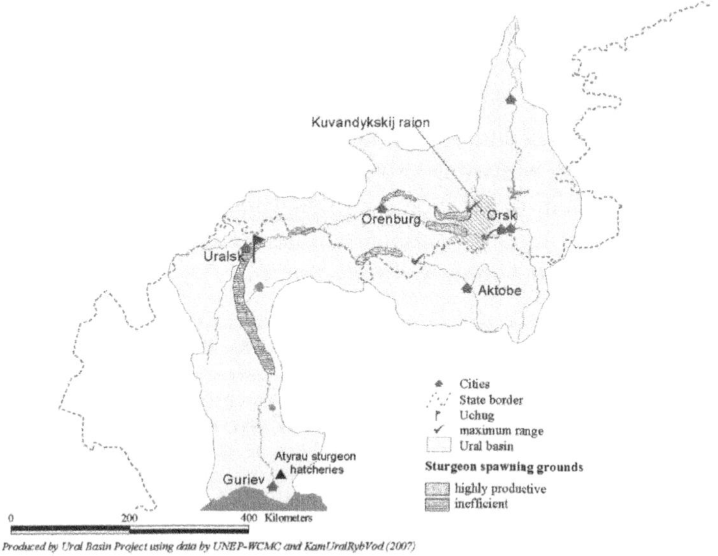

Figure 3. Sturgeon spawning grounds in the Ural river basin. There are evidences on spawning taking place up to Kuvandykskij raion (Orenburg Oblast, Russia)

Due to the high-level floods and preserved self-purification capacities of the Ural River, the precise location of the sturgeon grounds was always changing. In some years historic grounds were silted while other areas appropriate for spawning appear. Systematic monitoring of the spawning grounds has not been conducted for at least last two decades, thus only approximate ranges and upper limits for spawners can be indicated. Most of the available data on the location of spawning habitats is based upon the outdated results of the field research or observations conducted from the 1930s–1970s. Regular monitoring utilizing modern equipment (i.e. spawner tagging) is urgently required to obtain reliable information.

Currently, depending on favorable conditions, only beluga and Russian sturgeon appear occasionally in the spawning places in the middle course of the Ural River (Orenburg Oblast, Russia). The Sevryuga and ship do not reach spawning grounds in Russia. The Sterliad appears rarely in the middle Ural course and has a very small body size.

According to the observations made by the Federal State Fishery Department, the number of sturgeon specimens arriving at spawning grounds in Orenburg in 2004–2005 was around 100. Not a single beluga was seen at the spawning grounds despite the high water levels during the spring flood.

In 2006 the spring flood was characterized by low water volume and not a single sturgeon arrived in the spawning grounds in the middle Ural. In 2007 high water volumes were discharged from the Iriklinskoe reservoir and 10–20 belugas and 50–100 Russian sturgeons were observed in the middle Ural (Dmitriev and Vasilenko 2007). This year spawning has occurred in the Ilek River and its tributaries.

The efficiency of spawning grounds in the Ural in the 1970s was estimated at 11 thousand tons,[6] including 0.3–1.95 thousand tons for beluga, 0.16–0.36 thousand tons for Russian sturgeon, 2.4–8.3 thousand tons for sevryuga and 0.002–0.6 thousand tons for ship (KazNIRH,1999 cited by KaspNIRH 1999).

Sturgeon population as an indicator of the sustainability of watershed management

Apart from its high economic value, sturgeon is a perfect indicator (an umbrella) species for the river basin it inhabits (Lagutov 1995, 1996, 1997; Uralbas 2007a). The presence and well being of the sturgeon population in a river network indicates the "good quality" of a river ecosystem's health.

First of all, sturgeons utilize a variety of habitat types throughout their life cycles: rivers for spawning; rivers, lakes, estuaries, or the sea for feeding and wintering. Depending on the life stage the sturgeon habitats are spread through the whole river network, estuaries and adjacent marine areas. Living in the Caspian Sea and regularly migrating for spawning to the upper river branches in Russia through the territory of Kazakhstan, the Ural sturgeon population links together the marine and riverine ecosystems.

Figure 4 depicts the general Ural sturgeon life cycle with sea and river based stages distinguished. Some factors influencing sturgeon well-being are also indicated. The most influential factors for the Ural sturgeon populations are over-fishing, including all types of fish removal: commercial fishery, scientific, poaching, etc.; changes in river water regime; and certain aspects of habitat degradation. Each of these factors depends on both environmental and anthropogenic factors.

---

[6] These estimations correspond to the highest catch in the history of the sturgeon fishery in the region. As a result sturgeon stocks were overexploited and have never restored afterwards.

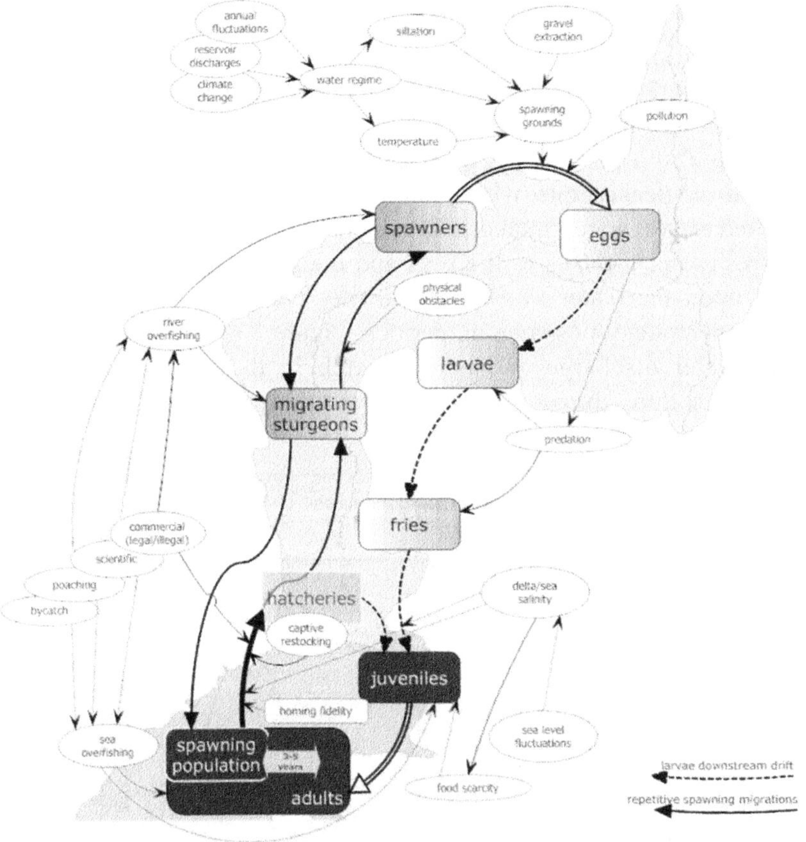

Figure 4. General sturgeon life cycle in the Ural Basin

Second, there is no natural predation for mature sturgeons, so apart from fishing efforts the sturgeon population is a function only of river environmental conditions, which can to a great extent be controlled by Integrated Water Resource Management.

Next, the sturgeon life cycle lasts up to 100 years which is comparable to the expected life duration of a human being. Actively migrating and feeding through all these years sturgeon presents a good subject for bio-accumulation. Taking into account its top position in the food chain (like human beings) sturgeon is a good *integral* indicator of water quality over a long period of time. In case of river contamination the river stream can be self-purified quickly (e.g. Baia Mare case (UNEP 2004)) and water quality tests will not indicate any problems, while living organisms (e.g. sturgeon and human beings) are subjects for the accumulation of harmful substances in their tissues.

Then, similar to a human being, sturgeon is a late maturing species, having an age-structured population. The reproductive age is reached depending on species at 10, 15, 20 years old. By that time harmful substances accumulated in the organism can affect reproductive abilities (Kajiwara et al. 2003; KaspNIRH 1999; Pourkazemi 2007) causing population decline as well.

There is also a positive relationship between sturgeon presence and a river's hydrological regime, which can be altered by damming, channelization or excessive water intakes. Spawning migrations are triggered by spring freshwater influxes to the seas and the entire success of spawning depends upon the water availability in the river, in other words water management strategy in spawning periods. Figure 5 shows the relationship between water discharge and beluga catch in the Ural river over 19 years when mature sturgeons returning for spawning (after Peseridi and Chertikhina 1967).

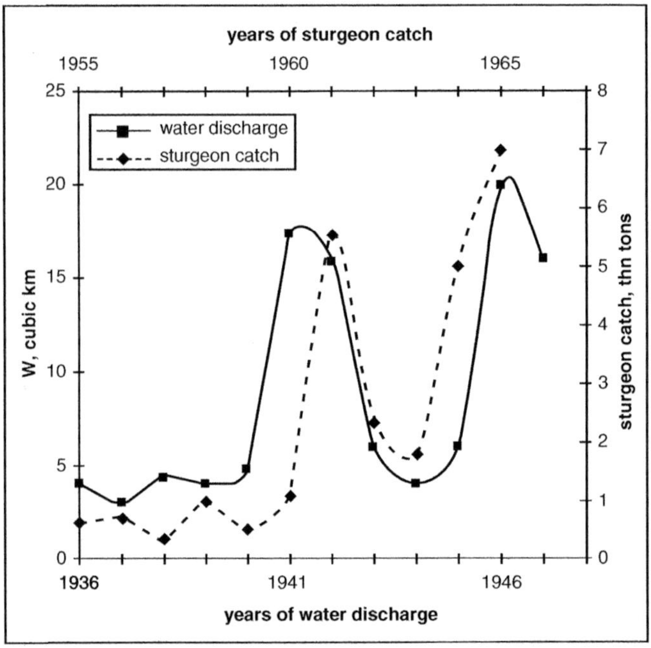

Figure 5. The relationship between river discharge and beluga catch in the Ural river 19 years later (After Peseridi and Chertikhina 1967)

Sturgeon presence in the river indicates the natural character of the hydrological regime, including regular floods and river self-purification service (Figure 6).

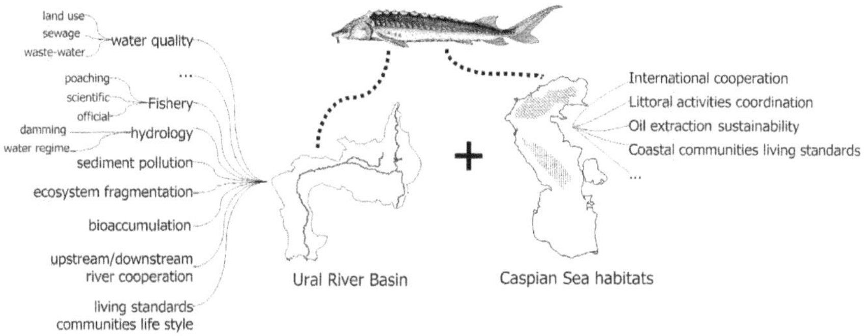

Figure 6. Some sturgeon functions as an indicator species

Apart from that, sturgeon is an indicator of other river physical characteristics: blockage of migratory routes, habitat degradation and fragmentation, siltation, pollution, water quality, etc. Some of these factors directly depend on the land use patterns in the river basin due to water runoff from the catchment area. In this way the terrestrial aspects of human activities are also brought into consideration.

Sturgeons also represents regional economic development and social structure, as poaching and illegal fishing which reduce sturgeon populations develop in areas with a poor unemployed population.

It is obvious that securing of natural sturgeon reproduction, protection and sustainable management of sturgeon stock is directly linked to integrated water resources management in the river basin and sustainable watershed development. These activities influence each other and should be considered only in an integrated manner.

Preserving sturgeon in the region would not only be of pure environmental benefit, but would also greatly contribute to economic and social stability in the region as well as food and water security. Traditionally, sturgeon harvesting was not only a major source of living for local communities but also an essential food resource. Thus, the measures aimed at preservation and sustainable management of the Ural sturgeon population can bring together environmental and socio-economic aspects of sustainable development and underpin the strategies for sustainable watershed development.

Cooperation in transboundary shared international river basins is complicated by the lack of incentives for cooperation. Upstream countries are not interested in securing environmental flows on the territory of downstream neighbors. There is no effective feedback from downstream regions suffering from water pollution or excessive upstream intakes to upper countries. Attempts to introduce feedback on the basis of hydrological cycles are often ineffective due to the large scale of the hydrological cycle and lack of evidence about causality (RAMSAR 2006). Sturgeons can provide such a feedback mechanism, and due to its high economic value all basin countries are interested in sturgeon stock rehabilitation and transboundary cooperation.

The role of anadromous species in general and sturgeon species in particular in integrated watershed management or regional sustainable development is a new concept in the basin-wide sustainability of environment society relations (Lagutov 1995, 1999). However, this idea is becoming increasingly recognized worldwide (Kliot et al. 2001; RAMSAR 2002).

For example, sturgeon species were suggested for the development and implementation of the Basin-based Concept of Regional Sustainable Development in the Don river and Azov Sea basins (AzovBas 2002; Russian State Duma 1995).

The European Freshwater Programme developed by the Worldwide Fund for Nature (WWF) also allocates a special role to sturgeon species. It introduces species classification to be used to secure success of environmental campaigns: *Flagship species, Species of special concern* and *Indicator species* (WWF 2002a, b).

According to this classification,

"**Flagship Species** act as a symbol and 'spokesperson' for their habitat. … major ecosystem programmes can be built around them….

Species of Special Concern are usually threatened species and their protection promotes conservation by safeguarding biological diversity and ecological processes.

Indicator Species are "markers" which help to measure changes or trends within a particular environment" (WWF 2002a, b).

All these functions perfectly suit sturgeon species, while the Ural sturgeon fits them the most (Lagutov 1995, 1996, 1997). The Ural River in general and sturgeon in particular were the main source of living for the Ural Cossack communities living along the Ural River. The sturgeon was depicted on their banners and coat of arms. Moreover, the Caspian sturgeon is world-renowned thanks to its caviar. The Ural River is the only spawning grounds for the "caviar carriers". The *flagship* function is fulfilled much better then in any other region.

Due to the high demand for caviar the Caspian sturgeon has almost disappeared. Its preservation is a matter of a special concern on both national and international levels.

The indicator function of the Ural sturgeon is also well defined and was discussed above.

Sturgeon was also included in the European Union Water Framework Directive, adopted in 2000, as an indicator of "a good status of surface waters" (WFD 2000). However, the situation in European rivers is much worse; all valuable sturgeons have been extirpated, habitats have mostly altered or degraded and now enormous efforts would have to be undertaken simply to try to restore sturgeons. By contrast, Ural River habitats are preserved and all historically available sturgeon populations are still present, though not for much longer if *business-as-usual* continues.

□istory of sturgeon fishery in the □ral region and catch statistics considerations

Periods in Ural fishery

Several distinctive periods in Ural fishing history can be specified. It should be noted that it is different from any other Caspian river basin, which can probably explain the higher river productivity through the 20th century. Though this history is unique and worth separate detailed investigation it is not well described in the available literature.

Historically, the low streams of the Ural River were populated by Ural Cossack communities, a self-governing paramilitary ethnic group. Cossack troops were traditionally involved in various State services in the Russian Empire. They were either protecting Russia's borders in their areas or serving as combatants during military campaigns. In exchange for military service they enjoyed exclusive rights to control natural resources on their territory (e.g. fish and water) and paid no taxes (Brockhaus and Efron 1898; Semple 1907; Von Harthausen 1972). The Ural Cossack community controlled the entire territory and resources of the lower Ural basin and adjacent sea area.

Living in harsh environmental conditions characterized by low soil fertility the Ural Cossacks had to fully rely on the river ecosystem, in particular sturgeon fishery, to support their communities.

Consequently, all the aspects of water usage and fishery were very carefully described, regulated and enforced. Fishery was limited to specific times in the winter, spring, and autumn. There were fishery and water laws. Out of two elected commanders (atamans) one was a military commander, while the other one was solely responsible for river-related issues (e.g. fishery). Fishing out of season was severely punished and the fisherman-violator lost his right to fish for the whole year. Special troops used to guard the river streams during spawning migrations. Another characteristic feature of the Ural fishery was *uchug*, the metallic or wooden fence constructed through the river stream near the city of Uralsk. The fence prevented sturgeons from going upstream out of Ural Cossack territory.

During this period until the Russian Civil War in 1917 when they were deprived of their privileges the entire water course of the lower and middle Ural was used exclusively for fishery (Brockhaus and Efron 1898). No other kind of activity was allowed, including navigation. Ferriage through the Ural was allowed only in a couple of places through the whole territory in order not to frighten the fish.

Fishery in the Ural was precisely organized and controlled (Borodin 1901). Any sturgeon fishing in the river was prohibited except for a couple of days in winter. During these days Cossacks equipped with special spears took sturgeons out of their wintering holes in the river bed through ice-holes. The catch in the sea was carried out with *okhans*, nets with coarse, more then 0.5 m, mesh (Malecha 2002). Fishing with coarse-meshed nets was allowed only upstream of uchug, and only Cossacks were allowed to fish.

Only three sturgeon species were considered as food fish: the beluga, Russian sturgeon and ship. The targeted species was mainly the beluga, 10% of the weight of which was caviar. Other species were used for fat rendering (Brockhaus and Efron 1898). Fish was used as a food supply for the local population and for trade.

The precise catch size can be estimated through the 19th century and early 20th century. The maximum catch conducted by Ural Cossacks for all purposes was 2.5–3 thousand tons (Lagutov 1995). Annually, the Ural Cossacks Land was exporting 128 tons of caviar, 1 thousand tons of sturgeons and 3.75 caviar in tons of balyk (smoked sturgeons) (Brockhaus and Efron 1898).

The First World War, Revolution and Civil War significantly decreased the pressure on the sturgeon stock due to the fact that most of the Cossacks participated in military campaigns.

After that the trends in fishing history and efforts in the Ural river basin till the late 1950s repeats the general Caspian pattern.

The 1930s was the period of collectivization in the USSR. Before this period fishing was mainly based on fisherman-individuals or small groups joined together (*artel*), but in the 1930s collective fishing artels (kolkhoz) were established. The state intensified its efforts in fishery, took all fishing activities under its own control and opened many dockyards to create and repair fishing boats. By that time the Caspian fishing fleet was extremely old and outdated. Up until 1931 most of the ships were wooden made in the beginning of the 20th century before the First World War. In 1930–1931 the Caspian fishing fleet was actively renovated – 2,305 new wooden made boats were created in the 1930 alone (APU 2000).

In 1928 the Caspian fishing fleet contained 19 trawlers and no seiners, while four years later in 1932 there were already 78 trawlers and 34 seiners. The efficiency of fishery also drastically increased. In 1931 for the first time new nets and fishing strategies were put into practice (APU 2000). According to some estimates the efficiency of fisherman in kolkhoz was two–three times higher than that of individual fishermen. Moreover, individual fishermen were persecuted by the authorities. In 1930s fishing efforts in the Caspian Sea became a subject for the established planned economy. In this way all, either successful or faulty, management strategies and policies in Caspian fishery were pre-developed, approved and controlled by the State (APU 2000).

The Great Patriotic war in 1941–1945 also demanded a lot of efforts and human resources from the local population and resulted in a tremendous decrease in catch. This fact is also considered to be beneficial for the partial restoration of sturgeon population, or rather the short delay in its total expiration. This drop is considered to be the only one (except that during the Revolution and Civil War) caused by reduced fishing efforts. During all other periods fishing efforts (in contrast to catch) were constantly increasing either by introducing new technologies, strategies or fleet increase. Moreover, the Second World War years were characterized by high water availability beneficial for spawning.

In 1951 it was decided to discontinue targeted sturgeon catch with okhans (coarse mesh nets) and harvest sturgeon as a by-catch from net-based fishing of usual species ("*chastik*"). New technologies (nylon fishing nets) were introduced in the region. Such fishing resulted in a high catch of young sturgeon of non-productive age. Despite new technologies a steadily decrease trend in the sea sturgeon catch can be observed through the 1950s in all Caspian regions, except the areas adjacent to the Ural River basin.

In 1955 the sturgeon hatcheries began their activities, gradually increasing the rate of larvae release. Up to 12 million beluga larvae were released every year. The idea of turning the Caspian sea into a big aquaculture fish pond was wide spread (Lagutov 1995), and the feeding grounds in the sea were called "pastures".

The 1950s were also characterized by the simultaneous introduction into practice of the dam complexes in the Caspian tributaries. As a matter of fact impoundment of the Caspian rivers started in the 1930s. Upon its continuation in 1950s the process of river damming continued for the next 20 years. The Volgograd dam was finalized in 1958, blocking the main spawning sites in the Volga River. Before that the Volga River was the main site for natural sturgeon reproduction, though comparable with the Ural River in terms of absolute yield to fishery.

Unlike in all other basin rivers the only hydraulic construction built up at the main Ural stream is located in its upper course, 1,810 km from the river delta. In this way the Ural River became the only river where natural spawning grounds were preserved. Because of that, no sturgeon hatcheries were installed in this river. Numerous sturgeon hatcheries were constructed in the deltas of the dammed rivers to compensate for the loss of spawning grounds and to sustain in this artificial way harvesting of the Caspian sturgeon. This fact alone makes the Ural River sturgeon a unique population relieved from the influence of hatchery-based specimens.

At the start of the 1960s (1962) significant changes in sturgeon fishery occurred. A ban on fishing with nets at sea was introduced and the fishery was transferred to rivers' mouth and streams (CEP 2002a). The reasoning behind this change was to protect juvenile sturgeon. Undoubtedly, this policy resulted in a drastic increase in catches. For example, the sevryuga catch in one year doubled in both the Ural and Volga rivers. The river itself and river mouth in particular is a bottleneck for the survival of any anadromous population. Fishing efforts in the river mouth are much more effective. However, annual systematic decrease targeting the spawning reproductive part of the population undermines sturgeon restoration and threatens the species' survival. Nevertheless, some believe that "a ban on sea fishing from 1962 to 1991 positively impacted the number and total biomass of commercial stocks" (Khodorevskaya et al. 1997).

Though not explicitly indicated, this measure seems to prioritize artificial sturgeon hatching over natural reproduction. The primary fishing pressure was put on the self-sustainable viable wild populations instead of the commercial ones fattening in the sea.

While it was claimed to be aimed at fish protection, the change in policy in 1962 could be caused by the decrease in catch and the need to secure food supply to the blooming Soviet economy. Figure 7 shows the gradual decrease in sturgeon catch from the 1950s to 1960s. The drastic catch increase in 1962 indicates not a stock restoration as hypothesized by some authors (KaspNIRH 1999), but the shift in fishery strategies towards a more aggressive river-based system imposing higher pressure on the migratory fish populations. Really, the generation hatched in low fishing pressure years of WWII had to reach maturation and fishing age in 1955–1960. However, these years are characterized by a decrease in total catch.

Also, sturgeon fishery in a river basin can maximize caviar production, the primary source of income from sturgeon, by targeting spawners directly. Moreover, only this fishing strategy can guarantee an industrial scale of caviar production.

This approach to fishery lasted till the collapse of the Soviet Union in 1991. The peak in sturgeon catch was observed in 1970. The peak in "fingerlings release into the basin rivers occurred in the 1980s. During this period fishing efforts were constantly increasing.

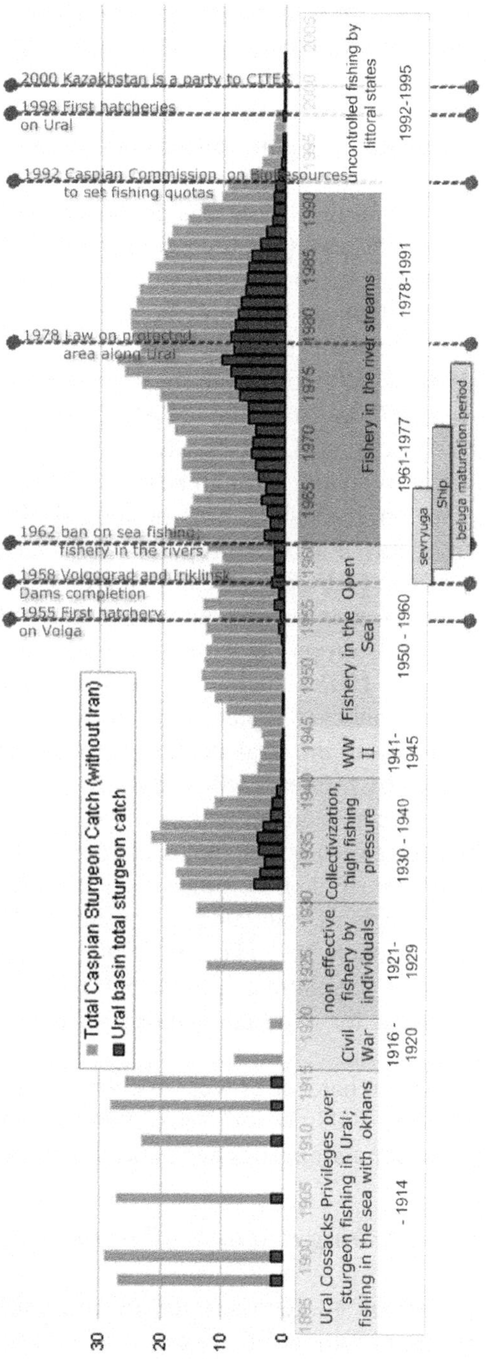

Figure 7. Timeline of Ural sturgeon fishing efforts

In general, Caspian fishery in the Caspian Sea was characterized by a high level of regulation and central planning with approved fish quotas, seasonal closures and gear restrictions.

By the end of this period the Ural sturgeon stock was already exhausted, but contributed a significant proportion of the whole Caspian catch.

The regional fishery in the 1990s was characterized by the collapse of centralized control over fishery (resulting in uncontrolled fishing efforts by littoral states), an outbreak of unemployment and, consequently, an increase in poaching and illegal fishery. By now, the sturgeon has almost vanished from the region. The Beluga catch dropped 750 fold from 1,500 t in 1932 to 2 t only in 2005 (FAO 2007b), while the sevryuga catch dropped 2,450 fold from 9,800 t in 1977 to 4 t in 2005 (FAO 2007b). Nevertheless, fishing efforts are actively continued. The Newly Independent Countries manage their fishing efforts individually through gear, catch, seasonal and regional regulations. The ban on sea fishing was prolonged, though according to some observations it has not been properly implemented (Pitikch et al. 2005). The Ural-Caspian Fishing zone came fully under Kazakh control.

The Caspian Sea sturgeon fishing quotas are distributed during regular meetings of the Commission on the Biological Resources of Caspian littoral states, established in 1992. The quotas are distributed according to the contribution each state makes to replenishing stocks. Kazakhstan's quota is based on the exploitation of the Ural stock and in 2007 it was only 18% of the total Caspian catch by former USSR countries.

In 1997 all commercial regional sturgeon species were included in the Appendix II of the Convention on International Trade in Endangered Species of Wild Fauna and Flora (CITES). Since then the international trade of sturgeon and its products is regulated according to CITES provisions. Kazakhstan became a party to CITES in 2000. Unfortunately, CITES' ability not only to stop population decline, but even to provide scientifically sound justifications for still high export rates are challenged by independent researchers.

The first two sturgeon hatcheries were opened on the Ural river in 1998 in Guriev (Atyrau) (RK 2003; World Bank 2004b), though their ability to sustain wild sturgeon populations has been questioned as discussed below.

Data availability on fishing and total allowable catch

Three out of six Caspian sturgeon species are recognized as commercial fish in the Caspian Sea basin and its rivers: the beluga, Russian sturgeon, and sevryuga. There is no standard commonly accepted methodology for estimating total sea fish stocks and commercial stocks in particular (Seijo et al. 1998). For instance, the international techniques are different from the ones used by the former USSR earlier or the littoral Caspian countries now (Lagutov 1995, 1996; Uralbas 2007a).The four former CIS countries use sample trawling to derive total annual catch quotas or total allowable catch (TAC), while the Islamic Republic of Iran uses a catch-per-unit-effort (CPUE) to determine fish abundance (CITES 2004b).

During the USSR Caspian Sturgeon TACs were allocated by the State Fisheries Committee using calculations by scientific agencies such as the Research Institute of Fisheries and Oceanography ("VNIRO") and Caspian Fisheries Research Institute (KaspNIRH) (CITES 2001). Nowadays the annual commercial catch quotas are allocated to Kazakhstan by the Inter-governmental Commission for Caspian Biological Resources which meets annually in Astrakhan. Any fishing activities, such as commercial catch, scientific catch and the catch of mature spawners for reproduction in hatcheries, are included in the TAC (CITES 2001).

Official statistics from specialized institutions are often contradictory. For example, the Caspian Fisheries Research Institute (KaspNIRH) is responsible, as follows from its title, for the research on Caspian fishery, stock estimations, quota establishment, etc. According to the field study results on sturgeon stock evaluation published by this Institute in one source (KaspNIRH 1999) the total abundance of beluga, Russian sturgeon and stellate sturgeon (sevryuga) in 1998 in the Caspian Sea was 42.2 million specimen. Surprisingly, this parameter for the same species next year 1999 was already ten million higher – 52.3 specimens (KaspNIRH 1999). As a matter of fact, the work of this particular Institute and other fishery-affiliated institutions in USSR, i.e. Azov Sea Fisheries Research Institute (AzNIRH), have been criticized by many authors for decades (Lagutov 1995).

The quantitative assessments of fish stocks conducted by USSR fishery institutes and later by littoral newly independent countries are often flawed and have been proved as biased or lacking scientific grounds (Crownover 2004b; Kirby 2002; Lagutov 1995; Morgan 2007; Pala 2004b; Raymakers and Hoover 2002; TRAFFIC 2007a, b; Uralbas 2007b).

There are different reasons to keep this situation running, including political and economic benefits as well as prestige of scientific schools in fishery. However, the discrepancies in estimation techniques provide good background for speculation and TAC establishment depending on the countries' or involved elite groups' interests.

As a result, there is no current commonly recognized Caspian sturgeon population assessment (Pitikch et al. 2005).

In this situation the data supporting population and stock analysis should come from official catch data, rather than from periodic quantitative assessments of fish stock (CEP 2002a; FAO 2007a; Seijo et al. 1998).

On the other hand even if a reliable universally accepted methodology were used, estimating the sturgeon stock would still be a very challenging task. Any methodology is based on the data catch statistics. However, historical statistical data often varies depending on the source.

For example, Figure 8 shows two different datasets for Ural beluga catches. The first one is based on the materials form KaspNIRH (KaspNIRH 1999), while the second one presents data by the Caspian Environment Program (CEP 2002b). Until 1993 both lines match since datasets were based on the same initial database from joint USSR statistics. However, with the Soviet Union's disintegration and the collapse of united basin fishery management, alternative sources for data appeared and discrepancies started to develop.

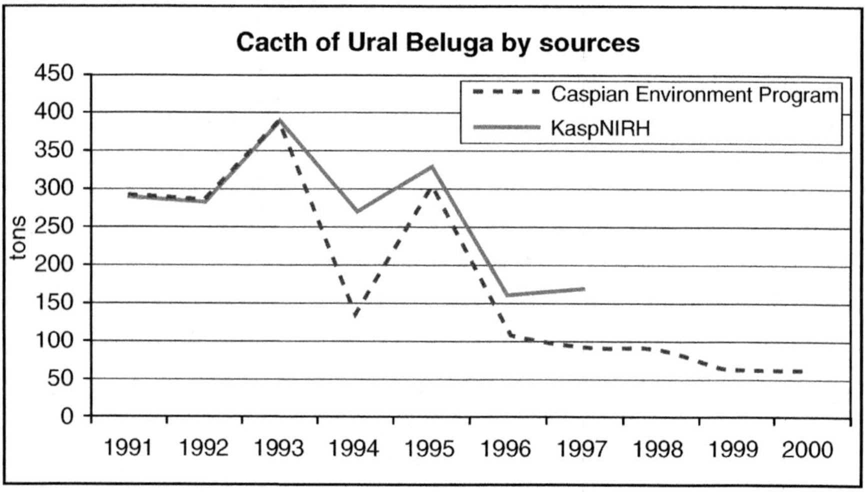

Figure 8. The discrepancy in reported sturgeon catch (KaspNIRH 1999)

Apart from these methodological problems, fishing zones and Fishery Departments were constantly changing and reorganizing.

The Ural Caspian Basin Fishery Department was created after the ban imposed on sea fishery in 1962 and fishing activities were relocated to the river basins. The statistics on fishery in the river basin, delta and adjacent sea area were collected in one center. After the disintegration of the Soviet Union the two basin countries organized independent fishery departments with their own statistical datasets. Furthermore, the fishery in Orenburg oblast was not considered to be part of the Caspian basin any longer and the successor of the Ural Basin Fishery Department on the Russian side was moved under the authority of KamUralRybVod, another basin fishery department in Russia. In this way the statistics on fishery in the Russian part of the Ural basin were excluded from the Caspian statistics. At the same time sturgeon commercial fishery in the 1990s in Orenburg oblast was negligible (KamUralRybVod 2007) and can be omitted from the preliminary analysis of the Ural catch. The analysis of fish stocks in the Ural basin can be based on the historical data before USSR disintegration and data provided by Kazakhstan for the later period.

Having said that it should be noted that the Ural-Caspian fishing zone in addition to the river Ural includes also the river Kigach. Though the catch in the Kigach River is not significant its possible influence should be taken into account while analyzing the Ural sturgeon population dynamics.

The recent problems and discrepancies with historic sturgeon catch in the Caspian Sea can be to some extent explained by the directives to operate only with the percentages of "socialist obligation", the planned level of catch in the centralized economy, but not with absolute catch values (Lagutov 1995). The latest statistics in the USSR were considered to be confidential information and were not distributed by fishery agencies (Figure 9).

As a result the proper analysis of the sturgeon population through the 20th century is complicated.

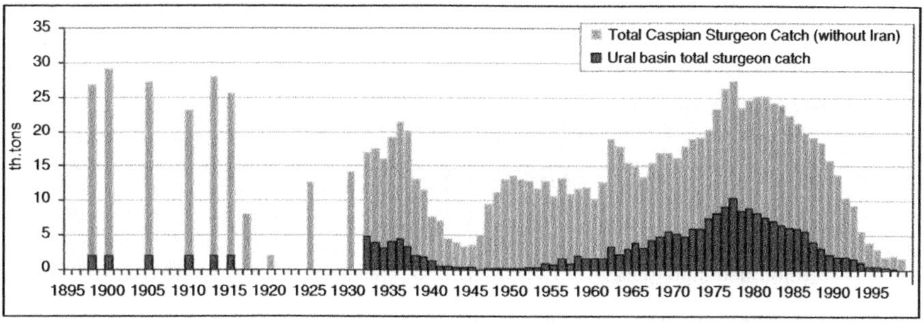

Figure 9. The share of Ural sturgeon catch in the total catch of f.USSR (KaspNIRH 1999)

Ural sturgeon species population dynamics and removal rates

Sturgeon species are late-maturing, slow-growing, long-lived fish and are able to withstand only light levels of harvest pressure. (Lagutov 1995; Uralbas 2007b).

Basic ecological theories claim that maximum harvest removal for this kind of fish cannot be more then 10% (Lagutov 1995). Nowadays the Russian secretariat of CITES claims to use the same principles for quota establishment (CITES 2004b). According to these regulations in the case of beluga, allowable removal is 9.4% of the stock. For Russian sturgeon the allowable removal is 13.7% and for stellate sturgeon 16.7% may be removed.

This approach is based on an estimate that the natural mortality rate for these species is about 10%, allowing the conclusion to be drawn that a reduction of 10% is harmless. The important feature of this approach is that due to the lack of natural predation this natural mortality in the sturgeon's case applies mainly to the old non-productive specimens. The same suggested removal rate in fishery is applied solely to the reproductive sturgeons prior to their spawning, in other words this is additional pressure on the stock, beyond the natural 10% mortality rate. At the same time annual fishing quotas allowed and scientifically approved by USSR fishing agencies and institutes in the 1970–1980s were 30–40% for some rivers in the south of Russia (Lagutov 1995).

The ratio between total species abundance and the catches in the Ural River shows much higher removal rate for some periods. Table 2 shows the official statistics on spawning sturgeon populations entering the Ural and the number of fish reaching spawning grounds. Though the methodology of such precise estimations of fish stocks on a non-regulated river is not described, these results are produced and disseminated by the Caspian Fishery Research Institute responsible for fishery planning and management in Caspian Sea (KaspNIRH 1999). The calculations of the removal rates are made by the authors.

Table 2. Spawning migrations and removal rates for the main sturgeon species harvested in the Ural fishing zone (Authors calculations on the base of materials from CEP 2002a)

		1971–1975	1976–1980	1981–1985	1986–1990	1991–1995	1996–1998
Sevryuga	Spawning migrants (thousand individuals)	1178.6	1227.5	884.3	463.1	184.4	98.7
	Reaching grounds (thousand individuals)	390.4	243.6	173.2	137.4	64.3	53.1
	Removal rate (%)	66.88%	80.15%	80.41%	70.33%	65.13%	46.20%
Beluga	Spawning migrants (thousand individuals)	6	13.1	18.1	16.2	7.7	3.5
	Reaching grounds (thousand individuals)	2.01	6.42	11.1	9.15	3.6	1.7
	Removal rate (%)	66.50%	50.99%	38.67%	43.52%	53.25%	51.43%
Russian sturgeon	Spawning migrants (thousand individuals)	18.14	33.6	37.1	43.5	28.5	10.2
	Reaching grounds (thousand individuals)	15.5	27.6	29.3	38.7	21.5	4.9
	Removal rate (%)	14.55%	17.86%	21.02%	11.03%	24.56%	51.96%
Ship	Spawning migrants (thousand individuals)	3.9	6.1	3.7	11	9.9	5.2
	Reaching grounds (thousand individuals)	2.3	2.64	3.2	9.9	6.5	2.6
	Removal rate (%)	41.03%	56.72%	13.51%	10.00%	34.34%	50.00%

These removal rates are applied to the spawning population annually not taking into account sturgeon life cycle features. Taking into account repetitive spawning periods in 2–5 years, exponential decay in reproductive sturgeon stock should be observed.

The fact that fish are extracted before the spawning occurred is also important for the understanding of population dynamics. Lately almost all of the harvested sturgeons were going for their first spawning, which was not completed. Correspondingly, the size of new species generation would be reduced accordingly.

Not surprisingly, a significant decrease in natural sturgeon reproduction can be observed recently. The decrease in the population with highest removal rates, sevryuga, clearly indicates a rapid exponential decay pattern.

According to other sources (Pala 2004b) the number of spawning belugas in the Ural river was only 3,900 in 1994 and 2,500 in 2002. These specimens were underweight (two times lower then weight needed for effective spawning) and premature, yielding eggs of poor quality.

In 1990s the situation regarding percentages of spawners removal worsened further. Table 3 shows annual removal rates for the Ural beluga and ship for 1991–2000 calculated using official statistics from Caspian Environment Program (after KaspNIRH) (CEP 2002b).

Table 3. Spawning migrations and removal rates for beluga and ship harvested in the Ural fishing zone (Authors calculations on the base of materials from CEP 2002b)

Years	BELUGA			SHIP		
	Total abundance of spawning population (thousand individuals)	Catch (thousand individuals)	Removal rate (%)	Total abundance of spawning population (thousand individuals)	Catch (thousand individuals)	Removal rate (%)
1991	7.5	3.6	48.0	13.6	0.5	3.7
1992	6.2	3.1	50.0	15.1	7.6	50.3
1993	13.2	6.9	52.3	8.06	2.96	36.7
1994	3.9	1.7	43.6	2.7	1.2	44.4
1996	3.2	1.4	43.8	5.6	1.3	23.2
1997	4.3	1.1	25.6	5.6	1.4	25.0
1998	3.1	1.2	38.7	4.4	2.8	63.6
1999	2.1	0.7	33.3	6.55	1.598	24.4
2000	2.66	0.67	25.2	6.28	1.268	20.2
Average	–	–	40	–	–	32.4

The official removal rate by legal catch for the Beluga reaches 50%, and for the ship it is even higher, around 60%. The average values for beluga and ship for 1991–2000 are 40% and 32% correspondingly. Taking into account such phenomena as scientific catch and poaching, the actual removal rate will be even higher. Late-maturing species cannot sustain such a high harvest rate. These facts explain the drastic sturgeon decrease in the river under the condition of available spawning grounds and undisturbed migration routes.

According to the statements by the same Caspian Fishery Institute the ship is not a commercial species (KaspNIRH 1999). Surprisingly, the official removal rate by state fishery for a non-commercial species in the Ural River was as big as 60% of the spawning population during the 1990s. Moreover, the ship is the only sturgeon species listed in both National Red Books as an endangered and protected species. It is somewhat surprising to see such a high level of official exploitation of a protected species.

Sturgeon species composition in the Ural

Insight into catch species and regional composition is important for understanding the dynamics of sturgeon populations and restoration programs' development. Figure 10 represents the Ural basin sturgeon catch by species.

Statistical data on sturgeon catch in the Caspian basin and its tributaries is often available as lumped amounts for the total sturgeon catch without separating statistics by the harvested species. This obstacle undermines proper analysis of individual species populations.

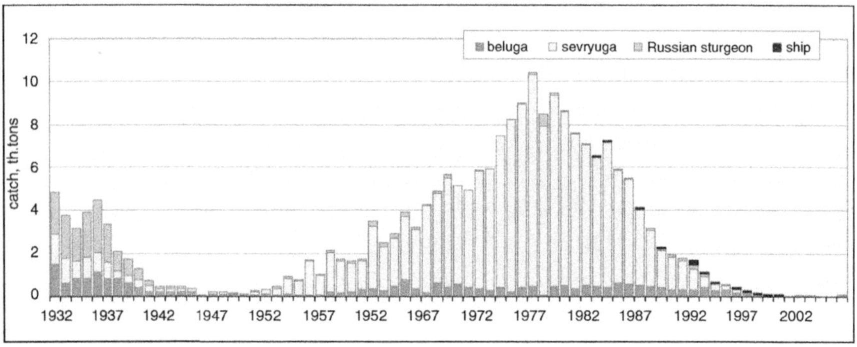

Figure 10. Species composition in total sturgeon catch in the Ural fishing zone (KaspNIRH 1999)

Another feature of catch statistics complicating the population analysis is data on total weight in contrast to number of specimens. For instance, the primary target for fishing efforts in the 19th and first half of the 20th centuries was the beluga (Brockhaus and Efron 1898). Correspondingly, the beluga constituted the biggest share in the catches. In the early 20th century, the beluga accounted for nearly 40% of the sturgeon catch (CEP 2002a).

The body weight of the beluga is much higher than that of all other sturgeon species. Official records indicate up to 1.5–2 t per fish with an average weight around 350 kg, while the average weight for other species varies from 5 to 50 kg depending on species (Table 1). During the 20th century the targeted sturgeon populations were changing: after the beluga stock's depletion the Russian sturgeon, Sevryuga or Ship was subsequently actively harvested. All these species have different body weights, which definitely should distort population analysis on the base of total sturgeon catch in tons. Operating with catch weight statistics might be a good approach for commercial fishery to estimate food production and other important living standards, but it is not very useful to evaluate population viability and extinction risk. Unfortunately, most Caspian sturgeon stock estimates are presented in total tons.

The Caspian sturgeon catch over the 20th century seems to be very stable up until the 1990s. In particular, the decline in catch at the end of the 1980s was often explained by usual multiyear fluctuations in catch and treated as a normal natural phenomenon (KaspNIRH 1999). However, the analysis of the species regional dynamics suggests a different explanation. The stable total catch seems to be the result of sequential overexploitation of various sturgeon populations. For example, Figure 11 presents the shares of Ural sevryuga catch in the total USSR sevryuga catch and total USSR sturgeon species catch.

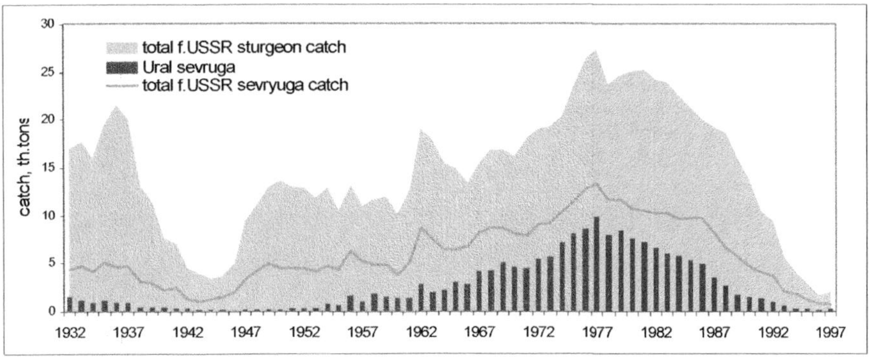

Figure 11. The dynamics of sevryuga catch in the Ural river against total USSR catch (KaspNIRH 1999)

The Ural sevryuga was not the primary subject for harvesting till the mid-1950s. Its contribution to the total Caspian sturgeon catch was only about 5%. In 1953 the fishing pressure on this population started to grow. After that time, the Ural sevryuga population, to be more precise the sevryuga catch in the Ural River alone, in some years constituted up to 75% towards the total sevryuga catch in the Caspian Sea. In the 1970s the share of the Ural sevryuga in total USSR sturgeon catch was up to 40%. The maximum sevryuga catch in the Ural occurred in 1977 and was equal to approximately 9,800 t, while total USSR sevryuga catch that year was 13.35 thousand tons (KaspNIRH 1999).

After this short term maximum the catch of sevryuga in the Ural River showed a steady decline. Unlike the 1930s when (1) the sevryuga was not a primary subject for fishing and (2) fishing in the Ural River stream was limited, the situation in the 1980s is characterized by active fishing of all sturgeon species in the river basin. The Ural sevryuga population was exploited until its total depletion in one approach. In 2005 the catch dropped to 4 t only (by 2,500 times) (FAO 2007a). No fluctuations or stock restoration periods can be observed.

This observation suggests that the total more or less steady sturgeon catch through the 20th century in the Caspian basin consists of a sequence of similar one peak total exploitation patterns for a particular sturgeon population in a particular region.

The analysis of catch in different sturgeon species in different Caspian basins supports this idea, as the same dynamics are repeated in other populations. For example, Figure 12 shows the catch in Russian Sturgeon by region. The regions/republics/countries correspond to the river basins and fishery areas: the river Kura – Azerbaijan, Volga – Astrakhan, Dagestan – Terek and Sulak, etc. Thus, the dynamics of sturgeon populations endemic

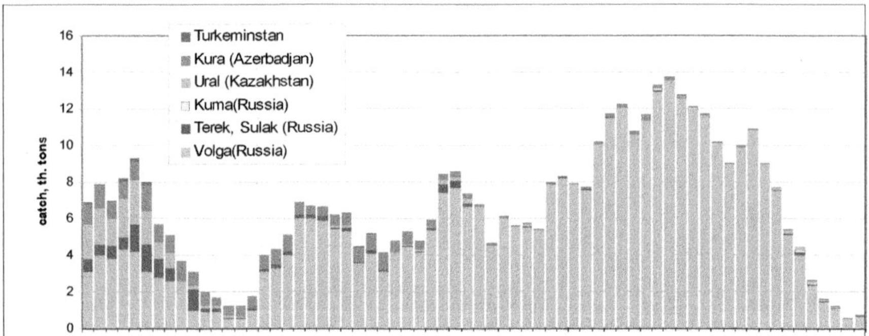

Figure 12. Russian Sturgeon catch in Caspian basin by regions (KaspNIRH 1999)

to the particular river basin can be estimated on the basis of the regional statistics. In the beginning of the observed period in the 1930s Russian Sturgeon populations were available in all fishery areas. Most of these populations vanished before dam construction, migration routes blockage or other factors could play a role. The scale of pollution, habitat degradation, sea level fluctuations and massive water intakes in rivers was negligible. The single factor which played an important role up to that time was over-fishing. A Russian Sturgeon population remained only in the Volga, due to the excessive initial fish abundance in it. This single population was exposed to increasing fishing efforts and overexploitation and also collapsed in due time.[7]

In the Ural the catch of Russian Sturgeon was most intensive in the 1930s. At its peak the maximum catch was 2.5 thousand tons (KaspNIRH 1999). Since that time it has never recovered. Only occasional specimens of Russian sturgeon have entered the Ural for spawning in recent years, while in other rivers – the Terek, Kura, and Sulak – spawning has not been observed since 1983 (KaspNIRH 1999).

Next, the current trend in Caspian sturgeon catch towards younger and lighter individuals should be taken into account. The average age in the commercial catch for every sturgeon population has been steadily decreasing. This fact means that in order to sustain the same level of reported catch in tons a higher number of specimens should be collected. In sum, on the one hand, by observing the stable total sturgeon catch in tons the conclusion of stock exploitation sustainability can be drawn, but on the other the real pressure on the stock has increased many folds. From this perspective the analysis of catch statistics in tons for the discussion of population sustainability should be applied with reservations.

The sturgeon catch itself, even in absolute numbers, is not a sufficient and adequate indicator of the real sturgeon stock size. To be closer to the real situation the sturgeon catch should be compared against fishing activities.[8] This aspect is often missing from sturgeon population analysis.

□opulation structure

The population structure of all sturgeon species in both the Volga and Ural River basins has changed, causing additional concern over population sustainability. Over the past 30–40 years the average age for all commercial species has decreased by more than 10 years: the beluga's average age has

[7] Another important factor for the Volga stock was lack of recruitment and natural stock replenishment due to the damming of major sturgeon spawning habitat – the Volga river.

[8] The fishing efforts in the Ural basin were discussed earlier.

declined from 40 to 20, the Russian sturgeon's from 33 to 20, the sevryuga's from 28 to 11–12 (Baimukanov 2007; Khodorevskaya et al. 2000). Not a single beluga older than 50 years has been recorded in catch lately. The predominant age of spawning fish has also decreased from more then 26 years to 11–17 years (Khodorevskaya et al. 2000). Female Sevryuga specimens older then 25 years males over 21 years cannot be found in the catch in recent years. Often, no specimens of reproductive age could be found (Khodorevskaya et al. 2000). Some authors believe that this is an indicator of maturing of hatchery-originated sturgeons and proof of success in sturgeon stock rehabilitation programs (KaspNIRH 1999). In reality, this fact can be better explained by total depletion of older age groups by systematic fishing of spawners in the river basins.

An additional indicator of the significant changes in the beluga population is the changes in female proportion in spawners in Ural from 50% in 1980s to 21–24% in the early 1990s (CEP 2002a; KaspNIRH 1999). This phenomenon is usually explained by targeting of productive sturgeon females for caviar harvesting and traditionally attributed to poachers' activities (EPA 2004). However, in the late 1980s poachers (as opposed to state fishery companies) were not the significant problem for the region. These facts rather characterize fishing efforts (legal and illegal) by state companies in the 1980s prior to the collapse of the Soviet Union.

It should also be noted that the CITES version of statistical data on fishery in Kazakhstan the ratio of males and females caught is not regulated or monitored (CITES 2001).

Beluga

A maximum catch of the Ural beluga population of 1.4 thousand tons was registered in the 1930s. Since then the river-based fishery has yielded only 0.4–0.6 thousand tons and a steady decrease in Ural Beluga population can be observed from 1985. The number of spawning belugas going up to the Ural river steadily declined to 2,500 individuals in 2002 (Pala 2004b) from 3,900 in 1994 (CEP 2002b).

Figure 13 depicts the beluga catch in the Ural-Caspian fishing zone in absolute values and as a proportion of the total f.USSR beluga catch. As can be seen from the figure, the Ural catch gradually grew from 5–10% in the 1950s to 50–70% in the 1980s–1990s. Comparing this result with absolute values it can be concluded that the Ural River sustained more or less equal catch through the 20th century and was exposed to smaller fluctuations than in other Caspian regions.

The catch in the Volga and Ural fishing zones, combined in one graph (Figure 14), supports this statement and suggests the influence of the Volgograd Dam construction on the catch levels in both rivers. The manifold increase in beluga catch in the Ural fishing zone can be observed during and after the completion of the Volgograd dam and is accompanied by a two fold decrease in the Volga catch. It might be explained by massive beluga migrations to the undisturbed landings in the Ural River.[9] It also could be argued that this rapid increase can be explained by the introduction of the aggressive fishing strategy due to fishery reallocation to the rivers. However, this change occurred only in 1961, several years later. The effect of that policy can be very well traced in the case of the sevryuga catch (Figure 11).

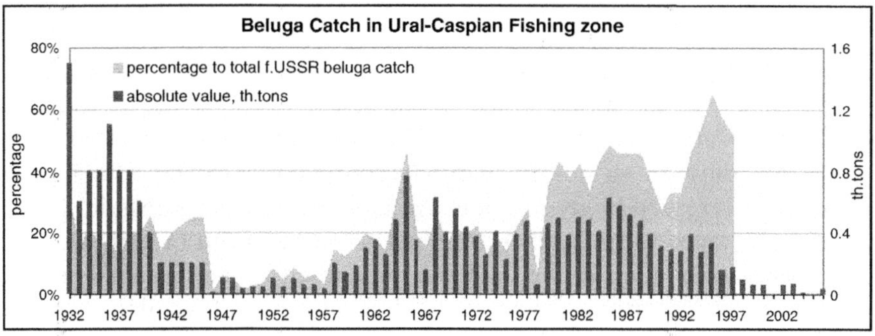

Figure 13. Ural Beluga catch in absolute values and as percentage in total USSR Beluga catch (KaspNIRH 1999)

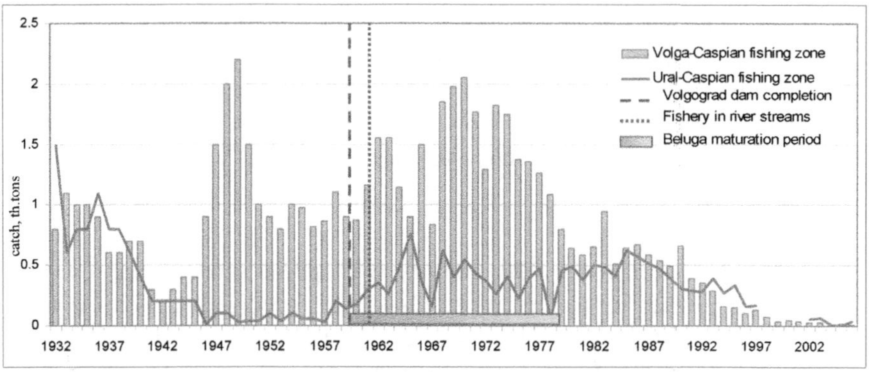

Figure 14. Ural and Volga beluga catch (KaspNIRH 1999)

[9] This speculation is a subject for biologists investigation. However, there are some indirect evidences to support this idea. For instance, molecular analysis often indicates cases of mislabeling of products from the different sturgeon species (Birstein et al. 1998).

Another interesting conclusion can be made by superimposing the beluga maturation period, the time before the beluga goes back to the rivers for first spawning (~20 years), over the catch graph. The abrupt substantial decline in the Volga beluga catch can be observed starting one generation after the damming, at the end of the 1970s. At the same time the beluga catch in the non-regulated Ural River stayed more or less stable till the end of the 1980s. Thus hatchery-based restocking programs, intended to mitigate the consequences of Volga impoundment, failed to substitute natural reproduction and to sustain beluga population.

Some researchers believe that the beluga is no longer naturally reproduced in the Caspian Basin (Birstein et al. 1997). As of 1997 the beluga population in the Volga region was considered by some authors to consist of 96.3% hatchery-reared fish (Khodorevskaya et al. 1997). This situation is explained by the fact that 100% of beluga spawning grounds were cut off by the Volgograd dam in 1958. The grounding assumption was that since that time no successful natural reproduction for the Beluga has occurred. Correspondingly, the few available beluga specimens in the region are believed to be hatchery originated. At the same time in the Ural river beluga spawning in the wild was monitored by the Russian Federal State Fishery Department even in 2007 (Dmitriev and Vasilenko 2007). According to the statements by the Russian Fishery Inspections, beluga spawning occurs in the Ural tributaries at the territory of Orenburg Oblast occasionally during the high flood years even now when total abundance is negligible. Twenty years ago, when the current adult beluga population would have hatched, spawning in the Ural occurred regularly.

The brief analysis suggests the close relationship and interlinkages in the Ural and Volga ecosystem and sturgeon populations. This line of reasoning justifies the point of view that the Northern Caspian should be considered and treated as one ecosystem (Dmitriev and Vasilenko 2007).

Ship

The ship in the Ural in the 20th century was spread through the low and middle river courses up to the city of Orsk. This species was not a fishing target in the Ural Cossacks Land, while its catch upstream was 16.4 tons annually (ORB 1998). The Caspian Fishery Research Institute, where data on sturgeon catch was collected and analyzed to develop further fishing strategies in the USSR, reports the ship stock's decrease in the early 1960s (KaspNIRH 1999). On the basis of this conclusion its fishing, according to KaspNIRH, was forbidden until 1994. Surprisingly, official statistics, including the very same source and others (CEP 2002a; Dmitriev and Vasilenko 2007; KamUralRybVod 2007), on its catch in the Ural river, exist starting from 1978 to 2000, while catch data on other sturgeon species is available from 1932. After the ban was removed in the 1990s the ship catch was only about 20–30 t. Moreover, by that time the ship was protected by national Red Books in basin countries. It should be noted that ship was available only in the Ural and Kura River. The catch in the Kura even in the 1980s was only 4–5 t per year. Now it has vanished from the South Caspian and is available only in the Ural.

According to the official statements by the Caspian Fisheries Research Institute (KaspNIRH) the catch of Persian Sturgeon was never officially monitored due to the low level of catch and no such statistics are available (CEP 2002b; KaspNIRH 1999). Nevertheless, Persian sturgeon contributed up to 23% of the experimental catch in the area downstream the Volgograd Dam (Artiukhin 1979). According to CEP data in both Volga and Ural the Persian sturgeon comprised around 5% of total catch in the 1980s (CEP 2002b). The total sturgeon catch in both regions in the 1980s was more then 20 thousand tons, which makes Persian sturgeon catch in that period equal to 1,000 t. This catch is more or less equal to the beluga catch for the same period. This fact can be used for various speculations over commercial catch statistics, e.g. Persian sturgeon, having characteristics similar to Russian sturgeon, might be accounted in Russian sturgeon catch in Volga or Ural-Caspian fishing zones. Despite the official statements on the absence of statistics on Persian sturgeon there are some claims about its stock increase (Pitikch et al. 2005) reflected in the contemporary trade quota increase for this species (CITES 2007).[10]

Having observed this evidences of the drastic decrease in Caspian sturgeon stock, the statement by the Sturgeon Management Authority of Russia that by 2004 "sturgeon stocks in the Caspian Sea as a whole appear to have stabilized or are beginning to increase" (CITES 2004b) sounds strange to say the least. The Authority's document, prepared for the justification of higher export quotas distributed by CITES, indicates that the estimate of the stock size is done based on the number of released fingerlings one generation ago and the trawl surveys in the open sea. The trawl surveys conducted by joint efforts of littoral states through the Caspian Sea were able to catch only 56 mostly pre-mature belugas (CITES 2004b). Based on this result, the Russian Sturgeon Management Authorities believed that the belugas are abundant in the sea and demanded higher export quotas of sturgeon products under CITES. To do so they announced that numbers of beluga sturgeon in 2002 rose to 11.6 million from 9.3 million in 2001, 25% in one year. In other words, the total sturgeon abundance in 11.6 millions was derived from 56 specimen (Pala 2004b). These conclusions were

[10] The increase in Persian Sturgeon is to be attributed to Persian sturgeon populations endemic to Iranian rivers thanks to high efficiency restocking programs and precise regulation in fishing.

challenged by many specialists claiming the annual 25% rate of increase is biologically impossible for late-maturing species such as sturgeon (Uralbas 2007b). The "miracle" is probably better explained by the corruption in Russian (as well as other former Soviet Union) fishery-affiliated authorities and research institutes.

Such data provision deficit and discrepancies often result in a situation when decision–makers operate with outdated or falsified data. For instance, the TRAFFIC secretariat, the wildlife trade monitoring network, in one of its publications states that "the Caspian Sea sturgeon population has been reduced by 40%" by 2007. In other words more than half of the historic sturgeon stock is still available for further exploitation (TRAFFIC 2007b).

Factors affecting the Ural sturgeon population

Its high economic value, the characteristic features of the sturgeon life cycle and the low priority of environmental issues and habitat preservation measures caused a situation where the significance of the problems related to sturgeon stocks were greatly underestimated not only within former Soviet Union countries (Lagutov 1995), but also in European countries and the USA (Bachmann 2000).

According to Reid and Miller (1989), threatened species are often characterized by one or more of the following: large body size, high trophic level, small population size, restricted geographic distribution, poor dispersal and colonizing abilities, colonial breeding habits, dependence on specialized habitats or ecosystems, migratory life history, dependence on unreliable resources, and inability to respond to environmental change or disturbances (Reid and Miller 1989).

Almost all of these risk factors are applicable to sturgeon species and can cause sturgeon extinction. All of them are migratory, large and at the top of food webs. Due to their bony exterior sturgeons do not have non-human predators in nature (Williamson 2003). Sturgeon species were distributed over the Northern Hemisphere, but local populations occupy restricted areas (river basins) and may be strongly isolated (Bachmann 2000; Waldman and Wirgin 1998; Williamson 2003).

Both anthropogenic and natural reasons can trigger the negative influence of these factors and affect sturgeons. Natural Caspian Sea fluctuations, climate changes or natural spread of invasive species in the Ural River courses may cause unfavorable conditions for the sturgeon population. Having lived in the neighborhood for more then 200 million years sturgeons have proved themselves to be highly resilient species which are resistant to various natural disturbances; nevertheless, during the last few decades sturgeons have been brought to the edge of extinction.

Historically, the sturgeon species' extirpation is attributed to overharvesting of sturgeon species worldwide (Cohen 1997; Hensel and Holchik 1997; Pourkazemi 2007; Qiwei et al. 1997; Zhuang et al. 2002). As a matter of fact, the absolute values of catch in the second half of the century do not exceed the harvest levels in the 19th century. On the contrary, according to the Caspian Fishery Institute (KaspNIRH 1999) the total Caspian sturgeon catch in the beginning of the 20th century was 39.4 thousand tons. The highest catch in the second half of the century was 27.4 thousand tons, followed by immediate and abrupt decline. Facing higher fishing pressure the sturgeon population did not collapse till the end of the 20th century, when other important factors started to play a major role. Overfishing as a total catch cannot be the primary reason for the sturgeon's extirpation from the Caspian region, but rather a combination of negative factors played a crucial role.

All researchers agree on the list of the factors causing sturgeon extinction, though the order of the impact magnitude for a particular factor is still actively discussed (Williot et al. 2002b).

The traditional list of negative anthropogenic factors includes blockage of migration routes, overfishing, pollution, habitat degradation, loss of spawning grounds, siltation, changes in hydrological regimes, sea salinity changes etc. The importance of these factors varies for different sturgeon species; they are considered in turn below.

Spawning migration blockage

Blockage of migration routes is the most significant anthropogenic impact on the sturgeon population (AzovBas 2002; Craig 2000; Lagutov 1995, 1997; McAllister et al. 2000).

Dozens of dams were constructed on the Caspian tributaries in the 20th century from the beginning of the 1930s to the 1970s (Figure 1). The dams have blocked the migration routes for both anadromous and semi-migratory fish types. Being deprived of their spawning grounds sturgeon populations became absolutely sterile, incapable of any reproduction and doomed to extinction in 1–2 generations even without any influence from other factors such as overfishing.

Figure 15 depicts the remaining spawning grounds in the Caspian rivers after impoundment (CEP 2002a). After the construction of the Volgograd Dam 100% of beluga spawning grounds were lost, 80% for the Russian Sturgeon and 40% for the Sevryuga. It is estimated that the total area

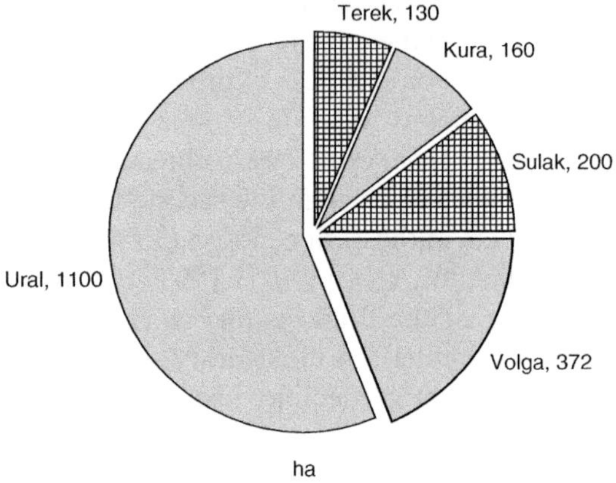

Figure 15. Remaining sturgeon (total for all species) spawning grounds in the Caspian basin. Checked slices represents rivers were no spawning was observed since 1980s

of spawning grounds in the Volga river decreased by 85% (AzovBas 2002; Russian State Duma 1995; Lagutov 1995). The beluga spawning grounds in the rivers Terek, Kura and Sulak were also lost completely.

Most of the constructed dams are high pressure dams constructed for hydropower generation purposes. The high level water drop does not allow the dam to be equipped with effective fish passing facilities. Though most of the dams are equipped with fish-passing devices of various designs, aimed at enabling migratory fish (in particular sturgeons) to pass through the dams, the overall efficiency of fish passage is extremely low due to a combination of factors (AzovBas 2002; Russian State Duma 1995; Lagutov 1995). As a rule fish-passages are costly, massive constructions requiring substantial operational and maintenance costs and resources. Yet despite their presence no sturgeon species spawning upstream of dams in the Caspian has been observed or described in the literature. Ichthyologists and fishery specialists agree that dams have "effectively cut off the spawning grounds upstream" and do not contribute to sturgeon population reproduction (Lagutov 1995). Some researchers deem that even in the case of successful sturgeon transfer to the dam's upper water sturgeons will not be able to find their way through to the spawning grounds upstream, if any is available in the impounded river segment. The former spawning habitats were either permanently flooded, silted or do not have proper environmental conditions (i.e. stream velocity and temperature regime).

At the same time "forced" spawning grounds downstream of the dams are believed to play some role in sturgeon spawning.[11] These grounds are located out of the historical spawning range and have a "forced" character

[11] See AzovBas 2002; Lagutov 1995 for the discussion on this topic.

of functioning. The migrants which are not able to overcome the dam to reach the upper river branches accumulate under the dam. If not collected by fishery, poachers and for scientific or hatching purposes different sturgeon species spawn jointly on any available substrate when the time of spawning comes. The survival rates for this kind of embryos and larvae are questioned by some authors due to many factors (Lagutov 1995, 1996). For example, the proximity to the sea might result in high losses of larvae/juveniles migrating downstream due to exposure to brackish sea water at early stages of the life cycle (Peseridi et al. 1979).

In addition, the environmental conditions in the habitats below dams are often unsuitable for spawning. The spawning grounds in the major sturgeon habitat, the Volga, were flooded only 13 times during 40 years after the Volgograd dam's completion in 1958 (Dubinina and Kozlitina 2000).

In any case, no natural reproduction was observed downstream the dams in most of the impounded rivers (except the Volga) during the 1990s (KaspNIRH 1999), 1–2 sturgeon generations after river damming.

The migration routes in the Ural River are still not obstructed. *Uchug*, used by Cossacks in the 19th and beginning of the 20th century to prevent big sturgeon from migrating upstream the Uralsk city, was an obstacle only for sizeable mature specimens. Moreover, every spring during the spawning migrations it was dismounted. The entire historic range of sturgeon habitats in the Ural is available for migrants and spawning with no reservations.

It is believed by some fishery-affiliated officials that one of the reasons for the decline in sturgeon appearance at the effective spawning areas in the Orenburg oblast was sinking of a barge in the middle of the Ural river close to the lake Indera (around 200 km from the Ural delta). There is some speculation that this occasion was used, or even intentionally created, to prevent sturgeons from going upstream and to maximize the catch in the area. Surprisingly, after the barge was lifted and evacuated exactly several pontoon bridges were constructed at the same location. Their removal due to obstructing the sturgeon migration routes was a matter for discussion between Russian and Kazakhstan regional authorities (Korina 2006). Being bottom-feeders sturgeons always swim near the river bottom, so a sunken ship across a medium sized river can be as an effective obstacle for sturgeon migration as a permanent *uchug*, and far downstream of the latter's historical location.

Habitat degradation, loss of spawning grounds

Traditionally, loss of spawning grounds for sturgeon species is understood as a consequence of river habitat fragmentation by the construction of dams and the blockage of migratory routes (Marmulla et al. 2001).

This is a major problem for all Caspian sturgeon stocks, apart from those in the Ural River. The Ural sturgeon spawning grounds are historically located up to the territory of the Orenburg Oblast and branches of the rivers Ilek and Sakmara. The results of the field studies conducted by Kazakhstan Fishery Institute (Guriev, Kazakhstan) in the 1980s showed that the most productive and viable sturgeon juveniles appear at spawning grounds in the middle Ural course close to Orenburg (Dmitriev and Vasilenko 2007). Moreover, historically the spawning grounds of the valuable sturgeon species (beluga, Russian sturgeon, Persian sturgeon, sevryuga) were located more then 1,000 km upstream the Ural delta (CEP 2002b; Peseridi 1971, 1986; Peseridi and Chertikhina 1967). Lack of a barrier complex on the Ural guarantees free access to spawning grounds for a hypothetical spawner.

However, due to the general decrease of sturgeon stock and active fishing efforts on the territory of Kazakhstan only a few sturgeons have been observed in this area lately.

At the same time changes in hydrological regime due to water intakes, climate change or water discharge regulation during the flood period may cause the spawning grounds to be unavailable for spawning even if migration routes are not obstructed. For instance, more then half of the spawning grounds in the Ural River are temporarily flooded, and to secure their proper functioning certain environmental conditions are required. Some mention the level of irrevocable water consumption from the Ural River as being 50–60% of the annual flow, resulting in 90% of larvae and young sturgeon perishing on their way to the sea (Fashchevsky 2003). However, this level of water intake seems to be overstated,[12] and the survival rate for juveniles from the deposited eggs to the sea may even be higher than natural levels.

Siltation, cover of spawning grounds with mud, is often mentioned as a problem for sturgeon spawning grounds' destruction. As a result of siltation the survival rate for sturgeon eggs will be low, because (1) eggs do not stick to the rocky bottom and (2) eggs are suffocated by silt/sand at the bottom of the river. There are some claims that from 1970 to 1994 a third of historical spawning grounds in the Ural was covered with mud, a sign of habitat degradation.[13] Siltation of some river intervals and cleaning of others is a natural dynamic process in the free-flowing steppe rivers depending on the water discharges in the river. The Ural's natural hydrological regime with high level floods maintains river self-purification services. High water

[12] See the article on the Ural river hydrology in this volume. The hydrological regime of the Ural river did not have drastic changes over the last century.

[13] According to other estimations 50% of the Ural's spawning grounds are lost due to the habitat degradation and pollution (cited by Pitikch et al. 2005).

flows, occurring in the Ural once 3–5 years in general, clean the potential spawning sites easily or create new ones. Since it flows through a wide valley the Ural River has a dynamically changing river bed shape with a high number of meanders and old river beds. In comparison to other European rivers, located in highly developed areas and limited by artificial channels and dams, the Ural River is a "living" water course, exposed to constant natural changes, including siltation and vegetation growth.

On the other hand higher then usual siltation rates can be caused by dredging works for navigation and extraction of sand and gravel conducted lately in the lower courses of the Ural on the territory of Kazakhstan. But these works obviously can affect only spawning grounds downstream and not productive sites upstream at the Orenburg oblast.

At the same time the very nature of dredging works suggests the erasure of gravel and pebble-formed rifts, where sturgeon spawning sites are located, causing direct irreversible destruction of spawning habitats.

Besides this, the mining of sand-gravel results in habitat degradation, loss of feeding grounds, siltation, and alterations in hydrological river regime. According to USSR Fishery Regulations in the Caspian Sea the mining of sand-gravel was prohibited in the Ural River stream up to the village Borodinsk in the Orenburg Oblast. However, these regulations are no longer enforced. For example, since 2000 Kazakhstan has been conducting sand-gravel mining in the watercourse of the Ural River near the village Priuralnoe, where many sturgeon wintering and spawning grounds are located. The specialists in the Orenburg Oblast claim that this has a strong negative effect on the sturgeon population.

Siltation cannot significantly affect spawning sites in the temporarily flooded areas, which is a substantial proportion of all available spawning grounds.

In this way, spawning habitats in the Ural River are abundant and underutilized and in case of the producers' availability can sustain numerous sturgeon populations. Unfortunately, due to the fact that the Ural River is outside the scope of many Caspian sturgeon restoration programs a systematic specialized study of the river's conditions has not been conducted. However, the underexploitation, or lack of any exploitation, of the Ural spawning grounds is well documented by the Orenburg Branch of the Russian State Fishery Department (KamUralRybVod 2007).

River's hydrological regime

Changes in the river's hydrological regime altering the volume and timing of the river flow have substantial direct and indirect impacts on successful sturgeon spawning.

The optimal conditions for sturgeon reproduction in the Ural river are created when the total annual flow is more then 9 km^3 (KaspNIRH 1999). Figure 16 presents the total Ural flow for the observation period 1915–2000. The optimal value of total flow in 9 km^3 is slightly less than the mean total flow for the period of observations. However, frequency of the favorable floods is approximately once per three years.

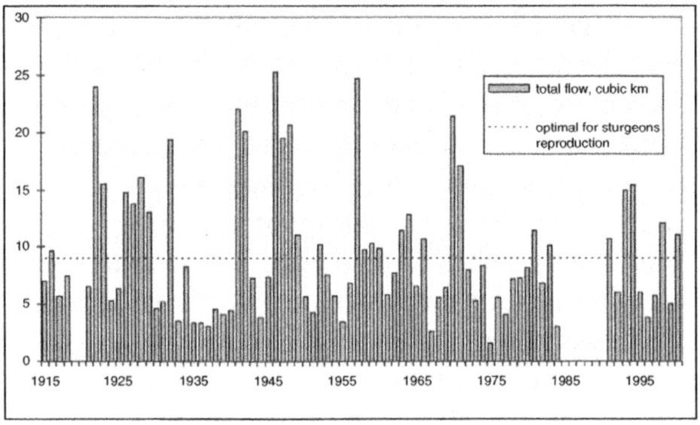

Figure 16. Total Ural flow for 1915–2000 (KaspNIRH 1999)

The comparison of sturgeon abundance/catches and total annual flow shows a very good correlation taking into account the time delay needed for sturgeon to mature in the sea before returning to their rivers for spawning. Figure 5 depicts a combined graph for total flow for 1936–1947 and beluga catch for 1955–1965. It should be emphasized that the higher total flow under the conditions of a non-regulated snowmelt-fed river, such as the Ural, means mainly higher spring floods, taking up to 80% of total flow. As has been indicated by several authors, spring (vernal) sturgeon race plays an important role in reproduction in the river Ural (Peseridi and Chertikhina 1967).

Unfortunately, proper statistical analysis of this correlation on the basis of the later data is not possible due to the intensive fishery in the river and sturgeon disappearance. Though the precise relationship between the total Ural flow and sturgeon spawning is hard to establish, the causal links between river flow and certain aspects of the spawning process are well known:

- First, ichthyologists claim that water salinity in the river delta changed by spring flood is one of the triggers for the sturgeon spawning migrations (Dmitriev and Vasilenko 2007).
- Next, the water salinity in the Northern Caspian directly depends upon the Ural River's hydrological regime.[14] The water salinity influences juvenile survival rates and food composition and availability.

[14] See article on the Ural river hydrology in this volume.

- Third, environmental conditions at the spawning grounds (current velocities) and their availability (water depth at temporarily flooded spawning grounds) are defined by water discharges in the river.
- Temperature regime in the river is also a function of the water level.
- Another factor of hydrological regime influence on the sturgeon population is the higher exposure of the migrating spawners to fishing efforts in low waters in the river stream. The record-breaking sevryuga catch in the low water years of the 1970s prove this statement (Figure 18).
- A number of other factors depending on the river's hydrological flow and influencing sturgeon population can be mentioned. Among them are fish-kill (oxygen-deficit), river self-purification service and fish exposure to pollution.

The existing water reservoirs in the upper branches of the Ural River do not have significant influence on the river's hydrological regime.[15] However, an appropriate management scheme of water discharges can improve the spawning conditions downstream the dams. The facilitation of sturgeon stock restoration was one of the main reasons for the creation of the Iriklinskoe reservoir, the biggest water reservoir in the Ural River. Unfortunately, nowadays this reservoir is also used as a pond for inland fisheries. The favorable conditions for inland fishery often contradict the interests of sturgeon migration. Taking into account the low number of specimens reaching Russian territory nowadays and the related lack of financial motives, the Russian fishery managers are not interested in providing good environmental conditions for hypothetical sturgeon migrations and spawning at the expense of inland fisheries' stable financial profit.

Sea salinity

Though sturgeons are euryhaline (salinity tolerant) species, the sea salinity level is an important factor in sturgeon population dynamics.

Sturgeons utilize a number of distinct habitats through their life cycle, but most of the time they spend in the sea for growing, feeding, fattening and maturing.

The well-being and survival rate of most species of Caspian sturgeon depend upon the conditions in three basin ecosystems:

- Rivers (freshwater)
- Estuaries with a salinity level of 0–4‰ and desalinated shallow waters of the Northern Caspian region (4–7‰)
- Northern Caspian Sea ecosystem (5–7‰ to 10‰ salinity)

[15] This influence is analyzed in the article on the Ural river hydrology in this volume.

According to the regional fishery officials the current salinity changes are having a dramatic impact on the sturgeon species' population spawning in the Ural river (KamUralRybVod 2007; Uralbas 2007b).

Sea salinity affects the Ural sturgeon population in several distinct ways:

- Feeding grounds are shrinking. Due to high salinity highly productive benthos and small fish are disappearing from the region.
- Non-freezing sea water of high salt concentration with negative temperatures can result in severe damage to fish.
- Survival of juveniles entering the sea for the first time after hatching rapidly decreases with salinity increase.

While all sturgeons are euryhaline species, larvae and juveniles are less tolerant to a saline environment than adults; water salinity of 8‰ is lethal to larvae at early stages of development (CEP 2002b; KaspNIRH 1999; Lagutov 1995, 1996). Fry and larvae need freshwater or brackish waters during the first few weeks. Depending on the water amounts delivered by rivers, the North Caspian estuaries' salinity can cause high mortality in sturgeon larvae.

Historically, before the creation of the Volgograd Dam, the juveniles in the Volga river used to stay in the river freshwaters after hatching for up to three months and on entering brackish salted water had an average weight of 171 g and length of 36 cm. Currently, larvae reach only a weight of 4.2 g and length of 5–9 cm (KaspNIRH 1999). Russian Sturgeon juveniles also often stayed in the river freshwaters for 3–4 years after hatching (Chugunov 1968). Current trends towards salinity increase in the Northern Caspian suggest the need for the usage of historical spawning places located in upper river branches. Due to the construction of high pressure dams on most of the Caspian basin rivers the Ural river is the only river stream with spawning habitats in their historical range.

Even in the Ural river 95% of Russian sturgeon, 98% of sevryuga and 65% of beluga juveniles appear in the delta at an age sensitive to high salinity exposure (Peseridi et al. 1979). In the case of water salinity close to 8‰, most of the new sturgeon generation will be lost. Such high salinity occurs in the Ural River delta during years with low water availability. Also the changes in the Volga river's annual stock have caused a rapid increase in the salinity of the Northern Caspian. The closer the spawning grounds to the river delta, the higher the risk of significant larvae and fry losses and a further decrease in Caspian sturgeon stock.

Overfishing

Commercial fishery

The problem of overfishing in the Caspian fishery is a very interesting phenomenon. Traditionally, the sturgeon, once abundant in all European rivers, was harvested by many different states, local communities, kingdoms or dukedoms. Though there were some attempts to regulate sturgeon fishing (Keysler 1762) at the beginning of the 20th century sturgeons have since disappeared from European rivers in commercial quantities. The second half of the century was characterized by a drastic increase of environmental awareness, international conventions and scientific approach to natural resources exploitation, which presumably should have helped sturgeons to survive. However, the Caspian sturgeon stock vanished exactly at this time. Though severely overexploited in the 1930s–1940s the Caspian sturgeons still inhabited the sea and rivers in great numbers.[16] In order to preserve available stocks several institutions were created to give scientific-sounding grounds for the establishment of fishing quotas. The Caspian sturgeon was driven to extinction despite all the activities aimed at its preservation and the new scientific approach (CEP 2002a).

According to the official records (FAO 2007b; KaspNIRH 1999) the legal beluga catch in the Ural basin dropped by a factor of 750 (from 1,500 t in 1932 to 2 t in 2005), while the sevryuga catch decreased by 2,500 (from 9,870 t in 1977 to 4 t in 2005). It can be argued that this tremendous decrease in catch is caused by introduction of quotas and thorough compliance with these regulations by fisheries.

As it is known, quotas are calculated as a percentage of the available fish stock (CITES 2004b; Seijo et al. 1998) to limit the catch with the purpose of securing sustainable stock reproduction. From this perspective the official statistics of the quotas/catches in the Ural River basin reveals very interesting dynamics. Figure 17 plots sturgeon fishing quotas and reported catches in the Ural-Caspian fishing zone by Kazakhstan. During 15 years from 1992 to 2007 the quotas gradually decreased by a factor of 10. However, even these small quotas cannot be utilized. So, the sturgeon fishing quota in Kazakhstan in 2007 was 184 t and it was only 70% completed (Uralbas 2007b). For the same period of 1992–2007 the reported catch in the Ural dropped by a factor of 15, which exceeds the drop in quotas by 50%.

[16] There are some sources stating that the peak in sturgeon catches in the Caspian Sea was 50 thousand tons (Pitikch et al. 2005). In this case the Caspian sturgeon stock was already overexploited in 20th century.

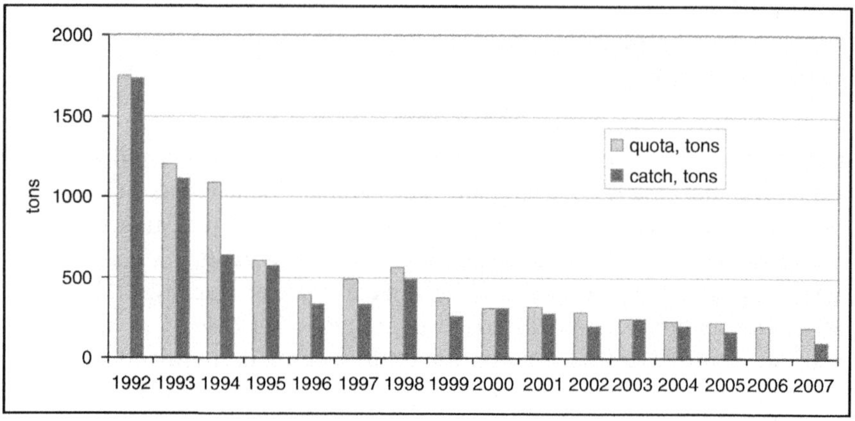

Figure 17. Sturgeon quotas and reported catch in Kazakhstan (RK 2003. Reported catch for Kazakhstan in 2006 was not found by authors)

The failure to utilize the sturgeon harvesting quotas is often treated as kind of sturgeon-protection measure. At the same time it should be taken into account that efforts to utilize the quota are more or less the same as decades ago, when the catch was 10–20 times higher. The only credible reason for the failure is the diminishing number of the specimen in the region. Nevertheless, official – constantly shrinking – quotas are still distributed. The reasoning to explain the existence for more than a decade of quotas which are higher than the maximum possible catch can be hardly found. Moreover, according to the Sturgeon Management Authority of Russia (CITES 2004b) the estimated sturgeon abundance in the Ural river for the period 1998–2001 was 197.6, 183, 226.7 and 226.7 t correspondingly. This is a much smaller than the quota or the real catch in this period. In 1998 the total estimated abundance was almost three times smaller then quota.

Another interesting observation concerning these statistics is connected to the fact that sturgeons are long-lived late-maturing species and considerable time period is required for the population rehabilitation. Surprisingly, there are significant annual fluctuations in the official quotas, including an increase in 1996–1998. Sturgeon stock cannot be restored in 1–2 years to satisfy higher fishing pressure.

As already discussed above, by fishing in the river basins the state fishery, as well as poachers, has for decades targeted the spawners in order to secure caviar production. For decades the fish of reproductive age were systematically removed from the stock. As a result, fishery nowadays aims at sturgeons returning to the Ural spawning grounds for the first time (Dmitriev and Vasilenko 2007). The average age of species in the sturgeon spawning flock in the Ural river during 2001–2006 was around 20 years for the beluga (Huso Huso) and 11–12 years for the Sevryuga (Acipenser stellatus) (Baimukanov 2007). Taking into account sturgeon species' age of maturity in fact suggests that these sturgeons were first-time spawners. As noted above, the reported catches of the state fishing companies are considered to be understated by 2–3 times (World Bank 2004b). Thus, the scale of the first time spawners' removal is much higher than claimed. If caught they could not complete even one spawning cycle. If even these spawners are removed then the sturgeon population is doomed to total extinction within a few years.

Taking into account the complex long-term sturgeon life cycle, a minimum level of population needed for reproduction should be established. Until reaching this level a total ban on fishing should be imposed (Jonsson et al. 1999; Lagutov 1995, 1996; Uralbas 2007a). There are numerous indications that the Ural and Caspian sturgeon have crossed the threshold after which population recovery is hardly possible. Furthermore, the hatcheries' inability to find enough producers to carry out artificial propagation is another indicator of the species' extermination from the region.

Nevertheless, the constantly decreasing quotas are still granted officially. The inefficiency of the sturgeon fishing quota system to revive dwindling sturgeon populations was also confirmed by the analysis made within the framework of the Caspian Environment Program (CEP 2002a).

The high intensity of open-sea fishing in the 1950s is proclaimed as one of the biggest overfishing-related causes for the sturgeon's decrease. The official statements by the Caspian Fishery Research Institute indicate this fishing strategy as one of the main reasons for the sturgeon stock's depletion in the 1990s (KaspNIRH 1999). According to these sources "high intensity" sea-based sturgeon fishery Sturgeon catch in this period is characterized by high number of young fish of non-productive age and small body weight (Marti 1972).

At the same time, reallocating of fishery to the rivers (i.e. Ural) resulted in a two fold increase in catch within a year. This strategy focused fishing efforts exclusively on spawners entering the rivers with a removal rate up to 80% on some species prior to spawning. These estimates do not take into account illegal fishing, poaching and removal for scientific or reproduction needs. Though claimed to be aimed at sturgeon stock preservation and protection, the limitation of sturgeon spawning in rivers should cause significant decrease in spawning and stock replenishment. Such a fishing strategy should be scientifically grounded, precisely regulated and controlled. Instead, the fishing strategy focused on annual systematic removal of spawners has undermined natural sturgeon reproduction and caused the drastic stock decline which can be observed recently.

By now the total removal of the productive spawning population (repetitive spawners) is confirmed by the same authors. Surprisingly, this statement was recently also supported by the Caspian Fisheries Institute (KaspNIRH) working on justification of fishing in deltas and rivers earlier. In 1999 the KaspNIRH report on the state of Caspian sturgeon stock and reasons for its decline states "the fishery in … delta and river was extremely intensive during the spawning period… The most valuable and productive part of spawning population was annually extracted…" (KaspNIRH 1999). However, the shift towards fishery in the rivers is still called a fish-protection measure in comparison to open sea fishing, even in the face of the Caspian sturgeon's extinction and the collapse of sturgeon fishery.

A deeper insight into the problem can be gained by observing fishing efforts coupled with other factors. For instance, the decrease in total sturgeon numbers in the 1990s in the Ural could be mainly caused by such a combination of several factors. In particular, according to the observations by the Caspian Fishery Research Institute the period from 1973–1979 is characterized by a *drastic* decrease in natural spawning in the Ural river (KaspNIRH 1999). Unfortunately, the authors do not pay proper attention to this fact. Nevertheless, this period is exactly the period when the generation of the 1990s was supposed to be incubated. Depending on species sturgeons have 10, 15, 20 years to reach reproductive (and commercial) age. In other words, the drastic decline in abundance and low catches of sturgeons in the 1990s is the result of the low level of spawning one generation ago.

The low level of spawning in the 1970s seems to be a combination of both environmental and anthropogenic factors. Figure 18 depicts a combined graph of the total Ural flow and sevryuga catch.

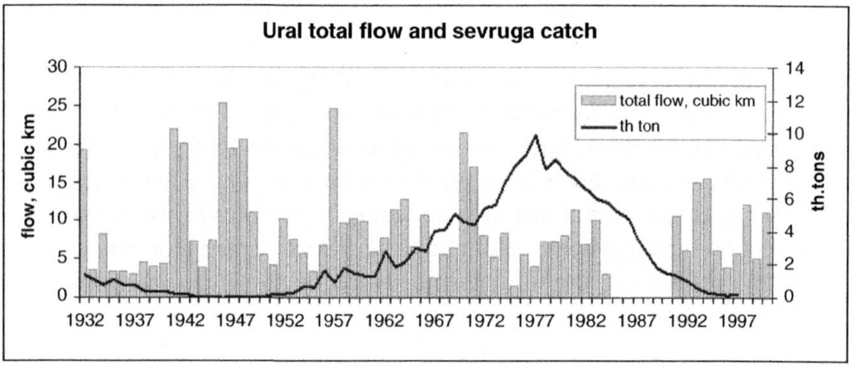

Figure 18. Sevryuga catch in the Ural River and total annual flow of the Ural River. There is no data on the Ural River flow available for the period 1985–1990

As can be observed from the graph an extremely low level of total water flow in the Ural River coincides with the highest catch during the whole history of fishery in the region. The catch of 10 thousand tons is the maximum catch observed for a single sturgeon species in the 20th century (Pitikch et al. 2005). It should also be noted that this enormous catch was removed out of only one spawning population in one river in one year. Most probably this high level of catch is the result of low level water when the fishing efforts are more efficient. As a result just a few spawners managed to pass to the spawning grounds. The reiteration of the same situation during several consecutive years resulted in total extermination of the reproductive population fraction accompanied by the lack of incubation of a new generation. Indeed, the sevryuga population could not recover after such a pressure.

Analysis of the population dynamics based on the catches without consideration of other factors such as scale of fishing efforts can also produce a distorted picture. In particular, the gradual increase in catches in the 1950s is often explained by population restoration during the years of the Second World War, characterized by lower pressure on the stock. These conclusions are true for fish populations with a short life span, but not for the sturgeon, which is a long-lived late maturing fish. Overexploited stock of this kind of fish cannot be restored in 4–5 years (Jonsson et al. 1999). Instead, the increase in catches in the post-war period should be explained by increasing fishing efforts. During this period more efficient and aggressive fishing technologies and equipment were introduced (KaspNIRH 1999).

This line of reasoning suggests that official "legal" fishery and fishery-affiliated institutions played a leading role in the sturgeon population's decline in the non-regulated Ural River.

□oaching

Poachers are traditionally recognized as unemployed local population, collecting sturgeons out of fishing zones and seasons with banned gears, processing their products and selling them on the black market. Some authors suggest that poaching is the primary factor for sturgeon species' (in particular the beluga's) decline in the Ural river in the 1990s (DeMeulenaer and Raymakers 1996; EPA 2004). The most significant role in this process is attributed to large-scale organized poaching mafia believed to arise after the collapse of the Soviet Union. Coastal population in Newly Independent Countries in the beginning of 1990s after the collapse of regional economy desperately needed new sources of income. In many coastal areas the fisheries, including poaching, appears to be the major source of income and jobs even now. Poacher communities are making their source of living out of fishery using available equipment (boats, nets, ammunition), often better than that in the possession of state inspectors.

Undoubtedly, poaching has significantly worsened the situation with sturgeon stock. The scale of poaching and its strength in the 1990s in Dagestan (Russia) was high enough to undertake military campaigns against state border guards, and called for sturgeon stock protection. There were numerous reports on machinegun attacks and exchange of fire between fishery inspectors/border guards on one side and poachers on another. The same problems were reported by the Kazakhstan Fishery Inspectors in 2007 during the First Ural Basin Sturgeon Workshop (Uralbas 2007b). The poachers sometimes even attack official sturgeon warehouses on the sea coast to take away the official catch (Uralbas 2007b).

However, the drastic decline in catch in the Ural basin started from the beginning of the 1980s (Figure 7), ten years before the disintegration of the Soviet Union. At that time poaching was severely punished by the authorities and was not a large scale problem.

A significant change in Ural sturgeon populations occurred in the 1990s, namely that the male-female proportion in spawning populations drastically changed. The historical proportion in the flock entering the Ural River for the migration upstream was 55:45. According to the latest observations, this proportion shifted to 75:25 (CEP 2002a; Dmitriev and Vasilenko 2007). Such a population structure results in lower number of new larvae to be hatched at the spawning grounds.

This feature is usually assigned to poachers hunting only for the caviar. They often capture only sturgeon females running for spawning, cut them open right on the boats, remove the caviar and throw the sturgeon bodies back to the sea. The caviar has much higher market value and does not burden the boat much in case of chasing by the state fishery inspectors.

By capturing the spawners the poachers cause a decrease in the abundance of future generation numbers. If the poachers significantly damaged the spawning population in the 1990s the effect would be revealed 10–15 years later. However, according to the official story poaching has only bloomed after the collapse of Soviet Union in the 1990s, which was already characterized by a tremendous decrease in sturgeon abundance and catch. Moreover, there are some suggestions that official fishery itself targets the reproductively mature females (EPA 2004).

In any case, poachers cannot compete with state fisheries in catch size and cannot significantly undermine their efforts and drastically decrease their catch. The influence of this kind of poaching by local communities on sturgeon stock decline may therefore be overstated.

According to the most widely spread estimation of poaching activities it takes up to 11–12 times the volume of the official catch (CEP 2002a; ZIN 2006) In other words, having the legal catch in 1995 in approximately 550 t the total catch (without scientific, productive and official unaccounted ones) in the Ural should be 7 thousand tons (550*12 + 550). Such a high level of catch corresponding to the maximum catch in the 1970s and at least four times as big as the catch by the Cossacks in the beginning of the 20th century, when sturgeon numbers in the Ural were plentiful. This obviously contradicts the situation when enough producers cannot be found even to perform captive breeding in hatcheries (Khodorevskaya et al. 2000).

This estimation migrates from one report on Caspian sturgeon to another without explanation how the calculations were made and referring to the source as "some Russian experts"(CEP 2002a; ZIN 2006). It seems that the origin for this estimation is the Caspian Research Fishery Institute (KaspNIRH 1999). In the report prepared by KaspNIRH within the framework of Caspian Environment Program the methodology for this estimation is described. The authors state that these poaching rates are calculated using mathematical models based on the difference between the expected level of catch and the real catch. This difference is then somehow distributed between poaching and "illegal" (unaccounted) official catch (KaspNIRH 1999). The expected level is derived using a set of assumptions, which in fact might not be correct. In particular, one of the main assumptions suggests the maturation of millions of the released fingerlings from the Russian hatcheries since 1955. The survival rates (if any) for these fingerlings are unknown since no proper estimations were carried out and no tagging technology used. Next, if there are any survivals they are not expected to appear in the rivers, the fishing zones, due to the peculiarities of release technology.[17] Consequently, they will not contribute to the legal catch in the rivers upon their maturation. On the other hand, the announced size of the real catch itself is influenced by the value of the "illegal" unaccounted official catch. These and other founding principles of the poaching estimation methodology are questioned by experts and provide wide opportunities for manipulations depending on the experts' beliefs and biases.

These considerations suggests that the well accepted rates of poaching during the last decades as calculated now are very unreliable figures and require careful examination and revision.

At the same time, new poaching technologies were detected lately in the Ural river (Dmitriev, personal communication, June 14, 2007). Some individual poachers use electric rods powered by portable generators,

[17] According to the USSR hatching technology fingerlings were supposed to be delivered by the ship to the "pastures" in the brackish waters and released there.

paralyzing and killing fish. The fish that survive electrocution are believed to become sterile, reducing future fish populations. Though infringers are severely persecuted by both fishery inspectors and local communities this way of fishing is believed to cause serious damage to fish stocks in the shallow Ural tributaries.

With regards to poaching activities in the open sea, adjacent to the Ural delta, numerous international poaching groups are actively hunting for the sturgeon here, utilizing modern equipment and ammunition (Uralbas 2007b). The numerous poaching boats from Dagestan, Kalmikiya and Azerbaijan fishing in this region might be an indication of the greater sturgeon availability in this area in comparison to other Caspian regions.

Indeed, poaching does exist in the region and causes serious damage to the vanishing population. However, this is rather a social phenomenon which is hard to solve by prohibitive acts and occasional patrolling. There are numerous reports on close cooperation between poachers and fishery inspectors. For instance, selling of confiscated poacher's production through official shops seems to be an excellent loophole for such cooperation. Both sides, poachers and fishery inspectors, benefit from this situation. Local communities should become interested in long-term sturgeon stock preservation. Significant changes in society are required as well as technical solutions to secure sturgeon preservation. In case of the Ural River the reviving Cossacks communities, which have a high regard for sturgeon and the Ural River, can serve as a foundation for grassroots anti-poaching campaigns.

It should also be noted that drastic declines in sturgeon stock lead to greater fishing efforts to make poaching in the sea profitable. This should result in a decrease of regional poaching activities.

Catch for scientific and reproductive purposes

While commercial fishing catch is monitored in one way or another and some, though sometimes controversial, statistics are available, so called scientific catch and removal for reproductive purposes are not properly counted (Lagutov 1995). Nevertheless, uncontrolled removals of reproductive sturgeons for these purposes have contributed considerably to stock decline, especially in the situation when sturgeon populations are already threatened and fewer specimens are available.

Although the return rate of hatchery-reared sturgeons is dubious, high number of producers have been collected for hatcheries in an uncontrolled manner (Lagutov 1995), diminishing the already depleted stock. This activity is not under CITES or any other kind of agreement and gives ground for various data manipulation and unreasonably high producer collection.

The same considerations are applied to scientific fishing, the catch intended to supply researchers with study materials.

In Russia the commercial fishery of the beluga has been closed since 2000. However, according to a resolution of the Russian State Committee on Fisheries this species can be caught for scientific and reproduction purposes and the meat and caviar can be sold afterwards (RF 2000b). In accordance with this regulation in 1998 in the Volga river alone 266 belugas were caught using drag seines, while the overall total allowable catch limit (TAC)[18] was only 710 specimen for the whole of Russia (CITES 2004b). Figure 19 shows the ratio between announced scientific catch, official TAC and commercial catch. The commercial catch was obtained from FAO fish database (FAO 2007b) in tons and converted to number of specimen using average beluga commercial weight (75 kg) provided in the very same CITES document (CITES 2004b) and KaspNIRH report (KaspNIRH 1999). The beluga catch for scientific purpose alone contributed up to 52% of total allowable catch even according to the official KaspNIRH data. Taking into account the low number of spawners and high number of hatcheries in the region the scientific and reproduction catches can exceed TAC even without considering impacts from poaching, commercial or illegal fishing. However, legal commercial fish reported to FAO is even higher then TAC. In any case, all these values are of the same order of magnitude. The catch announced as scientific is comparable to the legal commercial catch.

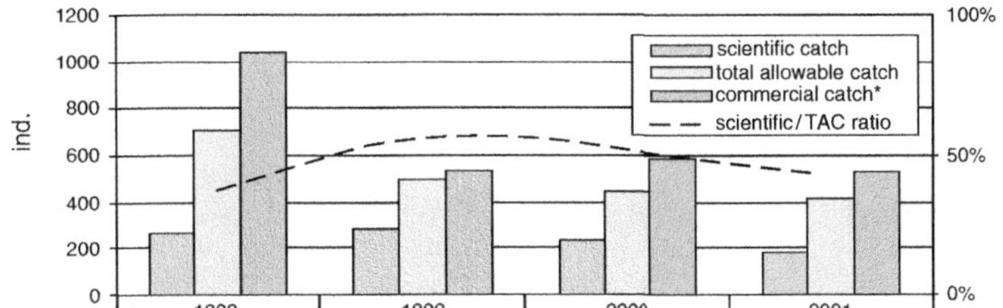

	1998	1999	2000	2001
scientific catch	266	281	236	180
total allowable catch	710	500	450	419
commercial catch*	1040	533	586	533
scientific/TAC ratio	37%	56%	52%	43%

Figure 19. The proportion between scientific catch and total allowable catch (CEP 2006; KaspNIRH 1999; Uralbas 2007b)

In Kazakhstan fishing for scientific purposes is allowed not only in the river, but also in the sea. It is limited by a certain annual quota. The scientific quota for Russian sturgeon alone is 20 t (CITES 2001), while commercial catch quota for the same species was 52 t for 2000 and 41.3 for 2001. In other words, the scientific catch adds almost 50% to the commercial catch quota.

[18] As it was already discussed the total allowable catch (TAC) is calculated based on questionable assumptions and should be considered as overestimated value.

The trend is also well represented by the scientific catch in the Volga River. Figure 19 shows the ratio between the number of beluga specimens caught for scientific purposes in the Volga (by using drag seines in the delta) and Russian Total Allowable Catch (CITES 2004b). According to this data, presented by the CITES Management Authority for Sturgeon of the Russian Federation, scientific catch comprised more then 50% of the TAC. Special attention should be paid to the scale of catch. Such a high scientific catch occurred in the situation when TAC, whatever reasoning is used for its calculation, is only 500 individual beluga specimens.

Despite the high level of scientific catch most of the available data on high-profile sturgeons is a compilation based on studies conducted a long time ago (1930s, 1960s–1970s) under different conditions. These statements are especially true for the river basin aspects of sturgeon life cycle.

Hybridization

The regulation of rivers resulted in spawners of all sturgeon species accumulating in the areas downstream the obstacles. The changes in water temperature regime force fish to spawn in the same areas as other species. This results in the appearance of hybrids with unknown characteristics. This also applies for the hatchery-reared sturgeons, often using producers from different populations resulting in mutations (Brown 2002; Kirby 2002).

Pollution

Due to the life cycle characteristics and long time span sturgeons are subject to *bioaccumulation* and *biomagnification* processes. During bioaccumulation a sturgeon absorbs some toxic substances at a rate higher than the substance is lost. The longer the organism's life span the greater the risk of chronic poisoning, even if environmental levels of the toxin are very low. Biomagnification is the process of the increase in toxin concentration in the organisms on higher trophic levels that occurs through a food chain. In this way low concentrations or occasional high level pollutants can be accumulated in sturgeon tissues, affecting its health and reproductive abilities.

Fortunately, the water pollution level in the Ural River is not high due to low population and industry density in the region. Nevertheless, some pollutants concentration and impacts on sturgeon health (i.e. reproductive behavior) were detected in sturgeon tissues (CEP 2002a, b; KaspNIRH 1999).

Some researchers indicate concern at the increasing rates of oil extraction in the Northern Caspian in general and the sea areas adjacent to the Ural delta in particular (CEP 2002a, 2006).

Other

Though there is no natural predation on adult sturgeons, sturgeon eggs are subject to predation by some river fish, such as catfish, pike, or bream (CEP 2002b).

The influence from invasive and introduced species on sturgeon populations should be insignificant due to the lack of natural predation on sturgeons (except for earlier life stages in the rivers), the wide range of food resources and underexploited food abundance in the historical sturgeon habitats. There are indications of some changes in the sturgeon food chain caused by some invasive species (CEP 2002b).

The reduction in the Ural water level has resulted in changes in water temperature regime. The sturgeon spawning behavior, dates, duration and larvae survival strongly depends on water temperature regime (Dmitriev and Vasilenko 2007).

Changes in population characteristics were detected during the last decades. In particular, mean individual weight for spawning beluga population was 110 kg in 1970s, while in the 1990s it has decreased to 75 kg (KaspNIRH 1999). Though according to KaspNIRH the relationship between population size and food availability was not discovered, such loss in weight is often explained by food scarcity due to climatic changes and sea level fluctuations. However, nowadays the total abundance of beluga and other sturgeon species is negligible in comparison to the prior size of the stock on the same grounds. Furthermore, even if there is a link between sea level fluctuations and loss of weight in sturgeon species, changes in food availability are of a much smaller scale than the catastrophic decrease in fish utilizing this resource. Fewer sturgeon individuals cannot compete for the available food sources. On the contrary, the food resources in the Northern Caspian Sea are underutilized by fish stocks (CEP 2002b). This fact is often used by the Fishery Institutes to justify usage of Caspian Sea as a fishing pond for commercial sturgeon harvesting by establishing sturgeon hatcheries in the river mouths (KaspNIRH 1999).

There are different plausible explanations for the decrease in individual weight. First of all, many researchers indicate decrease in average age of the sturgeon population. No belugas older then 25 years old have been caught lately, while they are reported to live for more than 100 years. Another possible reason in weight decline is sturgeon hybridization and influence of artificially hatched sturgeons.

Climate change does not have a direct affect on sturgeon populations. During more then 250 million years sturgeons proved themselves to possess a high level of flexibility with regards to changing environmental conditions.

In case of extremely low population size any, otherwise insignificant, factor may play a crucial role. So, according to the representatives of fishery agencies (Dmitriev, personal communication, June 14, 2007) massive beluga specimen death was detected in the 1990s in one of the traditional beluga wintering habitats in one of the Ilek river meanders (aged river bed linked to the main river course) due to the complete freezing up of the entire water body right to the bottom. Though the reasons for this phenomenon are unknown, some practitioners link these occasions with underground explosions conducted nearby in the 1980s to create gas storage reservoirs. At the same time this phenomenon can be the result of natural geomorphological changes of the river bed. If the beluga population in the river were abundant it would spread through numerous wintering habitats and escape the negative effects of the changes in one particular habitat.

Another environmental factor limiting sturgeon population is a fish-kill (oxygen deficit) in lower reaches of the river during the winter period, when sturgeon are hibernating in river depressions (Uralbas 2007b).

As one of the factors limiting sturgeon spawning migrations some authors indicate shallowing of the river delta due to siltation and sea fluctuations (Caviaremptor 2004; EPA 2004). As the Ural river delta has become shallower, fish cannot enter the stream for spawning. Several internationally-funded projects were launched aiming at dredging of the channels through the Ural river delta to facilitate sturgeon movements to hatcheries (World Bank 2004a). On the other side, the Caspian Sea level has been constantly fluctuating. In the last 15 thousand years it has varied from -20 to +50 m relative to current levels (Asarin 1997). These sea level fluctuations, far more significant than can be observed recently, and related changes in the sea ecosystem did not cause sturgeon extinction. By contrast, dredging for so-called sturgeon passage purposes can increase access of salted sea water to the estuary and increase mortality rate for the fingerlings and larvae sensitive to salinity.

A combination of these factors can also result in decrease of sturgeon feeding grounds, such as siltation of the stony substrate, low water level in the river, disappearance of temporary spawning grounds, change in food availability and composition, etc.

Restoration activities

The terrible situation with regard to sturgeon stock and the galloping price of caviar caused international discussion of the ways to restore the Caspian sturgeon (Williot et al. 2002b). Different measures are suggested to preserve sturgeon species: from “an absolute ban on uncontrolled fishing for sturgeon in the sea” (Luk’yanenko et al. 1999) to avoid buying caviar in the shops (WWF 2004). These recommendations often depend upon the perception of the problem: unique ancient species extinction or decrease in the stock of a valuable delicacy source. Moreover, the suggested strategies are often biased by experts’ vision of the problem’s roots and their professional affiliation.

Two primary alternative strategies are considered for restoring sturgeon population:

- Stop harvesting and allow natural rehabilitation and recolonization.
- Hatchery-based re-stocking in parallel with commercial exploitation of the resource.

The first approach requires a long time till population restoration occurs, if any, depending on the current population status. The second one is risky due to the possibilities of genetic modifications and other factors.[19]

The second one is advocated by fishery-affiliated institutions trying to maximize short-term food production through catch.

Both approaches can be accompanied by other activities assisting the main strategies: fishery limitations, quotas introduction, spawning grounds monitoring, establishment of protected areas, etc.

Combination of these two approaches is possible as well as their total incompatibility, when artificial re-stocking has negative effects on natural restoration.

Until now the second approach has prevailed in the Caspian basin in general and in the Ural river in particular. The best proof for this statement is the location of the sturgeon hatcheries in the Ural river delta, while the entire historic extent of sturgeon migrations is not only freely available and natural spawning habitats are accessible, but also has the status of protected territory according to the national legislature. If the aim of the restocking program is to restore wild population and to secure natural reproduction abilities the hatcheries should at least be placed close to the historic spawning grounds.

The most productive sturgeon spawning grounds are located in the Ural’s upper branches on the territory of Russia, while the migration

[19] The complications of hatchery-based sturgeon restocking are discussed below.

routes, nursing and feeding habitats are in Kazakhstan. Thus, in both alternative sturgeon restoration strategies only joint efforts of the basin countries can secure success. Thus, the sturgeon can be preserved only by joint efforts and transboundary cooperation in river basin management. Taking into account the high economic value and worldwide demand for both sturgeon products and gene pool for restoration programs, maintaining its natural reproduction and sustainable extraction is a genuine interest of the basin countries. In order to secure this possibility integrated sustainable management of water resources in the basin should be ensured.

Sturgeons are high on the international political agenda nowadays and this region increasingly attracts attention from international and national institutions. For example, from August 1, 2007 Russia has introduced a total ban on sturgeon caviar production to facilitate sturgeon restoration programs. In August 2007 a Russian State Council presidium took place in the Caspian region and focused mainly on fishery and sturgeon restoration. Special attention in these efforts has been paid to cooperation with neighboring countries, in particular Kazakhstan.

Though the importance of the Ural river basin sturgeon habitats for the conservation of the Caspian Sturgeon population is increasingly recognized, practical measures which have been undertaken so far in this area are not satisfactory. For instance, the Russian National Action Plan developed within the framework of the Caspian Environmental Program (RF 2002) does not mention the river Ural even once, even though the restoration of the spawning habitats is one of the Caspian Strategic Action Programme's primary objectives.

During the last decade a number of bilateral summits devoted to Russian-Kazakhstan cooperation in the Ural river basin have been conducted. Unfortunately, transboundary cooperation on sturgeon species conservation in the Ural river was not an issue for the discussion until the First Ural River Basin Workshop conducted in Orenburg in 2007 (Uralbas 2007a). The basin countries are trying to undertake sturgeon-protection measures, if any, independently – a strategy which is unlikely to be effective.

Endangered status and ban on fishery

Formally, the territory of the Northern Caspian adjacent to the Ural river Della was recognized as unique ecosystem from the biological and sturgeon commercial point of view in the 1970s. A protected area including the Ural delta and adjacent sea was established in 1974 in accordance with Resolution N 352 by the Government of Kazakhstan Soviet Socialist Republic "On the establishment of the protected area in the Northern Caspian Sea". After four years (by the Kazakh Government Resolution N284) the protected area was extended to the Ural river floodplain from the river delta to the mouth of the river Barbastau (near the city of Uralsk next to the border with Russia). The current status and anthropogenic activities in the protected area are defined by the Law of the Republic of Kazakhstan "On Protected Areas" (from 15.07.97), article 48. According to this Law, the main function of this protected zone is the preservation and conservation of the sturgeon species (RK 2002). The economic activities are limited within these protected areas. The extent of application and status of this zone is not clear since not only sturgeon fishery takes place in this area, but many other dangerous anthropogenic activities, such as extraction of sand and gravel from the river bed and oil production. In particular, drilling for oil extraction has been conducted in the areas adjacent to the Ural delta since 1993 (Bolshov 2000). The Northern Caspian protected zone is basically represented by sparse patches of small reserves with limited economic activity.

Paradoxically, despite the formal existence of the specially designated zones aimed at sturgeon protection the sturgeon species themselves are not protected under the national legislatures.

Listing of sturgeon on national endangered species lists (Red Books), ban on their catch and preventing trade and export of their products is considered to be a crucial step if not to restore, then at least to conserve the vanishing species. Any restoration activities should start by providing the species with protected status.

The Table 4 shows recognition of Ural sturgeon status by main lists of endangered species: IUCN Red List, National Red Books of basin countries

Table 4. Status of sturgeon species in national and international Red Lists (IUCN 2007; ORB 1998; RF 2000a; RK 1996)

	IUCN	Red Book of Kazakhstan	Russian Red Book	Orenburg Red Book
Ship *Acipenser nudiventris (Lovetsky, 1828)*	Endangered	Protected (only Aral Sea population)	Protected	Protected
Sterlet *Acipenser ruthenus (Linnaeus, 1758)*	Vulnerable		Protected	Protected
Sevryuga *Acipenser stellatus (Pallas, 1771)*	Endangered			
Beluga *Huso huso (Linnaeus, 1758)*	Endangered			
Russian sturgeon *Acipenser gueldenstaedtii (Brandt, 1833)*	Endangered			
Persian sturgeon *Acipenser persicus (Borodin, 1897)*	Endangered			

and Orenburg Regional Red Book. The classification accepted in IUCN Red List distinguishes the following classes (in order of threat decrease):

EXTINCT – EXTINCT IN THE WILD – CRITICALLY ENDANGERED – ENDANGERED – VULNERABLE – LOWER RISK

The endangered and vulnerable statuses were assigned to Caspian sturgeons by IUCN in 1994 when the situation was not as catastrophic as now. Recently, IUCN Red Book (IUCN 2007) recognizes these classifications of sturgeons as outdated. Nevertheless, according to this classification almost all sturgeon species were enlisted as endangered, while only one, the Sterlet, is seen as vulnerable.

The appropriateness of these classifications to anadromous species, i.e. sturgeons, is questioned by many researchers. Measuring extinction threats is not a straightforward process. So, the common practice used, for example, by World Conservation Monitoring Centre, is to consider a species *extinct* if it has not been observed for 50 years. In accordance with this approach, WWF Factsheet on endangered species published in the framework of CITES convention claims that only 13 species of sturgeon are threatened, and two species located in Aral Sea are "close to extinction" (WWF 2002a).

It might be too late to restore sturgeon population in case a few specimens are seen in the wild on the occasional basis. The few remaining sturgeons are not capable of restoring the population even in case of immediate measures on their habitat restoration and total ban on fishing due to their life cycle characteristics. The sturgeon populations of the Sea of Azov are doomed to extinction with no chance for natural restoration (Lagutov 1997). The European Atlantic sturgeon has been extirpated from main European rivers (Birstein 1993; Birstein et al. 1997; Dulvy et al. 2003; Granado-Lorencio 1991). Some authors believe that the Caspian sturgeon species spawning in the Russian rivers are also not capable of recovering (CITES 2004a; Crownover 2004a). Although there are cases of rare accidental catches of some sturgeons in these rivers, unfortunately natural restoration of wild populations from these spawners is not possible (de Groot 2002; Williot et al. 2000, 2001).

In any case, Caspian Sturgeons have endangered status according to IUCN classification, which presumably should at least raise the regional awareness and facilitate restoration programs.

However, neither Russian nor Kazakhstan Red Books, created to enlist threatened species at the territory of the corresponding countries, list any of the valuable species. Only two species are included, the Ship and Sterlet, which, according to KaspNIRH, do not have commercial value. Moreover, the Red Book of Kazakhstan contains only the ship population of the Sea of Aral, considered to be extinct, and not the Caspian population.

While the compilers of the national Red Books might not be very familiar with the regional situation, local academia and fishery management should be aware of the stock status. The regional Red Book of the Orenburg Oblast, where all sturgeons (except sterlet) mostly disappeared in the 1990s, was compiled by local academia at the end of the 1990s (ORB 1998). According to this book, only two species are protected: Ship and Sterlet. It literally replicates the National Red Book in terms of sturgeon species. None of the formerly commercially valuable fish was included (Uralbas 2007b).

Furthermore, from the beginning of the 1960s until 1994 a ban on ship catch in the Ural river was imposed (KaspNIRH 1999). Surprisingly, though ship has been protected in a number of ways its commercial fishing continued. Official statistics on its commercial catch exist from 1978 (CEP 2002a, b; KamUralRybVod 2007; RK 2002, 2003). Moreover, in the 1990s the catch of the ship exceeded the catch of the Russian Sturgeon in the Ural. On top of that it should be mentioned that Kazakhstan was actively utilizing export quotas obtained from CITES on the Ural ship caviar and meat trade during 2001–2003.

The national and international efforts to limit or suspend sturgeon fishing faced active opposition from the Russian Fishery Authorities, CITES and Fishery Institutes. For example, the demand to impose a total ban on sturgeon fishing to secure population rehabilitation was formulated and announced by one of the authors in the mid 1990s on the highest national legislative level, in particular Hearings in Russian State Duma on the status and reasons for decline in Azov Sea sturgeon stock (AzovBas 2002; Russian State Duma 1995). Despite strong support by environmental experts, these efforts were not successful. In 2002 the US Fish and Wildlife Service proposed to enlist beluga as an endangered species under the US Endangered Species Act. The proposal could result in an outright ban on beluga caviar import to the world's biggest caviar consumer, US, which would decrease pressure on sturgeon populations. However, this initiative was opposed by the Caspian Fisheries Research Institute (KaspNIRH), claiming increasing status of Caspian beluga. The papers were signed by the Directors of KaspNIRH and the CITES deputy secretary general.

Unfortunately, the current total Russian ban on caviar production from 2007 does not have any affect on the Ural stock, since there is no commercial sturgeon fishing in the Ural river within Russian territory. In any case, the Russian ban on sturgeon caviar does not seem to be effective tool in sturgeon restoration either. 2–3 t of the caviar will be allowed to be produced for fishing farms, which will be permitted to do scientific and productive catch for their needs.[20]

---

[20] The possible amounts of the scientific catch were discussed above.

CITES, quotas and caviar business

Caviar, or Black Gold, is one of the most expensive products on a weight basis on the world commodity markets (CEP 2002a). The most valuable and expensive caviar is derived from beluga roe.[21] Other important species used for caviar production are Russian sturgeon, sevryuga and Persian sturgeon. The price increase from the region where it is produced to the consumers is more than 100 times.

The collapse of the Soviet Union, followed by the appearance of the Newly Independent States competing for the natural reserves of the former USSR caused uncontrolled and unregulated over-exploitation of fish stocks in international waters. With the caviar industry in the Caspian Sea facing possible collapse, connoisseurs turned to Northern American caviar. But North American fisheries alone cannot supply global demand (TRAFFIC 2003).

The global caviar trade is dominated by just a few nations. In 1998 about 99% of the supply came from seven countries, with more than 90% originating from the four sturgeons species in the Caspian Sea basin: beluga, sevryuga, Russian and Persian sturgeons (Pitikch et al. 2005). Almost 100% of the caviar was imported into 12 countries, with 95% going to the European Union (EU), Japan, Switzerland and the USA. (Raymakers and Hoover 2002; WWF 2002a). In 2000 the US alone imported about 15 t of beluga caviar only, Germany 1.8 t, Switzerland 1.2 t, and France 0.9 t (Speer et al. 2000). The statistical data reveals the obvious trend: the US import of caviar is constantly increasing, sometimes doubling every two years (!), and constitutes up to 60% of total world caviar imports (Speer et al. 2000). It should be mentioned that taking into account the internal caviar production at fish-farms in USA or Germany, the size of imports might not reflect the consumption adequately. However, it is widely acknowledged that demand in major caviar-consuming countries is far greater than the caviar supply which can be provided by the newly established commercial aquaculture industry (Williamson 2003).

However, in 2006, after numerous attempts, the import of beluga caviar was banned in the USA. According to TRAFFIC, in 2007 the EU became the biggest consumer of caviar with 591 t imported per annum compared to 300 t per annum by USA (TRAFFIC 2007b). Unfortunately, this ban concerns only beluga-originated caviar, while all other kinds of caviar can be freely imported.

---

[21] According to other estimations the caviar derived of the Persian sturgeon eggs are the most expensive (Pitikch et al. 2005).

Decreases in caviar supply and corresponding price increases in the 1990s made the worldwide caviar market more financially attractive than ever. To establish the rules of the trade in 1997 all commercially utilized sturgeon species worldwide were listed under the Convention on International Trade in Endangered Species of Wild Fauna and Flora (CITES) by the World Conservation Union [International Union for the Conservation of Nature (IUCN)]. Despite the catastrophic condition of sturgeon populations sturgeons were and still are listed on Annex II of Convention as a species "currently not necessarily threatened with extinction, but which may become so unless trade is closely controlled" (CITES 2004a).

In reality, the "close control" employed within the framework of CITES does not work towards sturgeon population rehabilitation. Failures in CITES efforts to facilitate sturgeon stock restoration are well known (Crownover 2004a, b; Pikitch and Lauck 2002).

For instance, before Kazakhstan became a party to CITES in 2000 trends in world export quotas revealed a 33% decrease in expected levels of exports for caviar from 1998 to 2001 and a 775% (!) increase of sturgeon meat from 1999 to 2001 (Raymakers and Hoover 2002). These trends are difficult to interpret as compatible with preserving sturgeon populations.

Quotas are calculated depending on a country's contribution to the preservation of the sturgeon stock. So, the approval for the high export quota for former Soviet Union countries in 2003 was based on the fact that these countries proclaimed "a new approach that gives them an economic incentive to reduce poaching, the main cause of a 90 percent decline in stocks of sturgeon over the past few decades" (Pala 2004a). Disregarding the dubious official claim that poaching is the main reason for the sturgeon's decline, the positive role of the hypothetical new approach should at least be proven before granting export quotas if species preservation is the final target.

Having the only natural spawning grounds in the Caspian Basin, Kazakhstan plays the primary role in the natural sturgeon's restoration. However, unlike other regional caviar exporters, it did not have sturgeon hatcheries before 1998. The release of hatchery – reared sturgeon fingerlings is considered to be a substantial contribution to the Caspian sturgeon's restoration and has been awarded by higher export quotas. The Kazakhstan export quotas under CITES for 2001–2007 are presented in Table 5. For most of the years it is explicitly stated that sturgeons for the export should originate from the Ural stock.

Table 5. Ural sturgeon caviar and meat quotas for Kazakhstan under CITES[22] (CITES 2007)

Species	Products	2001	2002	2003	2004	2005	2006	2007
		(in tons)						
	Catch	49.6	47.9	54				
Beluga	Meat	24.8	23.95	27	52.1	27		21.9
	Caviar	3.6	5.616	4.62	2.36	2.555		1.7
	Catch	37.4	41.9	38.5				
Russian sturgeon	Meat	16.5	21.5	19.25	30.35	20		20.25
	Caviar	2.8	4.2	3.41	3.204	2.969		3.07
	Catch	161.5	144.9	121.81				
Stellate	Meat	80.75	70.38	60.545	109.27	73		48.1
	Caviar	18.41	14.5	15.15	11.01	9		8.5
	Catch	26.5	3	3				
Ship	Meat	13.25						
	Caviar	2.1						
	Catch	275	237.7	217.31				
Total	Meat	135.3	115.83	106.795	191.72	120		90.25
	Caviar	26.91	24.316	23.18	16.574	14.524		15.668

In 2006 the CITES deprived Russia, Iran, Turkmenistan, Azerbaijan and Kazakhstan of their quotas for export of sturgeons and black caviar, since the countries did not provide enough information on sturgeon population condition.

However, in 2007 CITES lifted the ban on beluga caviar export. The way these countries could prove good sturgeon population status in order to lift the ban is not clear. This decision shocked environmental experts and agencies, such as TRAFFIC and WWF (TRAFFIC 2007a). These organizations believe that the re-introduced caviar export quotas are not based on catch quotas and do not have scientific and legal basis. For instance, export quotas for Russian sturgeon from Russia were increased from 14 t in 2005 to 20 t in 2007 while the catch for this species decreased from 230 t in 2005 to 11 t in 2007.

The same situation occurred with Kazakhstan export quotas. A comparison between export quotas on sturgeon meat and caviar for 2001 and 2007 (excluding amounts allocated for Turkmenistan) shows that the changes are insignificant, which presumably should indicate a stable population situation. So, in 2001 and 2007 the export quota on Beluga meat was 24.8 and 21.9 t respectively. The export quotas for Russian Sturgeon meat and caviar have even increased: from 16.5 and 2.8 t in 2001 to 20.25 and 2.07 t in 2007 respectively.

[22] The quotas allocated to Turkmenistan as well as unutilized quotas from previous years were deducted from Kazakhstan quotas.

Figure 20 shows the dynamics of the Kazakhstan export quotas under CITES for beluga caviar. As is known, maximum production of the caviar from beluga or any other sturgeon species cannot be more then 10% of the female weight. Currently, this ratio is much lower due to the high proportion of young and pre-mature beluga females in the catch. Taking into account that female share in Ural beluga catch is equal to only 20–25% (CEP 2002a; KaspNIRH 1999) the caviar proportion in CITES export quotas is eight times higher then the amount corresponding to the sturgeon meat export quota. For other species the caviar production per fish is even lower then 10% (Figure 21 and Figure 22).

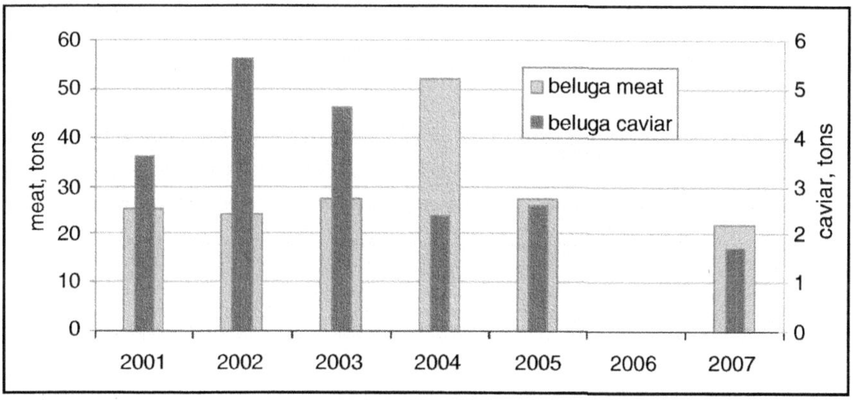

Figure 20. Meat and caviar export quotas under CITES for Ural-originated beluga

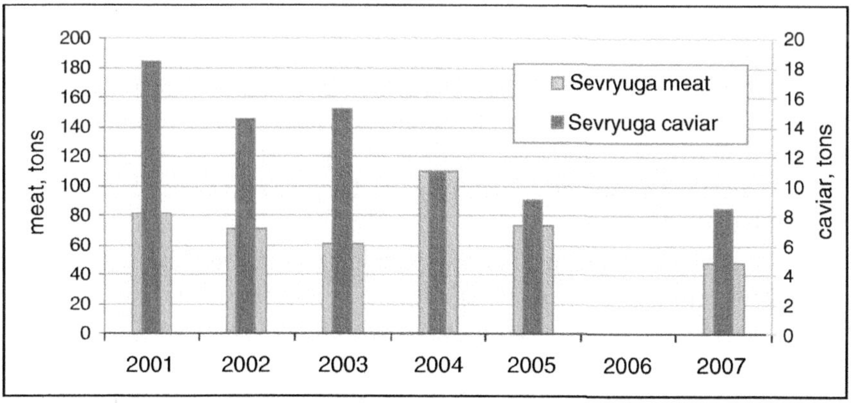

Figure 21. Meat and caviar export quotas under CITES for Ural-originated Sevryuga

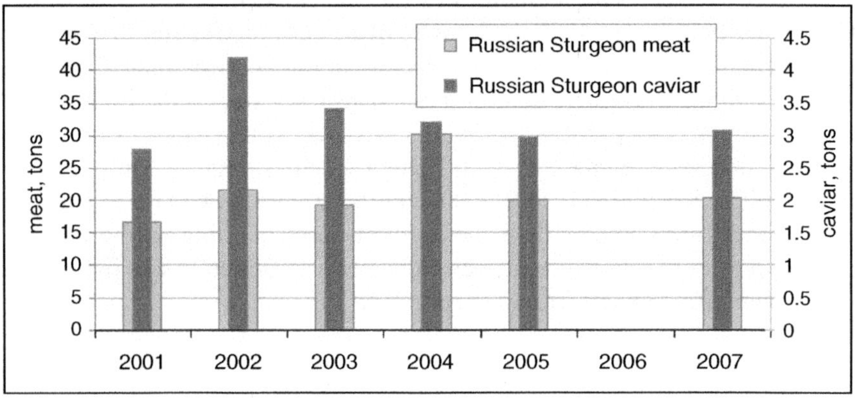

Figure 22. Meat and caviar export quotas under CITES for Ural-originated Russian sturgeon

Though having some regulatory effect on the international caviar trade, the work of National CITES Management Authorities for sturgeon in both Russia and Kazakhstan has been actively criticized. In particular, the management authorities are overstating their contribution to sturgeon restoration (i.e. fingerlings release), hiding the real catch as well as overestimating Caspian sturgeon stock (CITES 2004b; Kirby 2002; Pala 2004b; TRAFFIC 2007a; Uralbas 2007b). In this way the higher fishing and export quotas can be obtained providing a legal background for species extermination.

Finally, within the framework of CITES there are also some attempts to implement a standard labeling system for caviar exports, which to date is still not operational (TRAFFIC 2007a).

Summing up, the role of CITES in sturgeon population conservation and restoration is dubious. Nevertheless, the national export quotas have been assigned by the CITES Secretariat based on Parties' own estimations of stock populations and quota requests. The CITES secretariat only approves the export amounts demanded by Parties. Consequently, the national authorities should be considered the primary source for unreasonably high export levels.

On the other hand, the national fishing quotas, the basis for export quota calculations, are distributed by the Commission on the Biological Resources of Caspian Littoral States according to the country contribution to sturgeon stock replenishing. The hatchery-based sturgeon restocking is mainly counted as such a contribution. Despite the unique opportunities for natural restoration in the Ural river Kazakhstan's fishing quota based on the recently opened sturgeon hatcheries in Atyrau is only 18% of total f.USSR Caspian catch. Correspondingly, the existing quota distribution system prioritizes artificial over natural reproduction and complicates the restoration and conservation of natural spawning habitats.

As a result the quota system is considered to be a rather inefficient tool in sturgeon population conservation and restoration (CEP 2002a; ENS 2007; Uralbas 2007a). From this point of view one of the most common recommendations for sturgeon restoration suggesting "better calculation of national fishing quotas reflecting the real contribution of a particular state to overall sturgeon stocks" (Luk'yanenko et al. 1999) does not seem likely to contribute to sturgeon population sustainability.

Hatchery-based restocking

In order to maintain and restore the diminishing sturgeon's wild stock, intensive hatchery sturgeon production has been used since the mid-1950s (Secor et al. 2000).

Actually, the first trial on artificial sturgeon propagation started in the Volga basin in Russia at the beginning of the 20th century (Secor et al. 2000). However, this phenomenon began on a massive scale only after sturgeon hatcheries were put into operation starting from the 1950s. Thirteen hatcheries were constructed in the Caspian Basin during the Soviet period. Millions of fingerlings were released annually. In the 1980s the release rate was up to 101 millions fingerlings per year (Ivanov 2000). Even nowadays Caspian sturgeon propagation is the world's largest restocking program.

The primary goal for the introduction of artificial breeding and propagation was to support commercial fishery. These activities were carried out in accordance with the prevailing doctrine on converting the Caspian Sea into a fish pond for sturgeons. Numerous theoretical investigations and calculations were carried out to implement this strategy by fishery institutes. The shift of fishing efforts from the feeding grounds in the sea to the naturally spawning populations in the rivers in 1962 also indicates the priority of commercial sturgeon fattening and harvesting over natural reproduction.

There are some claims of an increase of the proportion of hatchery-reared sturgeons in total catch recently. By some estimations of KaspNIRH the share of artificially produced sturgeons in the catches is only 20–25% (KaspNIRH 1999). This is a very unexpected result taking into account the long history of juveniles release by numerous hatcheries. In fact, other estimations provide different shares of hatchery-based sturgeons in commercial catch. In 1997 the beluga, sturgeon and sevryuga had shares of

99.50% and 40% respectively (Khodorevskaya et al. 1997).[23] Unfortunately, the identification and counting methodology of hatcheries-originated sturgeons usually is not well described. No proper tagging system has been introduced even now, not mentioning 20 years ago when the currently caught sturgeons were released.

In this situation it is important to understand how these estimations were obtained. They are the results of mathematical calculations based upon the simple assumption that, after the Volga river's damming in 1958 cut off 100% of the beluga spawning grounds, all new generations of beluga are hatchery-originated (KaspNIRH 1999). The same approach was used for estimating Russian sturgeon and sevryuga stocks taking into account that some spawning grounds downstream of the Volgograd dam for each particular species are still available. Starting from this assumption a particular hypothetical sturgeon stock is calculated and used for deriving hatchery-reared shares as well as estimating poaching activities (11–12 times higher than official catch).

As a matter of fact, the yield to fishery from the millions of juveniles released starting from 1950s was supposed to be tens of thousands tons. The peak of release was observed in the mid 1980s: more then 100 millions were released annually. The sturgeons released in the 1980s should have reached their commercial age and size by the end of 1990s, yet the total sturgeon catch since the end of the 1990s is miserable (several hundred tons in Volga and 100 t in Ural for 2007 in comparison to 20 thousand tons in the 1970s–1980s). The impact from poaching is insignificant since these specimens could not reach maturity and commercial size until the end of the 1990s. According to the calculations, the yield of fishery in this period should have been tens of thousands. Poachers had to dump their catch to the markets. However, even on the black market the price of sturgeon products went up dramatically. Sturgeon products can hardly be found even on the internal market in the Caspian region.

On the one hand there is an increase in the virtual shares of hatchery-reared sturgeons in total sturgeon population, on the other the sturgeon fishing industry has collapsed due to the tremendous decline in sturgeon stock. Coupling these two facts challenges the efficiency of the hatchery-based re-stocking programs.

Though artificially reproduced sturgeon can to some extent be a substitute for the natural one in terms of gourmets' tables, the ability of these sturgeons to sustain a wild population is doubted by many researchers (Craig 2000; Lagutov 1995).

---

[23] It should be noted that this statement is coupled with beliefs that the sturgeon stock in 1990s is abundant due to the ban on open-sea fishery in 1962 and massive hatchery-based restocking program (Khodorevskaya et al. 1997).

Due to the free-flowing nature of its stream and preserved spawning grounds the hatcheries were not constructed in the Ural river basin until the sturgeon catch had collapsed in the region in the 1990s. In Kazakhstan first two sturgeon hatcheries, Ural-Atyrau and Atyrau hatcheries, were put into operation in 1998 in Guriev (Atyrau) in the Ural river delta (CITES 2001).

Figure 23 shows the fingerlings release by two Atyrau hatcheries to the Ural river. Significant variations in fingerlings release can be observed, which can be explained by various reasons. However, the superimposing of beluga catch in the Ural river basin over the release graph suggests the dependence of the beluga release on the catch. As it is known, artificial sturgeon propagation depends on the wild stock. The spawners (both males and females) are taken from the migrating population and bred in captivity. In other words beluga catch in the region was not high enough even to secure re-stocking. However, as follows from the previous chapter the Kazakhstan quota on beluga caviar export under CITES in 2005, characterized by a lack of beluga juveniles release, was 2.5 t (CITES 2007). Instead of using caught fish for restocking, roe was exported in full accordance with the provisions and quotas of the Convention on International Trade of Endangered Species, created to prevent such an export.

To summarize, the collapse in natural sturgeon reproduction put an end to hatchery-based stocking programs.

Though it might be too early to evaluate the efficiency of hatchery-based sturgeon restocking carried out in the Ural basin, the general trends and shortcomings of this process are identical to the sturgeon restoration

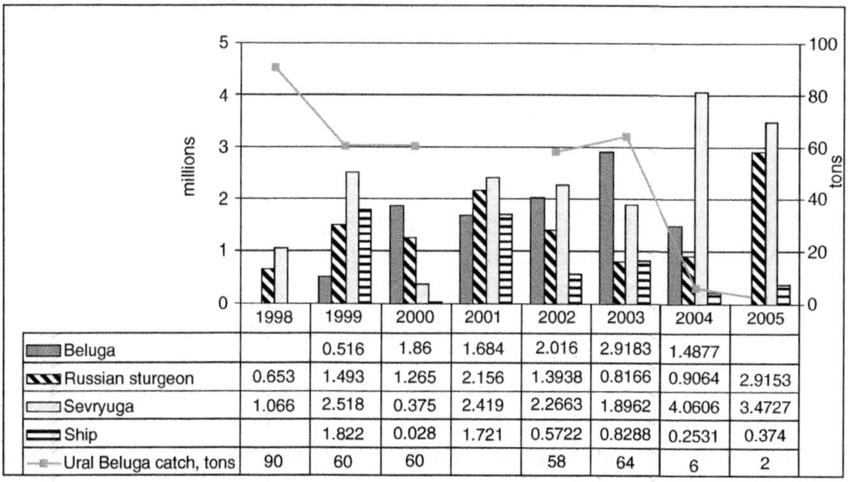

	1998	1999	2000	2001	2002	2003	2004	2005
Beluga		0.516	1.86	1.684	2.016	2.9183	1.4877	
Russian sturgeon	0.653	1.493	1.265	2.156	1.3938	0.8166	0.9064	2.9153
Sevryuga	1.066	2.518	0.375	2.419	2.2663	1.8962	4.0606	3.4727
Ship		1.822	0.028	1.721	0.5722	0.8288	0.2531	0.374
Ural Beluga catch, tons	90	60	60		58	64	6	2

Figure 23. The release of sturgeon fingerlings to the Ural river by two Atyrau hatcheries (CEP 2006)

problems throughout the former USSR countries. Moreover, these inherent problems are further complicated by other contemporary issues (i.e. trans-boundary sturgeon migrations, lack of producers, etc.). The various possible obstacles to the success of Ural sturgeon restoration are speculated on below.

Homing fidelity

There are some theories suggesting that sturgeons released from the hatcheries in the river deltas do not contribute to the wild sturgeon population's sustainability (Lagutov 1995). It is argued that hatchery-originated specimens, if they survive till reproductive age at all, do not have homing fidelity. Homing is needed not only to find the natal river but also to arrive at a particular spawning site at the proper time. The ability of the hatched specimens to do so is not proven. Homing fidelity for the Caspian sturgeons was demonstrated by different researchers (Ivanov et al. 2005). Thus, the above mentioned threat exists and should be carefully evaluated before drafting any restoration strategies. The influence of this factor increases many-fold upon consideration of fingerlings release technology. The fingerlings were deliberately delivered to "pastures" located in the brackish sea waters in the Northern Caspian by special boats (KaspNIRH 1999). In fact, this technology was employed during most of the hatcheries operation years. Only in the 1990s was this policy cancelled due to the lack of state financial support. If the believers in the inherent Caspian sturgeon homing fidelity are right these sturgeons have difficulties in returning to the rivers for spawning.

Rearing technologies

There are three main rearing and fingerlings release technologies utilized by hatcheries (Abdolhay 2004). Correspondingly, there are different associated official mortality rates for the fingerlings. Some technologies employ fingerlings rearing in ponds for 40–60 days before the release with due feeding and fertilizing (Abdolhay 2004). However, the approach utilized by Volga hatcheries was different with much lower survival rates (KaspNIRH 1999). Since the beginning of the restocking program in the 1950s the peculiarities and efficiency of the technological process was constantly changing in Soviet, then Russian and Kazakhstan, hatcheries due to technological improvements, lack of funding, equipment deterioration, lack of producers, etc. Apart from the release numbers the larvae survival rates should also be constantly fluctuating.

According to some sources (Kirby 2002) the Kazakhstan hatcheries release fingerlings when they are two months old and about 10 cm long. These statements contradict well established sturgeon rearing technology used in the USSR for a long time, when fingerlings were released with a weight of 2–3 g and 6–7 cm long, depending on species. According to this technology hatching is conducted without feeding (CITES 2004b, KaspNIRH 1999), while two month old fingerlings are feeding actively.

The release of fingerlings from two Atyrau hatcheries for 2002–2005 according to the Ministry of Environmental Protection of the Republic of Kazakhstan are presented in Table 6 (CEP 2006). Significant fluctuations in the weight of the released fingerlings can be spotted. The weight of beluga fingerlings varies from 12 to 3.5 g within four years. This irregularity suggests drastic changes in technological process and undermines the calculations of fingerlings survival and return rates to fishery. Total annual fingerlings release shows a steady increase and, presumably, indicates active actions towards sturgeon stock rehabilitation, while species composition analysis suggests significant problems of the artificial sturgeon propagation in the Ural river. The overall increase is secured at the cost of two species (Russian Sturgeon and Sevryuga), while hatching of others has been discontinued (Beluga and Ship). This consideration suggests the remaining two species are exposed to increasing fishing pressure for captive breeding.

Sea salinity

Next, as was discussed above, sturgeon fingerlings are very sensitive to sea salinity levels. High salinity is lethal for sturgeon fries. Historically, sturgeon juveniles stayed in the river freshwaters after hatching for up to three months and by entering brackish salted water had average weight of 171 g and length of 36 cm. According to the CITES report nowadays fingerlings are released into the sea brackish waters with a weight of only 2–5 g (CITES 2004b). Apparently, taking into account the evolutionary developed mechanisms for sturgeon life cycle stages, these embryos are exposed to high mortality rates to say the least.

With regards to later practice of fingerlings release into the river stream the same line of reasoning can be applied. Hatcheries are usually located in the rivers' deltas or close to them upstream. Nevertheless, according to the official statements by Caspian Fisheries Research Institute (KaspNIRH 1999) up to 70% of released larvae has perished already on the way to the sea.

Both Kazakhstan sturgeon hatcheries are located at the city of Atyrau (Guriev) in the river delta. The increase of the salinity in the estuary areas adjacent to the Ural river delta observed in recent decades could cause even higher mortality rates among the released fingerlings.

Table 6. Release of sturgeon fingerlings from two Atyrau hatcheries. AH stands for Atyrau Hatchery, while UAH is Ural Atyrau hatchery (CEP 2006)

Species	2002				2003				2004				2005			
	AH		UAH		AH		UAH		AH		UAH		AH		UAH	
	Release (thousand individuals)	Weight (g)	Release (thousand individuals)	Weight (g)	Release (thousand individuals)	Weight (g)	Release (thousand individuals)	Weight (g)	Release (thousand individuals)	Weight (g)	Release (thousand individuals)	Weight (g)	Release (thousand individuals)	Weight (g)	Release (thousand individuals)	Weight (g)
Beluga	1002.7	8.6	1013.3	12.0	1728.6	7.3	1189.7	4.1	932.0	4.1	555.7	3.5	–	–	–	–
Russian sturgeon	987.8	2.3	406.0	2.6	341.7	3.7	474.9	3.0	507.8	3.5	398.6	3.1	1762.3	3.4	1153.0	3.2
Ship	–	–	572.2	3.7	828.8	4.1	–	–	253.1	3.3	–	–	374.0	3.1	–	–
Sevryuga	1090.3	1.6	1176.0	3.5	297.5	3.2	1598.7	2.8	1748.6	3.5	2312.0	2.7	1245.7	3.3	2227.0	2.8
Sterlet	20.1	0.8	–	–	–	–	–	–	–	–	–	–	–	–	–	–
Subtotal	3101.0	–	3167.5	–	3196.6	–	3263.3	–	3441.5	–	3266.3	–	3382.0	–	3380.0	–
Total	6268.5				6459.9				6707.8				6762.0			

In this way the assumptions of the high efficiency of existing artificial stock rehabilitation and the high proportion of hatchery-originated sturgeon in river catches are severely undermined.

The survival rates for the released juveniles should be reconsidered and carefully estimated by independent experts.

Lack of "producers"

Starting from the beginning of this century the substantial decrease in release of sturgeon juveniles from hatcheries in the Caspian region was reported (Uralbas 2007b; ZapKaspRybVod 2008). Moreover, the quantity of producers (female beluga) was considered to be insufficient to support hatchery production efforts already in 1995 in the Volga River delta (Birstein et al. 1997). This statement on the decline in juveniles release from the mid 1990s was also confirmed during personal communication with hatchery managers. As the primary reason for the decline managers indicated the lack of "producers", wild sturgeons used for breeding. The number of spawners in the river is not sufficient for the hatcheries functioning. This can only be explained by the fact that only naturally reproduced sturgeons are returning to the rivers for spawning.

This is an amazing result considering all the proclaimed success of beluga hatchery rearing. It also presents an interesting point in the entire theory of hatchery-based restocking. If the sturgeon numbers in the sea are abundant as is stated by KaspNIRH (KaspNIRH 1999) and the Russian Management of CITES (CITES 2004b) the logical question is why there are no producers in the rivers. There are only two possible answers:

- The sturgeon stock is depleted and the efficiency of the present-day artificial restocking is miserable at least or a big scale fraud at most; in this case chronic deficit of spawners in the rivers even during high water years indicates population extinction or
- Sturgeons living in the sea cannot return to the rivers due to the lack of homing fidelity or some other reason. Whatever that reason is it jeopardizes not only the sturgeon population's natural reproduction, but also artificial restocking programs.

Both answers urge a review of the current restocking programs and suggest hatchery based restocking should be avoided until the reasons for its failure are clarified and dealt with.

Changes in reproductive behavior

Sturgeon reactions to *stress* and, in particular, the influence of stress on sturgeon reproduction are one of the main practical problems in fish management in general and aquaculture in particular. However, little is known of sturgeon physiological response to stress (Bayunova et al. 2002). Most researchers nowadays agree that general management practices (capture, handling, transportation) negatively affect sturgeon reproduction (Bayunova et al. 2002; Williot et al. 2002b). At the same the time long-term consequences of stress during artificial reproduction are not yet properly studied. Results of some research challenge the ability of artificially reproduced sturgeons to reproduce successfully in the wild (Lagutov 1995, 1997; Secor et al. 2000).

Genetic Problems

As a result of some studies *genetic fitness* and impact of the hatchery-reared on the wild sturgeon population is argued as well (Aprahamian et al. 2003; Ludwig 2006). In fact, analysis of the caviar composition from aquaculture shows significant variation from that of wild origin already within the first generation (Gessner et al. 2002). Hatchery-based sturgeon re-stocking is also endangered by genetic risks of interstock transfer and inbreeding depression (Firehammer and Scarnecchia 2007; Grunwald et al. 2007; Waldman et al. 2002). Due to the lack of spawners, producers from different populations are often used, which results in mutations (Arndt et al. 2002; Brown 2002; Kirby 2002).

These considerations question the hatchery efficiency for the rehabilitation of wild sturgeon populations. There are definitely certain risks associated with restocking and all restoration programs should undergo risk screening in order to minimize negative impact on ecosystems and wild populations.

The problems with artificial propagation of anadromous migratory fish were also approached by many authors (Altukhov and Evsyukov 2001; Aprahamian et al. 2003; Arndt et al. 2002; Bachmann 2000; Belanger et al. 2001; Brown and Day 2002; Chebanov and Galich 2002; Jonsson et al. 1991; Jonsson et al. 1999; Schreck et al. 2001; TRAFFIC 2003; Williamson 2003). There is a clear trend on growing scientific concern over the negative influence of fish farming and restocking on natural populations.

Nevertheless, the representatives of sturgeon hatcheries and affiliated institutions demand an increase in juveniles release from hatcheries up to 100–110 million individuals (KaspNIRH 1999). For comparison, in 2007 the plan on juveniles release from six Russian hatcheries was only 23 million, which was already impossible to fulfill (ZapKaspRybVod 2008). Apart from the lack of spawners the aging equipment aging and insufficient financial support are indicated as the reasons for this failure.

Despite the dubious character of hatcheries' influence on sturgeon populations and catch the multi-million release of fingerlings is still a very profitable business. The sturgeon fishing quotas in the Caspian Sea are defined according to the contribution each state makes to replenishing stocks (i.e. number of fingerlings release). Currently the quota of Russia is 70% of the total Caspian catch, while Kazakhstan's is only 18% (ENS 2007).

The short-term profit from hatcheries opening is obvious – the higher the announced release the bigger the quota. In reality, the announced figures often proved to be overstated in order to demand higher fishing quotas (Uralbas 2007a).

Despite all these problems, artificial sturgeon propagation might play a positive role in Ural sturgeon rehabilitation, if certain conditions are met.

First of all, hatcheries should be placed next to the historical spawning ground to minimize possible problems with homing fidelity and negative influence by sea salinity. Indeed, artificial propagation will not lead to recovery and sustainability of the sturgeon population unless the fundamental problems that caused the population declines are properly addressed.

Second, the commercial sturgeon propagation can satisfy the needs of the market to decrease the pressure on the wild stock. However, in this case the Caspian sea should not be used as a "pasture" for the fattening of sturgeons, to avoid disturbance to natural population (Lagutov 1995). There are certain risks and limitations associated with this role of artificial propagation (lack of producers), which should be addressed accordingly.

In conclusion, the following statement from KaspNIRH should be quoted (KaspNIRH 1999):

> Multiyear research conducted by KaspNIRH and other institutes showed that sturgeon population restoration using artificial propagation is not possible without conservation and restoration of natural sturgeon reproduction.

As a matter of fact, this statement acknowledges the failure of the previous fishery management strategies and misleading role of scientific recommendations of fishery institutions aimed at securing artificial sturgeon fattening and harvesting in sea pastures and extermination of the natural spawning population in the river basins.

Unfortunately, this official recognition of basic ecological concepts comes too late since it might already be impossible to restore Caspian sturgeon populations from the consequences of the previous management paradigm. What is much more important is that this discovery still contradicts the practical steps on sturgeon stock rehabilitation suggested in this document, submitted in the framework of the Caspian Environmental Program. The euphoria over the 99% population share of hatchery-originated beluga also obscures the necessity to maintain natural reproduction to secure artificial reproduction.

Priority of natural sturgeon reproduction for the Ural river

The positive contribution of the existing approach to the hatchery-based sturgeon restocking program is questioned. Extensive independent research is needed to confirm its usability and efficiency. Before its efficiency or harmless nature is confirmed primary efforts in sturgeon conservation and, if any, rehabilitation should be made to secure the sturgeon's natural reproduction (AzovBas 2002; Russian State Duma 1995; Lagutov 1995, 1996, 1997).

Artificial propagation and release might only be an auxiliary short-term tool for stock replenishment which should be used with due care and reservations (Birstein et al. 1997). Millions of juveniles released from the hatcheries might prolongate commercial fishing for a short period but they cannot sustain a population. Hatchery production is only one, not the primary, of many strategies required to protect and increase the levels of natural reproduction. The primary strategy towards sturgeon stock consumption should be to restore, maintain and secure its natural restoration mechanisms, created over hundreds of millions years of evolutionary development. Only in this case the sturgeon population, and its consumption upon full rehabilitation, if any, can be sustainable.

From this perspective the Ural river provides a unique opportunity to preserve the sturgeon gene pool and to restore the sturgeons through the entire Caspian basin. As indicated above, the Ural river contained the natural spawning habitats for every sturgeon species historically inhabiting the Caspian Sea. Though slowly decreasing in size due to habitat degradation, climate change and anthropogenic activities (i.e. dredging for navigation and sand-gravel extraction), the habitats supporting all sturgeon life cycles are still available throughout the entire historical species areal in the Ural river (Dmitriev and Vasilenko 2007; KamUralRybVod 2007). Currently they are underutilized, if utilized at all, for reproduction due to lack of producers. All possible measures should be employed to secure spawners' arrival to spawning grounds and their successful spawning. The priority of natural sturgeon reproduction in the region leads to the necessity to prioritize sturgeon conservation needs over other participants of the integrated water management process.

To promote the idea of preservation of the Ural sturgeon habitats and to facilitate international transboundary activities to secure natural sturgeon reproduction the Ural Basin Sturgeon Project, aiming at the Ural sturgeon's conservation and rehabilitation, was initiated by Central European University, the environmental NGO "DonEco", and a number of federal environmental agencies. The First International Ural Basin Workshop was conducted in Orenburg, Russia, in June 2007. The participants, including international and national experts, decision-makers and local community representatives, developed a set of recommendations[24] on Ural sturgeon restoration. The main recommendation was the establishment of a protected area along the Ural River stream (a so-called International Transboundary Ural Sturgeon Park) with a high level of local population and international community involvement.

Conclusions and recommendations

The sturgeon species in the Caspian basin nowadays have not only lost their former economic value, but also literally are on the brink of extinction.

Despite the high level of attention by international and national communities to the Caspian region, basin-wide regular biodiversity assessments in general and sturgeon-related aspects in particular have not been undertaken. Generally the sparse data on sturgeon abundance, catch and life characteristics are contradictory, flawed or biased. There are some attempts to create Caspian wide databases on Caspian biodiversity (CEP 2002b), but their usability is undermined by the lack of reliable information on a river-basin scale as well as biased and contradictory sources of information. Most of the available data is a compilation of the outdated results of field research or observations conducted in the 1930s–1970s. In addition, the Ural river basin was excluded from the few contemporary study projects. The results of the modern sturgeon population stock estimates by national fishery-affiliated agencies and institutions are significantly undermined by deliberate or unintended distortions and are doubted by international expert communities.

Though conservation and restoration of sturgeon stock is proclaimed as a priority target in national strategic action plans in Caspian littoral countries, the specific activities and policies in the region aim rather at short-term consumption of the resource until its total extinction.

Fishery management strategies and existing sturgeon stock restoration schemes have proved to be ineffective to say the least. Fishery-centric approaches to optimizing sustainable maximum yield are not adequate for the conservation and restoration of the sturgeon populations. The attempts to squeeze as much as possible from diminishing sturgeon stock (i.e. the current approach to CITES quotas) would result in its total degradation and extinction.

---

[24] The recommendations and resolution of the First Ural River Basin Workshop can be found in this volume.

At the moment, Ural sturgeon stocks are following the Caspian-wide trend. Fish population analysis, which is exposed to high level of uncertainty by default, is complicated by the above mentioned factors. However, it is obvious that the Ural sturgeon populations have been brought to the verge of extinction during the last few decades solely due to state (USSR and later Kazakhstan) regulated and organized overfishing, including commercial, illegal, scientific and productive catches. Unlike in other regions, poaching played a minor role in the stock decrease in comparison to other factors. All additional factors crucial for other regions, such as river regulation, or spawning grounds' loss did not a play significant role in the Ural sturgeon stock's depletion.

The sturgeon population in the Caspian Sea basin can be sustainable only in the case of preservation and restoration of natural sturgeon. The only remaining sturgeon spawning grounds for all Caspian sturgeon species are in the Ural River. Thus, the Ural River should become the center for Caspian-wide sturgeon conservation and rehabilitation programs.

The role of hatchery-reared restocking in wild sturgeon population restoration is dubious. With regards to the Ural basin with natural habitats available and unobstructed migration routes, the only place for hatcheries, if any, should be close to the historical spawning grounds. This requirement implies relocation of existing sturgeon hatcheries upstream the Ural river.

Since the historic Ural sturgeon areal spread runs through the territory of Russia and Kazakhstan only joint transboundary measures to preserve this unique ecosystem and its sturgeon population will be productive and meaningful.

Considering the Ural spawning grounds' underexploitation the very first and urgent step towards sturgeon population rescue should be to secure breeders' access upstream by imposing a ban on any kind of river sturgeon fishing (including scientific) and enforcing its implementation with a high degree of international involvement.

The cross-disciplinary multi-sectoral basin-wide approach should be utilized. The issue of sturgeon preservation goes far beyond fisheries management plans. Instead, the sturgeon populations should be considered as an indicator of sustainability in a basin-wide regional development strategy. All involved stakeholders and aspects of anthropogenic influence on the sturgeon population and riverine ecosystem should be taken into account. Being a perfect environmental bioindicator of the basin ecosystem conditions, sturgeon also allows environmental, social and economic aspects of regional sustainable development to be brought together.

In particular, water usage and land use patterns in the Ural watershed, especially in floodplain areas, should be closely monitored and regulated.

Though requiring some improvements, the national and international legislative basis for these activities already exists. However, closer attention to the enforcement of the existing national laws is required.

Considering the high economic and environmental importance of the sturgeon species and traditional biased estimates and study results by national sturgeon authorities this process should be closely monitored by the international community.

At the same time a high level of cooperation from local communities is required. In the case of the Ural river basin this can be easily achieved through the involvement of the reviving Cossack communities.

Faced with a lack of reliable information on sturgeon migrations and life cycle characteristics, regular independent monitoring utilizing modern equipment (i.e. using satellite tagging, satellite images analysis, GIS, modeling, etc.) is urgently required.

If the Ural sturgeon stock restoration is successful fishing efforts should be focused only on repeatedly spawning sturgeons, but the ban on catching first time spawners should remain.

The long-term economic benefits of restocking sturgeons can significantly outweigh the initial costs. Upon stock rehabilitation the Ural sturgeon can also serve as a gene pool for sturgeon restocking programs and aquaculture production in other regions.

To secure natural sturgeon reproduction it is recommended that an International Transboundary Ural Sturgeon Park should be established along the sturgeon migratory routes throughout the historic range of sturgeon areal in the Ural basin.

References

Abdolhay, H. 2004. Sturgeon stocking programme in the Caspian Sea with emphasis on Iran. Rome: FAO.

Altukhov, Y.P. and Evsyukov, A.N. 2001. Juvenile overproduction at fish hatcheries may explain the degradation of the Volga stock of the Russian sturgeon. Doklady Biological Sciences. 380. (1): 454.

Aprahamian, M.W., Martin Smith, K., et al. 2003. Restocking of salmonids–opportunities and limitations. Fisheries Research. 62. (2): 211–227.

APU. 2000. Istorija Astrakhanskogo Kraja. Astrakhan: Astrakhan Pedagogic University.

Arndt, G.-M., Gessner, J., et al. 2002. Trends in farming, trade and occurrence of native and exotic sturgeons in natural habitats in Central and Western Europe. Journal of Applied Ichthyology. 18. (4–6): 444–448.

Artiukhin, N.E. 1979. Persidskij osetr v rekah Severnogo Kavkaza i perspektivy ego i spolzovanija v osetrovom hozjastve In Biologicheskie osnovy razvitija osetrovogo hoziajstva v vodoemah SSSR ed., Moscow: Nauka.

Asarin, A. 1997. Problem of Caspian sea level fluctuations. Power Technology and Engineering (formerly Hydrotechnical Construction). 31. (11): 645.

AzovBas. 2002. Proceedings. Azov Sea Basin Workshop Novocherkassk, Russia, Green Don.

Bachmann, J. 2000. European Freshwater Species Strategy. WWF International European Freshwater Programme. WWF Austria.

Baimukanov, M. 2007. Change Dynamics of Some Biodiversity Representatives in the Ural-Caspian Region. Biodiversity & the Oil and Gas Industry in the Caspian Region, Atyrau, Kazakhstan IPIECA.

Bayunova, L., Barannikova, I., et al. 2002. Sturgeon stress reactions in aquaculture. Journal of Applied Ichthyology. 18. (4–6): 397–404.

Belanger, J.M., Son, J.H., et al. 2001. Effects of short-term management stress and ACTH injections on plasma cortisol levels in cultured white sturgeon, Acipenser transmontanus. Aquaculture. 203. (1–2): 165–176.

Birstein, V.J. 1993. Sturgeons and paddlefishes: threatened fishes in need of conservation. Conservation Biology. 7. (4): 773–787.

Birstein, V.J., Bemis, W.E., et al. 1997. The threatened status of acipenseriform species: a summary. Environmental Biology of Fishes. 48. (1–4): 427–435.

Birstein, V.J., Doukakis, P., et al. 1998. Population aggregation analysis of three caviar-producing species of sturgeons and implications for the species identification of black caviar. Conservation Biology. 12. (4): 766–775.

Bolshov, A. 2000. Otcenka suchshestvuiushih nacionalniuh okhraniaemiuh territorij i ih statusa zaschity, osobenno transgranichniuh i migriruiuschih vidov. Kazakhstan. Biodiversity CRTC, Second Regional Caspian Workshop, Alma-aty, Caspian Environment Program.

Borodin, N.A. 1901. Ural'skie Kazaki i ih rybolovstva. St. Petersburg. Izdanie "Vestnik Kazachih Voisk".

Brockhaus, F.A. and Efron, I.A. 1898. Brockhaus and Efron Encyclopedic Dictionary. St. Petersburg: Brockhaus & Efron Publishing House.

Brown, C. and Day, R.L. 2002. The future of stock enhancements: lessons for hatchery practice from conservation biology. Fish Fisheries. 3. (2): 79–94.

Brown, P. 2002. Giant beluga farm that could save the fish that lays the golden eggs. Guardian.

Buijse, A.D., Coops, H., et al. 2002. Restoration strategies for river floodplains along large lowland rivers in Europe. Freshwater Biology. 47. (4): 889–907.

Caviaremptor. 2004. Current Status of Beluga Sturgeon. Summary of Scientific Literature. [on-line] Caviaremptor. http://www.caviaremptor.org/current_status.html [cited April 2004].

CEP. 2002a. Transboundary Diagnostic Analysis for The Caspian Sea. Caspian Environment Programme. Baku, Azerbaijan.

CEP. 2002b. Caspian Biodiversity Database. Caspian Environment Programme. Baku, Azerbaijan.

CEP. 2006. Social Economic Review of Atyrau Oblast. Astana, Ministry of Environmental Protection of Republic of Kazakhstan.

Chebanov, M. and Galich, E. 2002. Conservation of biodiversity of sturgeon under conditions of the flow regulation of rivers: stock enhancement and living gene bank. Symposium on Inland Fisheries Management and the Aquatic Environment - The Effects of Fisheries Management on Freshwater Ecosystems, Windermere, UK, Food and Agriculture Organization of the United Nations.

Chivers, C.J. 2006. Caviar Exports Canceled in Move to Stop Sturgeon Extinction. New York Times.

Chugunov, N.L. 1968. Biologija molodi promislovih riub Volgo-Kaspiyskogo raiona Triudy Astrakhanskoj Ihtiologicheskoj Laboratorii T. 6. (Вып. 4):

CITES. 2001. Acipenser gueldenstaedtii. Species description.

CITES. 2004a. A brief history of sturgeons and CITES. [on-line] Steering Committee. Convention on International Trade in Endangered Species of Wild Fauna and Flora. http://www.cites.org/eng/prog/Sturgeon/history.shtml [cited April 2004].

CITES. 2004b. Total Allowable Catch (Tac) Estimation for Sturgeon Species in the Caspian Sea. [on-line] CITES UNEP. www.cites.org/eng/prog/Sturgeon/catch.pdf [cited April 2004].

CITES. 2007. Sturgeon Meat and Caviar Quotas. CITES.

Cohen, A. 1997. Sturgeon poaching and black market caviar: a case study. Environmental Biology of Fishes. 48. (1–4): 423–426.

Craig, J. 2000. Large Dams and Freshwater Fish Biodiversity. [on-line] Dams, ecosystem functions and environmental restoration. World Dam Commission. http://www.dams.org/docs/kbase/contrib/env246.pdf [cited 17 April 2004].

Crownover, S. 2004a. U.S. Government Acknowledges Beluga Sturgeon "Threatened with Extinction" But Takes No Action to Protect The Fish. [on-line] http://www.caviaremptor.org/alert.042104.html [cited April 20 2004].

Crownover, S. 2004b. Cites Fails to Suspend Caviar Trade Despite Caspian States' Lack of Progress on Sturgeon Conservation. [on-line] http://www.caviaremptor.org/alert.031804.html [cited March 18 2004].

de Groot, S.J. 2002. A review of the past and present status of anadromous fish species in the Netherlands: is restocking the Rhine feasible? Hydrobiologia. 478. (1–3): 205–218.

DeMeulenaer, T. and Raymakers, C. 1996. Sturgeons of the Caspian Sea and the international trade in caviar. Cambridge: TRAFFIC International.

Detlaf, T.A., Ginsburg, A.S., et al. 1981. Development of sturgeon fishes. Moscow: Nauka.

Dmitriev, A.N. and Vasilenko, A.V. 2007. The influence of the hydrological regime on sturgeon restoration in the Ural river. Ural Basin Workshop, Orenburg.

Dubinina, V.G. and Kozlitina, S.V. 2000. Water resources management of the southern rivers of Russia with reference to fisheries requirements. Fisheries Management. 7. (1): 157–165.

Dulvy, N.K., Sadovy, Y., et al. 2003. Extinction vulnerability in marine populations. Fish Fisheries. 4. (1): 25–64.

ENS. 2007. Caspian States to Revise Sturgeon Quotas. [on-line] Environment News Service http://www.traffic.org/seizures/2007/8/20/new-quota-system-for-sturgeon-fisheries.html [cited 21 September 2007].

EPA. 2004. Endangered and Threatened Wildlife and Plants; Listing the Beluga Sturgeon (Huso huso) as Endangered. [on-line] US Environmental Protection Agency. http://www.epa.gov/fedrgstr/EPA-SPECIES/2002/July/Day-31/e19250.htm [cited October 2007].

FAO. 2007a. The State of World Fisheries and Aquaculture 2006. Rome: FAO Fisheries Department.

FAO. 2007b. FishSTAT Plus. Capture Production. UN, FAO.

Fashchevsky, B. 2003. Human Impact on Rivers and Fish in the Ponto-Caspian Basin. Proceedings of the Second International Symposium on the Management of Large Rivers for Fisheries, Phnom Penh, Kingdom of Cambodia, FAO.

Firehammer, J. and Scarnecchia, D. 2007. The influence of discharge on duration, ascent distance, and fidelity of the spawning migration for paddlefish of the yellowstone-Sakakawea stock, Montana and North Dakota, USA. Environmental Biology of Fishes. 78. (1): 23.

Gessner, J., Wirth, M., et al. 2002. Caviar composition in wild and cultured sturgeons - impact of food sources on fatty acid composition and contaminant load. Journal of Applied Ichthyology. 18. (4–6): 665–672.

Granado-Lorencio, C. 1991. The effect of man on the fish fauna of the River Guadalquivir, Spain. Fisheries Research. 12. (2): 91–100.

Grunwald, C., Maceda, L., et al. 2008. Conservation of Atlantic sturgeon Acipenser oxyrinchus oxyrinchus: delineation of stock structure and distinct population segments. Conservation Genetics. 9. (5): 1111–1124.

Hensel, K. and Holchik, J. 1997. Past and current status of sturgeons in the upper and middle Danube River. Environmental Biology of Fishes. 48. (1–4): 184–200.

Itoh, Y., Tainaka, K., et al. 2004. Spatial enhancement of population uncertainty near the extinction threshold. Ecological Modelling. 174.

IUCN. 2007. The IUCN Red List of Threatened Species. International Union for Conservation of Nature and Natural Resources.

Ivanov, V.P. 2000. Biologicheskie resursy Kaspiiskogo Morya Astrakhan: Kasp NIRKH.

Ivanov, V.P., Nikonorov, S.I., et al. 2005. May caviar of Siberian sturgeon (Acipenser baerii) be present in sturgeon caviar supplied from the Volga River? Caspian Fisheries Research Institute, IWMC - World Conservation Trust.

Jonsson, B., Jonsson, N., et al. 1991. Differences in life history and migratory behaviour between wild and hatchery-reared Atlantic salmon in nature. Aquaculture. 98. (1–3): 69–78.

Jonsson, B., Waples, R.S., et al. 1999. Extinction considerations for diadromous fishes. ICES Journal of Marine Science. 56. (4): 405–409.

Kajiwara, N., Ueno, D., et al. 2003. Contamination by organochlorine compounds in sturgeons from Caspian Sea during 2001 and 2002. Marine Pollution Bulletin. 46. (6): 741–747.

KamUralRybVod. 2007. Vozdeistvie godovogo stoka na vosstanovlenie osetrovih v reke Ural. Orenburg, Orenburg Department of Federal State Agency "KamUralRybVod".

KaspNIRH. 1999. Osetrovie-sostoianie bioresursov, tendencii, prichiny istoschenija, predlojenija po vosstanovleniu. Astrakhan: Caspian Fisheries Research Institute.

Keysler, J.G. 1762. Travels Through Germany Bohemia Hungary Switzerland Italy and Lorrain.

Khodorevskaya, R.P., Dovgopol, G.F., et al. 1997. Present status of commercial stocks of sturgeons in the Caspian Sea basin. Environmental Biology of Fishes. 48. (1): 209.

Khodorevskaya, R.P. et al. 2000. Formation of the stock of the Caspian Acipenserids under present-day conditions. Journal of Ichthyology. 40. (8): 602–609.

Kirby, A. 2002. Farmed sturgeon 'only hope for caviar'. BBC News Online 2. 12. 2002.

Kliot, N., Shmueli, D., et al. 2001. Institutions for management of transboundary water resources: their nature, characteristics and shortcomings. Water Policy. 3. (3): 229–255.

Korina, L. 2006. CITES: Brakon'erstvo - eto bioterrorizm Kazakshtanskaja Pravda, Uralsk.

Lagutov, V. 1995. Mekhanizm unichtozheniya riubniuh zapasov Yuga Rossii i puti ih spaseniya [The mechanism of extermination of fish stocks in South Russia and possibilities of their restoration]. Novocherkassk: Donskaia Rech.

Lagutov, V. 1996. Imitacionnoe modelirovanie zhiznennogo tsikla osetroviuh v basseine reki Don. [Simulation of Sturgeon Life Cycle in the Don River Basin]. MSc thesis, Department of Applied Mathematics, South Russia State Technical University, Novocherkassk.

Lagutov, V. 1997. Simulation Model of Sturgeon Life Cycle as a Tool in Sustainable Water Management. MSc Thesis, Department of Environmental Sciences, Central European University, Budapest.

Lagutov, V. 1999. New Environmental Policy. Sustainable development: from myth to reality. 13th Global Biodiversity Forum, San Jose, Costa Rica.

Ludwig, A. 2006. A sturgeon view on conservation genetics. European Journal of Wildlife Research. 52. (1): 3.

Luk'yanenko, V.I., Vasil'ev, A.S., et al. 1999. On the increasing threat of extermination of the unique Caspian sturgeon populations and the urgent measures required to save them. Journal of Applied Ichthyology. 15. (4–5): 99–102.

Malecha, N.I. 2002. Slovar govorov uralskih (jaickih) kazakov. Orenburg: Orenburg Publishing.

Marmulla, G., Larinier, M., et al. 2001. Dams, fish and fisheries: opportunities, challenges and conflict resolution. FAO Fisheries Technical Paper 419, Fishery Resources Division. FAO Fisheries Department.

Marti, Y.Y. 1972. Voprosy razvitija osetrovogo hoziaystva v Kaspiyskom more In Osetrovie i problemy osetrovogo hoziaystva ed., Moscow: Pischepromizdat.

McAllister, D., Craig, J., et al. 2000. Biodiversity Impacts of Large Dams. [on-line] World Dam Commission. http://www.dams.org/docs/kbase/contrib/env244.pdf [cited 17 April 2004].

Morgan, D. 2007. Caspian: UN Lifts Embargo on Most Caviar Exports. [on-line] Radio Free Europe. http://www.rferl.org/featuresarticle/2007/01/97c22756-4ca0-45d6-b737-02fc5d57bb01.html [cited 2 November 2007].

ORB. 1998. Red Book of Orenburg Oblast. Orenburg. Orenburg Publishing.

Pala, C. 2004a. U.N. to Allow Caviar Exports, but Quota Is Cut. The New York Times, New York.

Pala, C. 2004b. Multiplication Problem Threatens Stock of Sturgeon. New York Times, New York.

Peseridi, N.E. 1971. Sezonnaja Dinamika khoda osetrovih v nizoviah reki Ural Osetrovie SSSR i ih vosproizvodstvo. 3. 355–358.

Peseridi, N.E. 1986. Fishes of Kazakhstan. Alma-Ata, Kazakhstan SFR: Nauka.

Peseridi, N.E. and Chertikhina, T.S. 1967. Some aspects of the effect of some factors on the migration, reproduction and yield of sturgeons in Ural River. Tr. tsentr. nauch-issled. Inst. osetro. khoz. 1.

Peseridi, N.E., Drido, E.A., et al. 1979. O vozmojnom neblagopriatnom vlianii povishennoj solenosti vod priustevogo vzmorija v Ural na vospoizvodstvo osetrovih. In Osetrovoe hoziajstvo vnutrennih vodoemov SSSR. ed., Astrakhan.

Pikitch, E.K. and Lauck, E. 2002. A Scientific Response to the CITES Justification for Setting the 2002 Total Allowable Catch of Beluga Sturgeon (Huso huso) in the Caspian Sea. Wildlife Conservation Society.

Pitikch, E.K., Doukakis, P., et al. 2005. Status, trends and management of sturgeon and paddlefish fisheries. Fish and Fishereis. 6. 233–265.

Pourkazemi, M. 2007. Caspian Sturgeon Extinction by 2021 - Official. BAZTAB News.

Qiwei, W., Fuen, K., et al. 1997. Biology, fisheries, and conservation of sturgeons and paddlefish in China. Environmental Biology of Fishes. 48. (1–4): 241–255.

RAMSAR. 2002. Sustainable Management of Water Resources: The Need for a Holistic Ecosystem Approach.

Running out of Freshwater or Maintaining Freshwater Through a Ecosystem Based Approach - An Easy Choice. 8th Meeting of the Conference of the Contracting Parties to the Convention on Wetlands (Ramsar, Iran, 1971), Valencia, Spain, the Ramsar Convention on Wetlands, 2005.

RAMSAR. 2006. River Basin Management. Ramsar.

Raymakers, C. and Hoover, C. 2002. Acipenseriformes: CITES implementation from Range States to consumer countries. Journal of Applied Ichthyology. 18. (4–6): 629–638.

Reid, W.V. and Miller, K.R. 1989. Keeping Options Alive: The Scientific Basis for Conserving Biodiversity. Washington, DC, World Resource Institute.

RF. 2000a. Red Book of Russian Federation. Moscow, Russia.

RF. 2000b. Russian Federation Committee of Fisheries. (In Russian).

RF. 2002. National Caspian Action Plan (NCAP) of the Russian Federation. Caspian Environment Program. Baku, Azerbaijan.

RK. 1996. Red Book of Republic of Kazakhstan. Institute for Zoology and Gene Pool of Animals of National Academy of Sciences of the Republic of Kazakhstan.

RK. 2002. National Action Plan for rehabilitation of the Caspian Sea environment for 2003-2012. Ministry of Environmental Protection, Republic of Kazakhstan.

RK. 2003. National Report on the State of the Environment in Kazakhstan 2003 Ministry of Environmental Protection of the Republic of Kazakhstan.

Russian State Duma, 1995. Committee on Ecology, Russian State Duma. Hearing on "Improvement of fishery policies in the Azov Sea Basin". Report by Dr. V. Lagutov.

Schreck, C.B., Contreras-Sanchez, W., et al. 2001. Effects of stress on fish reproduction, gamete quality, and progeny. Aquaculture. 197. (1–4): 3–24.

Secor, D.H., Arefjev, V., et al. 2000. Restoration of sturgeons: lessons from the Caspian Sea Sturgeon Ranching Programme. Fish Fisheries. 1. (3): 215–230.

Seijo, J.C., Defeo, O., et al. 1998. Fisheries bioeconomics - theory, modelling and management. FAO Fisheries Technical Paper 368, Rome, FAO. 108p.

Semple, E.C. 1907. Geographical boundaries. II. Bulletin of the American Geographical Society. 39. (8): 449–463.

Smith, F.D.M., May, R.M., et al. 1993. How much do we know about the current extinction rate? Trends in Ecology & Evolution. 8. (10): 375–378.

Speer, L., Lauck, L., et al. 2000. Roe to Ruin: The Decline of Sturgeon in the Caspian Sea and the Road to Recovery. Retrieved August, 2007, from http://www.caviaremptor.org/

Stephan, T. and Wissel, C. 1999. The extinction risk of a population exploiting a resource. Ecological Modelling. 115. (2–3): 217–225.

TRAFFIC. 2003. Bleak Outlook for World Caviar Trade: North America Sturgeon, Paddlefish Stocks at Risk. [on-line] 22 May 2003, http://www.traffic.org/news/press-releases/bleakoutlook.html [cited 20 April 2004].

TRAFFIC. 2007a. Caviar Figures Smell Fishy. [on-line] TRAFFIC. http://www.traffic.org/home/2007/2/8/caviar-figures-smell-fishy.html [cited June 2007].

TRAFFIC. 2007b. Stamp of Approval for Caviar. [on-line] TRAFFIC. http://www.traffic.org/home/2007/6/29/stamp-of-approval-for-caviar.html [cited 2007].

UNEP. 2004. Learning from Baia Mare. [on-line] The Environment Times. http://www.environmenttimes.net/article.cfm?pageID=112 [cited 10 December 2007].

Uralbas. 2007a. Workshop Results. First International Ural Basin Workshop "Rescue of Sturgeon Species by means of Transboundary Integrated Water Management in the Ural River Basin", Orenburg, Russia.

Uralbas. 2007b. Proceedings. First Ural River International Workshop Orenburg, Russia.

Von Harthausen, A. 1972. Studies on the Interior of Russia. Chicago/London: University of Chicago Press.

Waldman, J.R. and Wirgin, I.I. 1998. Status and restoration options for Atlantic sturgeon in North America. Conservation Biology. 12. (3): 631–638.

Waldman, J.R., Grunwald, C., et al. 2002. Impacts of life history and biogeography on the genetic stock structure of Atlantic sturgeon Acipenser oxyrinchus oxyrinchus, Gulf sturgeon A. oxyrinchus desotoi, and shortnose sturgeon A. brevirostrum. Journal of Applied Ichthyology. 18. (4–6): 509–518.

WFD, E. 2000. European Union Water Framework Directive.

Williamson, D.F. 2003. Caviar and Conservation: Status, Management and Trade of North American Sturgeon and Paddlefish. Washington, DC: TRAFFIC North America. World Wildlife Fund.

Williot, P., Brun, R., et al. 2000. Induced maturation and spawning in an incidentally caught adult pair of critically endangered European sturgeon, Acipenser sturio L. Journal of Applied Ichthyology. 16. (6): 279–281.

Williot, P., Sabeau, L., et al. 2001. Sturgeon farming in Western Europe: recent developments and perspectives. Aquatic Living Resources. 14. (6): 367–374.

Williot, P., Rochard, E., et al. 2002a. Biological characteristics of European Atlantic sturgeon, Acipenser sturio, as the basis for a restoration program in France. Environmental Biology of Fishes. 48: 359–370.

Williot, P., Arlati, G., et al. 2002b. Conservation and broodstock management status and management of Eurasian sturgeon: an overview. International Review of Hydrobiology. 87. (5–6): 483–506.

World Bank. 2004a. Biodiversity Preservation, Sturgeon Recovery Through Rehabilitation of Hatcheries, Reproduction and Protection of Sturgeon Stock. Project description. [on-line] http://lnweb18.worldbank.org/ECA/ceiforum.nsf/docbyid/837AE11C351640B485256EDE0069C487?Opendocument [cited July 2007].

World Bank. 2004b. Innovative Approaches in Fisheries Management in Kazakhstan. World-Bank, Kazakshtan Government.

WWF. 2002a. WWF Factsheet: Sturgeon. The materials of 12th Meeting of the Conference of the Parties to CITES, Santiago, 3–15 November 2002,

WWF. 2002b. Sturgeon 2020. Initiative for Providing Habitats for Sturgeons. Program despcription and schedule, Budapest, WWF.

WWF. 2004. Ten Things Not to Buy for Christmas. [on-line] WWF International. http://www.panda.org/about_wwf/what_we_do/marine/news/index.cfm?uNewsID=17351 [cited May 2007].

ZapKaspRybVod. 2008. Po osetram plan sorvan. fishnews.ru, Moscow.

Zhuang, P., Kynard, B., et al. 2002. Overview of biology and aquaculture of Amur sturgeon (Acipenser schrenckii) in China. Journal of Applied Ichthyology. 18. (4–6): 659–664.

ZIN. 2006. Report on Caspian Sea Biodiversity Project. St. Petersburg, Zoological Institute of the Russian Academy of Sciences.

FIELD AND GENETIC APPROACHES TO ENHANCE KNOWLEDGE OF URAL RIVER STURGEON BIOLOGY

PHAEDRA DOUKAKIS*

Pew Institute for Ocean Science at the University of Miami, 126 East 56th Street, Mezzanine, New York, New York 10022 USA

DANIEL ERICKSON

Pew Institute for Ocean Science at the University of Miami, 541 Willamette St., Suite 207A, Eugene, OR 97401 USA

MIRGALY BAIMUKHANOV

Hydrobiology and Ecology Institute Kirova 25 v, Irgeli village, Karassaisky rayon, Almaty, Republic of Kazakhstan

YELENA BOKOVA

Research and Production Center of Fish Industry, Ministry of Agriculture, Republic of Kazakhstan, Atyrau, Republic of Kazakhstan

SAGIDEN ERBULEKOV

Agip KCO, Atyaru, Republic of Kazakhstan

AKHAT NIMATOV

Atyrau Sturgeon Hatchery, Atyrau, Republic of Kazakhstan

ELLEN K. PIKITCH

Pew Institute for Ocean Science at the University of Miami, 126 East 56th Street, Mezzanine, New York, New York 10022 USA

Abstract Combined approaches utilizing tagging and genetic analysis can provide powerful insight into the biology and management of endangered sturgeons as described in a literature review herein. Since 2003, our team of USA and Republic of Kazakhstan researchers has attempted to use such techniques to study Ural River sturgeons. High-tech (satellite, acoustic) tagging methods were to be applied to examine movement and behavior of Ural River sturgeons within the River and the Caspian Sea and test hypotheses about sturgeon biology. In 2006, we successfully completed a satellite tagging project in which four adult sturgeons (three beluga, *Huso huso*, and one ship sturgeon *Acipenser nudiventris*) captured in the Ural River were equipped with Pop-up Archival Transmitting (PAT) satellite tags and released into the Caspian Sea. In 2007, efforts were made to initiate an acoustic tagging and tracking program within the Ural River to study sturgeon migratory behavior and locate

* To whom correspondence should be addressed: Phaedra Doukakis, Pew Institute for Ocean Science at the University of Miami, 126 East 56th Street, Mezzanine, New York, New York 10022 USA; pdoukakis@miami.edu

spawning grounds. Tagging has also been used to measure the contribution of hatchery-produced sturgeons to that of the wild population. These studies were to be combined with genetic approaches to study the uniqueness and diversity of Ural River sturgeons. Overall, our program produced mixed results: some projects could not be undertaken, others produced only preliminary results, and some initiated long-term monitoring. Our work has implications for the conservation of Ural River sturgeons.

Keywords: Sturgeon, beluga, Ural River, satellite tag, acoustic tag, genetics, mitochondrial DNA, caviar

Introduction

Valued for their eggs, commercially sold as black caviar, sturgeons (Order Acipenseriformes) have been the focus of fisheries for over a century (Pikitch et al. 2005). The endangered status of sturgeons is now well recognized, with these fishes protected by numerous international and domestic measures and international trade in sturgeon commodities regulated by the Convention on International Trade in Endangered Species (CITES) (Birstein 1993; Pikitch et al. 2005; Pourkazemi 2006). For some species, commercial exploitation has ceased after population declines prompted fisheries closure. Long lifespan, slow growth, infrequent reproduction, and late maturity make for lengthy rebound times after intensive harvest and population bottlenecks (Birstein 1993; Billard and Lecointre 2001). Habitat threats further affect recovery and population health (Birstein 1993; Billard and Lecointre 2001; Pikitch et al. 2005; Pourkazemi 2006). As all 25 species of true sturgeons depend upon freshwater for reproduction and survival, with about 15 species exhibiting an anadromous lifestyle, many species are prone to threats in freshwater, coastal, and offshore environments (Bemis and Kynard 1997).

In recent decades, the largest and most productive sturgeon fisheries have been in the Caspian Sea region, home to five anadromous (beluga (*Huso huso*), Persian (*Acipenser persicus*), Russian (*A. gueldenstaedtii*), ship (*A. nudiventris*), stellate (*A. stellatus*), and one strictly freshwater (sterlet (*A. ruthenus*)) species. Dam construction and water diversion has impacted the health of these sturgeons as has unrelenting fishing pressure (Birstein 1993; Pikitch et al. 2005; Pourkazemi 2006). Population declines prompted development of over ten hatcheries around the Caspian Sea that have accomplished seemingly little in changing downward population trends (Secor et al. 2000; Chebanov et al. 2002). Increasing concern over the future of sturgeons and their fisheries has resulted. The Ural River is the last river off the Caspian Sea used by sturgeons that does not have a dam impeding migration and thus may be home to the healthiest populations and greatest hope for the future.

Despite their endangerment and commercial, biological and cultural value, Caspian Sea sturgeons have been the focus of little contemporary life history research. Our work has focused on employing state-of-the-art techniques to characterize Ural River sturgeon biology. While we have had limited success, several important results have been obtained. As the beluga is the most endangered Caspian Sea sturgeon still subject to a lucrative fishery, our work has focused on this species. However, the lessons learned and studies undertaken could easily be expanded to the other species.

Tagging and genetics

Marking or "tagging" individual fish has been used in fisheries science since early in the discipline. Studies employing "simple" external and internal tags (e.g. spaghetti tags, Coded Wire Tags (CWT's)) in sturgeons have been used to examine population abundance and devise rebuilding strategies following intensive harvest (Whitlock and McAllister 2005), understand recruitment (Peterson et al. 2000) and study survival and spatial movement of hatchery-reared sturgeons (Smith et al. 2002; Yang et al. 2005). While simple tags have been used in the past to study adult migration and population abundance in the Caspian Sea, there has been no consistent application. As discussed later, the Caspian Sea nations have started a CWT program for hatchery reared and released individuals, a much needed program given that the impact of hatcheries is unknown.

In recent years, more sophisticated electronic tags have been applied, revealing new and important aspects of sturgeon biology. These tags are advantageous relative to "simple" tags because many more data points can be obtained in a short period of time (e.g., hundreds to thousands within a few years rather than less than 100 over decades) and results are fishing-independent (i.e. fish do not have to be recovered to collect information). Acoustic and radio tags are typically used to track migrations in and among rivers and near shore environments (Edwards et al. 2003; Benson et al. 2007; Erickson and Webb 2007; Lindley et al. 2008), identify concentration sites (Caron et al. 2002; Erickson et al. 2002; Hightower et al. 2002; Edwards et al. 2003; Heise et al. 2005; Benson et al. 2007), measure spawning periodicity (Erickson and Webb 2007), determine preferred habitats (e.g. depth and substrates; Erickson et al. 2002; Edwards et al. 2003; Hatin et al. 2007), identify vertical-migratory behavior (Paragamian and Duehr 2005), and understand timing of migrations to foraging and spawning grounds (Benson et al. 2007; Erickson and Webb 2007; Lindley et al. 2008). In some systems, like the Danube River, the application of telemetry revealed more about fishing pressure than life history, as many of the tagged fish were removed from the system (Kynard et al. 2002). Acoustic telemetry has also been used to study the behavior of captive-reared animals released into the wild (Bronzi et al. 2006) and of wild sturgeon transplanted above impassable barriers (Finney et al. 2006).

The newest technological application for tracking and identifying important habitats is the pop-up archival transmitting (PAT) satellite tags. These tags have been used to recover daily position estimates for pelagic fishes in clear water (Wilson et al. 2005). To date, these tags have not produced daily positional information for sturgeons because the species studied inhabited relatively deep and murky water. Daily depth and water temperature preference for up to eight months and precise, fishery-independent locations at the time of PAT detachment were, however, obtained (see Edwards et al. 2007, Erickson and Hightower 2007). Recent improvements in tags and analytical methods are permitting daily position estimates for sturgeons, as evidenced by an in-progress study of Atlantic sturgeon in the open ocean (Erickson et al. 2008).

Satellite tags are advantageous because fish do not have to be recaptured and position estimates are independent of receivers. Hence, satellite tags permit studies of migrations on a large, continuous oceanic scale. Sonic transmitters can be problematic in monitoring long-range coastal movements because the individual fish must be located (see Edwards et al. 2007) or fish positions are dependent on the deployment location of data-logging receivers (Lindley et al. 2008). Studies that employ a combination of sonic and satellite tagging may provide the most useful and complete dataset possible.

While tagging can aid understanding of movement on ecological time scales, genetic applications can reveal patterns on evolutionary time scales. Population structure, which can be difficult to understand in anadromous sturgeons with uncertain degrees of homing fidelity, can be more fully characterized through genetic study. Maternally inherited mitochondrial (mt) DNA markers are most often applied because of the polyploid nature of sturgeons and difficulty of using nuclear DNA (in Birstein et al. 1997). Advances are being made, however, in using nuclear markers (Rodzen and May 2002; Welsh and May 2006).

Most of the completed genetics population structure studies focus upon North American species. For the Atlantic sturgeon *Acipenser o. oxyrinchus*, an anadromous species spawning in over 20 different river systems, considerable structure is apparent on regional and river specific levels (Waldman et al. 1996; Wirgin et al. 2000, 2002; King et al. 2001; Grunwald et al.

2007). In the closely related Gulf sturgeon (*A. o. desotoi*), higher stock structure by natal river is evident due to rare movement of this species into marine waters (Waldman et al. 2002). In the shortnose sturgeon (*A. brevirostrum*), a species that rarely enters freshwater and occurs in about 20 river systems, fine-scale structure by river system is observed (Wirgin et al. 2005). For green sturgeon, genetic division supports northern (Rogue, Klamath, Eel, Umpqua Rivers) and southern (Sacramento River) population segments with some river-specific genetic structure (Israel et al. 2004; Adams et al. 2007).

Fewer genetic studies have examined species outside of North America (e.g. Zhang et al. 2003; Zhu et al. 2006). Research to differentiate Azov, Black and Caspian Sea populations of commercially important sturgeons failed to find strong genetic differences within mtDNA (Doukakis et al. 1998, 2005). Higher levels of genetic diversity exist in Russian sturgeon as compared to other species (*H. huso, A. stellatus*) suggesting that distinct populations of Russian sturgeon exist. Other Caspian Sea studies have focused on only small geographic areas within a basin (Pourkazemi et al. 1999; Rezvani Gilkolaei and Skibinski 1999; Rezvani Gilkolaei 2000) or small sample sizes (Kornienko et al. 2003). An effort is needed to identify genetically distinct units in Caspian Sea species so that management can be conducted on the level of the stock. The impact of dams and of hatchery restocking on population structure and genetic diversity is also an area needing investigation, possibly through studies examining museum specimens collected before dams and hatchery practices were instituted.

Genetic study can also aid fine-scale market traceability of commercial products. Trade in sturgeon products is regulated through CITES, with verification of CITES permits and species sometimes performed using genetic testing. These mtDNA–based tests have also been used in market surveys investigating labeling fraud, studies which further revealed cryptic species and questioned species status (e.g. for *A. persicus*) (DeSalle and Birstein 1996; Birstein et al. 1998, 2005). Traceability to the stock is not yet possible because studies of population structure are incomplete. In the future, genetic methods might also become useful in differentiating aquaculture from wild origin products and ensuring that aquaculture does not become an outlet for laundering illegally obtained wild-origin product.

Combined, tagging and genetic studies can yield results that challenge long-held assumptions about sturgeon biology and reveal new management needs. For example, green sturgeon (*A. medirostris*) was traditionally thought to be primarily marine, moving into freshwater for short periods to spawn. Telemetry work revealed that the species can spend up to six consecutive months in fresh water (Erickson et al. 2002). Green sturgeon also undertake long distance oceanic migrations moving from US rivers to

Canadian holding grounds. Transnational (i.e. US and Canada) conservation is therefore necessary as is management recognizing that animals from distinct US population segments intermingle in the ocean in areas where fishing occurs (Lindley et al. 2008). Similarly, Atlantic sturgeon from genetically distinct river systems aggregate as mixed stocks within oceanic environments and may be subject to fisheries as by-catch (Laney et al. 2007). Current satellite tagging studies are revealing the oceanic migratory pathways taken by Atlantic sturgeon (e.g. depth contour preferences) and providing the basis for developing management plans limiting fisheries interactions (Erickson et al. 2008).

Ural river sturgeon biology

There is considerable literature about the biology of Caspian Sea and Ural River sturgeons that provides hypotheses to be tested through tagging and genetics. As with many Caspian sturgeons, beluga are thought to harbor two spawning races in the Ural River, spring and winter. Levin (1997) suggested the winter run migrates into the river between August and October, overwintering and spawning the following spring, and the larger spring run enters between December and May, with a peak between February and March. Berg (1948) earlier suggested that the winter run enters at the end of spring and beginning of summer, forming winter aggregations in depressions in the river and migrating up the Ural the following spring. Peak spawning time is in May. Spawning occurs at water temperatures of 9–11°C and adult sturgeons are presumed to leave the river soon after spawning (Levin 1997). Beluga sturgeon undertake the longest in-river migration of any of the anadromous Caspian Sea sturgeon, spawning up to 800 km upstream from the Ural River delta. Males are thought to enter the river first, followed by females. In the spring fishery, males can be observed first entering the river in mid April.

After hatching and maturing, beluga fingerlings migrate downstream from June to September. Juveniles are known to congregate in the river delta and near-shore (Khodorevskaya and Krasikov 1999). The diet of beluga gradually shifts from invertebrates to fish as the fish becomes predatory (Khodorevskaya and Krasikov 1999). Beluga remain in the Caspian Sea for 11–20 years to mature, with males reaching maturity at 12–15 years and females at 12–13 years (Berg 1948; Mitrofanov et al. 1986; Mamina 1995). Spawning periodicity is thought to be three to four years in males and four to six years in females (Mitrofanov et al. 1986; Mamina 1995).

Much about the distribution and movements of beluga sturgeon within the Caspian Sea is known from trawl surveys. In general, beluga congregate in the northern Sea in the warmer spring and summer months and move to the central, southern and western parts of the Sea in winter (Khodorevskaya and Krasikov 1999). This migration pattern is thought to be the same for adult and juvenile beluga (Levin 1997). Juveniles move once water temperatures drop to 22–23°C to central and southern areas (Levin 1997). Trawl surveys suggest depth preference of beluga in different life stages and seasons but samples sizes are small (Khodorevskaya and Krasikov 1999). Mature beluga apparently tend to occur at depths of 10–50 m in the southern and middle Sea and 3–11 m in the north (Khodorevskaya et al. 2002). Statistical correlations between beluga occurrence and water temperature were found for mature beluga in spring (preference of 2–16°C) and immature fish in autumn (8–14°C) (Khodorevskaya et al. 2002). Animals appear to become more cold tolerant as they mature (Khodorevskaya et al. 2002). Beluga are thought not to congregate in groups within the Sea as other sturgeons do, remaining widely dispersed and occupying middle depths (Levin 1997). Once warmer temperatures return in spring, mature animals enter river systems to spawn (Levin 1997).

Ural river sturgeon research program

Through the projects detailed below, our multi-year program sought to better understand Ural River beluga sturgeon biology and test some of the hypotheses presented above.

Simple tagging: In 2005, we undertook a preliminary study using Passive Integrated Transponder (PIT) tags to tag juvenile sturgeons. Our goal was to lay the groundwork for an eventual large scale fingerling tagging program and examine the appropriateness of PIT tags. A total of 238 fingerlings (235 hatchery beluga, one wild beluga, one wild Russian sturgeon, one sterlet) were tagged. Tagged individuals were 5.5–55 g, with most between 8 and 13 g, and a mode of 9 g, and 10.6–27 cm in total length, with most between 13 and 16 cm, and a mode of 14 cm. One mortality occurred during tagging. Three fingerlings exhibited weak swimming behavior after tagging and an additional three individuals were bleeding following tagging, but all of these individuals survived. PIT tags were injected between the fourth and fifth dorsal scute. With the exception of 23 fingerlings retained (see below), tagged fingerlings were transported from the hatchery and released into the Sea (25 03 718 E 121 38 413). PIT tagging effect and loss was examined by keeping 23 hatchery beluga fingerlings at the hatchery for six days. Tag loss was confirmed in two individuals at the end of the six day period. Other tagged fingerlings appeared to be in good health. The fingerlings were released into the river in a standard release location. Subsequent recapture studies were not undertaken and would not have likely produced concrete results given the small number of sturgeons tagged.

Future studies using PIT tags would need to include larger sample sizes and additional work on tag loss, given that 10% loss was observed in our study in a short time period. Studies would also need to focus on tagging wild-produced juveniles to effectively understand hatchery vs. wild contribution. PIT tagging may not be the best method given the expense of PIT tags ($2 US/tag) and scale of an effective study.

In Kazakhstan and throughout that Caspian basin, current efforts are focusing on using a less expensive method (CWT). Iran has been the leader of the Caspian Sea nations in experimenting with CWT's, marking and releasing Persian and stellate sturgeons, performing field recapture surveys to examine tag retention, and undertaking tag retention studies at hatcheries (Fadaee et al. 2006). Although sample sizes were small, this study provides a good basis for future efforts and raises excellent points on study design and scale. Since 2006, Kazakhstan has also used CWT's for tagging hatchery reared juvenile sturgeon, with the number of individuals tagged increasing annually. This approach will likely slowly be adopted by all Caspian range States with sturgeon hatcheries. In the long term, once hatchery fish reach maturity, this program will yield information about the contribution of different hatcheries to adult populations.

We also attempted to initiate a tag (PIT and spaghetti tags) and release program for adult Ural River sturgeons, but efforts were hampered by the difficulty of establishing a catch and release program for such a valuable fish. Such a project could have investigated the number of spawners reaching the spawning ground (tagging at the river mouth and recapture studies upriver), outmigration timing (recapture study at presumed time of outmigration) and the occurrence of Ural River sturgeons in other river systems in the Caspian Sea (with establishment of a cooperative, Caspian Sea basin-wide monitoring program). Similar questions could be addressed using telemetry techniques.

Acoustic tagging: In the spring of 2007, we set out to complete an acoustic telemetry study to examine the migratory behavior of beluga in the Ural River. By tagging sturgeons with acoustic tags, installing stationary data-logging receivers at 50 km intervals from the river mouth to the presumed upriver migratory limit (800 km), and employing manual tracking using a boat and manual receiver, we hoped to accomplish similar results to those in published studies (see section 1.1). We sought to understand the true movement patterns of sturgeons in the Ural River and the timing of in and out migration, and to test the hypotheses about distinct seasonal races of beluga sturgeon. In the long term, information on spawning ground location and spawning periodicity could have been obtained.

Faced with the reality that beluga sturgeon are infrequently caught and are highly valued, we recognized that other sturgeon species might need to be included in our study. As we were relying on teams of fishermen at the last legal upriver fishing site, we knew that to catch and release a beluga, even for such valuable research, would be difficult and that catch and release of other species was a more reasonable approach. Our goal was thus to tag 27 sturgeons and track movements remotely and manually.

We could not undertake our planned research due to problems with importing equipment and an unsettled environment regarding fishing and scientific quotas. We did undertake an expedition to up the river to the presumed spawning grounds and observed a fairly pristine upriver system, free of industry and with a riparian corridor intact. Thus there is still likely habitat available for sturgeon spawning. Poaching was also apparent as evidenced by nets and boats. Should funding become available we may decide to undertake a telemetry project in the future.

Satellite tagging: The purpose of this project was to study the habitat use and movement of sturgeons after leaving the Ural River and entering the Caspian Sea and to use the data in managing fisheries interactions, protecting critical habitats and improving stock assessment trawl surveys. Wildlife Computers PAT tags were used in our study to record depth, temperature, light-level, date, and time data at programmed intervals. These tags detach on a programmed date and time, float to the surface, and transmit archived data to the research via the Argos satellite system. Our work marked the first time satellite-tagging technology has been used to study any Caspian Sea species.

Sturgeons were tagged with PAT tags at the Atyrau Sturgeon Hatchery in Kazakhstan. The animals used had been previously spawned in the hatchery. Table 1 describes the three beluga (*H. huso*) and one ship (*A. nudiventris*) used for this study. The ship sturgeon was used because sufficient numbers of healthy beluga were not available. Tags were attached to the dorsal fin of the study animals using procedures developed by Erickson and Hightower (2007). The tagged animals were transported to the Caspian Sea for release at a location several kilometers from the river delta (N46 47 067 E 051 27 892).

Table 1. Sturgeons used in Mk-10 PAT tag study in the Ural River, Kazakhstan, May 2006. Weight is post-hatchery use, TL is total length, release date is the day the tag is set to release from the sturgeon and is expressed as month/day/year

Species	Date tagged	Sex	Weight (kg)	Length (TL)	Tag #	Programmed-release date
Huso huso	5/19/06	Male	45	180	1	8/31/06
Huso huso	5/19/06	Female	60	192	2	11/1/06
Huso huso	5/19/06	Male	55	190	3	3/31/07
Acipenser nudiventris	5/19/06	Male	10	110	4	5/31/07

The satellite tagging project did not produce data on movement and habitat preference. Tag one (Table 1) released 26 days after the programmed release date, indicating possible tag malfunction. The tag then transmitted intermittently for 13 days ending October 31, 2006 but provided only four days of usable depth, temperature, and light data. The tag detached from the fish in the North Caspian close to the release area in approximately 8 m of water and the useable data illustrate that the study animal likely experienced mortality at some point before July 14, 2006. This combined with tag malfunction and possible satellite interference (see below) amounted to almost no usable information on movement and habitat use from this tag. Five geolocation estimates, depth, and temperature values were, however, recorded at depth. This is significant because positional data is usually inferred using light data. As such, if other problems associated with satellite tags in this region are overcome, the technology could prove useful in studying movements of sturgeons in the North Caspian.

The second tag released and transmitted on the programmed date but only transmitted unusable data. No location data were ever received and not enough data were transmitted to process. Tag three never transmitted. Tag four transmitted on the programmed date and provided good, clear transmissions for nine days. Unfortunately, the tag released or was removed from the fish on or before May 26, 2006, one week after the fish was tagged, and no information for this week was transmitted. It appears that neither the fish nor tag left the northeastern part of the Caspian Sea.

While we did not obtain information about migration patterns of Caspian Sea sturgeons, we did identify challenges to satellite tagging for sturgeons in the Caspian Sea as follows:

- Tags may not release in low salinity conditions. Satellite tags are designed for marine use, with the tag release mechanism optimized for saline conditions. The salinity of the North Caspian Sea is very low (0.1 ppt) and thus tag malfunctioned may have been due to incomplete release in low salinity. Tag release may function better in the middle and southern parts of the Sea where salinity can range from 10 to 13.5 ppt.

- The Caspian Sea region is affected by radio-frequency interference with the Argos system. This problem also occurs in the Mediterranean and Black Seas and is caused by an unknown source of interference that blocks tag to satellite transmission. As satellite tags transmit at relatively low power, the transmissions can be overwhelmed by other, higher-power transmissions. Increasing the transmission power requires sacrificing battery life and increasing tag size. At present under optimal conditions in the Caspian Sea, only 30% of the archived data transmitted will be received. The situation is worse in the west Caspian Sea. Remedying this could occur if the interference source is identified and eliminated.
- Sturgeons used in the hatchery procedure may be negatively affected by handling prior to release and more susceptible to capture or other sources of mortality. Some of our animals likely experienced mortality after tagging. Satellite tagging of other sturgeons has not directly resulted in mortality (e.g., Edwards et al. 2007; Erickson and Hightower 2007) and therefore, we do not believe this was the case here. Stress induced by holding prior to tagging, egg/sperm removal, and/or transport over land and boat to the release site could have made the tagged sturgeons more susceptible to fishing and disease. Our tagging project also coincided with a documented die-off of several hundred seals and several thousand sturgeons, with direct evidence of this apparent near the release site (i.e. dead sturgeons and birds floating on the Sea surface). Whatever caused this die-off could have affected the already stressed, tagged individuals.
- High rates of poaching occur in the Caspian Sea and thus our tagged sturgeons may have been captured. Sturgeon fishing is illegal in the Caspian Sea and is permitted only in rivers. Nonetheless, gill net density is high in the North Caspian and near the Ural River delta. Satellite tagging studies will thus require large sample sizes and might be more successful if animals are released in deeper regions of the Caspian where illegal nets are not found.

Although our sample size was small, there appears to be many problems with using satellite tags to study Caspian Sea sturgeon movement. Future studies should consider the results and recommendations presented.

Genetics: Our work has involved examining mtDNA of four sturgeon species from the Ural River (*H. huso*, *A. gueldenstaedtii*, *A. nudiventris* and *A. stellatus*) and comparing these profiles to an existing database of mtDNA profiles generated previously (Doukakis et al. 1999, 2005). Not enough results were available at the time if writing to present here. When this work is completed it will reveal whether Ural River sturgeons exhibit distinct genetic signatures, providing evidence of reproductive isolation and homing fidelity. This could then lead to improved management on the level of the genetically distinct stock as well as enhanced forensic identification of the stock origin of caviar and sturgeon products in trade.

Conclusions

Our study was designed with multiple, complementary components to better characterize Ural River beluga sturgeon biology and ultimately improve management and conservation. While efforts focused beluga, we hoped to eventually institute a multi-species program. Studying less commercially valuable species may prove easier. With the exception of the satellite tagging project those projects that did not move forward likely could have with sound political backing. Fully supported projects include hatchery tagging and genetics. With respect to hatchery work, the approaches taken should be expanded to devise good practices for maintaining genetic diversity of Ural River sturgeons. In Kazakhstan, hatcheries often must work with limited broodstock and the ramifications of releasing large numbers of hatchery-produced fingerlings from only a few females needs study. Cryo-banking of sturgeon reproductive products would also be a valuable avenue of research.

Although the work described herein is costly and must occur over a long time-scale, the benefits would be significant. This is well illustrated by the results of published studies in other species. For Ural River sturgeons, protected areas could be created around spawning grounds, better time and area closures could be designed to maintain the health of spawning runs and the effectiveness of hatcheries as a management tool could be elucidated. The significant controversy surrounding the design of stock assessment trawl surveys in the Caspian Sea could also be addressed, as survey design could be improved using the results of a satellite tagging study (i.e., characterization of habitat preference and migration in the Sea). Satellite tag technology would need improvement before such work could be undertaken. Perhaps the most controversial question to be addressed is that of how many individuals survive to spawn in the Ural River. Expansion of our work could uncover if the Caspian Sea contains one intermixing stock or if populations should be managed by individual river systems. If the latter were true, this would further aid in understanding where natural reproduction still occurs. Ideally a basin-wide program should be launched with coordinated research in different nations and rivers.

Successful implementation will require not only national and international commitment but also devotion of sufficient resources. Undertaking tag and release studies might further mean lower fishery yield in the short term but greater viability of the commercial fishery in the long term with improved fisheries management. Although not described here, our program also included outreach and educational activities in recognition of the fact that the public can shape policies and the future. Local and international journalists were involved in each of our field efforts, resulting in extensive media attention. We will also be completing educational brochures on Ural River sturgeon biology in the near future. With these efforts, we hope to create influence that will allow us to undertake many of the projects that described here. Our hope is that our work can be undertaken before the beluga, and other sturgeons, become too scarce to study.

References

Adams, P. B., Grimes, C., Hightower, J. E., Lindley, S. T., Moser, M. L., and Parsley, M. J., 2007, Population status of North American green sturgeon, *Acipenser medirostris*. *Environ. Biol. Fish.* **79**:339–356.

Bemis, W. E. and Kynard, B., 1997, Sturgeon rivers: an introduction to Acipenseriformes biogeography and life history. *Environ. Biol. Fishes* **48(1–4)**:167–183.

Benson, R. L., Turo, S., and McCovey, Jr. B. W., 2007, Migration and movement patterns of green sturgeon (*Acipenser medirostris*) in the Klamath and Trinity rivers, California, USA. *Environ. Biol. Fishes* **79(3–4)**:269–279.

Berg, L. S., 1948, *The Freshwater Fishes of the USSR and Adjacent Countries*. 4th ed. Akademiya Nauk USSR, Moscow and Leningrad. Part 1.

Billard, R. and Lecointre, G., 2001, Biology and conservation of sturgeon and paddlefish. *Rev. Fish Biol. Fish.* **10**:355–392.

Birstein, V. J., 1993, Sturgeons and paddlefishes: threatened fishes in need of conservation. *Conserv. Biol.* **7**:773–787.

Birstein, V. J., Hanner, R., and DeSalle, R., 1997, Phylogeny of the Acipenseriformes: cytogenetic and molecular approaches. *Environ. Biol. Fishes* **48**:127–155.

Birstein, V. J., Doukakis, P., Sorkin, B., and DeSalle, R., 1998, Population aggregation analysis of caviar producing species of sturgeon and implications for diagnosis of black caviar. *Conserv. Biol.* **12(4)**:766–775.

Birstein, V. J., Ruban, G., Ludwig, A., Doukakis, P., and DeSalle, R., 2005, The enigmatic Caspian Sea Russian sturgeon: how many cryptic forms does it contain? *System. Biodivers.* **3(2)**:203–218.

Bronzi, P., Garofalo, E., Dall Balle, Z., Battaglia, A., Ansferri, S., and Poggioli, C., 2006, Telemetry test trials on the behavior of sub-adults of *Acipenser naccarii* reared in captivity and released into the River Po (Italy). *J. Appl. Ichthyol.* **22(Suppl. 1)**:132–139.

Caron, F., Hatin, D., and Fortin, R., 2002, Movements and aggregation areas of adult Atlantic sturgeon (*Acipenser oxyrinchus*) in the St Lawrence River estuary, Quebec, *Can. J. Appl. Ichthyol.* **18(4–6)**:586–594.

Chebanov, M. S., Karnaukhov, G. I., Galich, E. V., and Chmir, Y. N., 2002, Hatchery stock enhancement and conservation of sturgeon, with an emphasis on the Azov Sea populations. *J. Appl. Ichthyol.* **18(4–6)**:463–469.

DeSalle, R., and Birstein, V. J., 1996, PCR identification of black caviar. *Nature* **381**:197–198.

Doukakis, P., Birstein, V. J., and DeSalle, R., 1999, Molecular genetic analysis among subspecies of two Eurasian sturgeon species, *Acipenser baerii*, and *A. stellatus*. *Mol. Ecol.* **8(S12)**:117–129.

Doukakis, P., Birstein, V. J., and DeSalle, R., 2005, Intraspecific structure within three caviar-producing sturgeons (*Acipenser gueldenstaedtii*, *A. stellatus*, and *Huso huso*) based on mitochondrial DNA analysis. *J. Appl. Ichthyol.* **21(6)**:457–460.

Edwards, R. E., Sulak, K. J., Randall, M. T., and Grimes, C. B., 2003, Movements of Gulf sturgeon (*Acipenser oxyrinchus desotoi*) in nearshore habitat as determined by acoustic telemetry. *Gulf Mex. Sci.* **21(1)**:59–70.

Edwards, R. E., Parauka, F. M., and Sulak, K. J., 2007, New insights into marine migration and winter habitat of Gulf sturgeon. *Am. Fish. Soc. Symp.* **56**:183–196.

Erickson, D. L. and Hightower, J. E., 2007, Oceanic distribution and behavior of green sturgeon. *Am. Fish. Soc. Symp.* **56**:197–211.

Erickson, D. L. and Webb, M. A. H., 2007, Spawning periodicity, spawning migration, and size at maturity of green sturgeon, *Acipenser medirostris*, in the Rogue River, Oregon. *Environ. Biol. Fishes* **79(3–4)**:255–268.

Erickson, D. L., North, J. A., Hightower, J. E., Weber, J., and Lauck, L., 2002, Movement and habitat use of green sturgeon in the Rogue River, Oregon, USA. *J. Appl. Ichthyol.* **18(4–6)**:565–569.

Erickson, D., Kahnle, A., Millard, M., Bryja, G., Higgs, A., Mohler, J., Sweka, J., Kenney, G., Difour, M., and Pikitch, E., 2008. Use of pop-up satellite archival tags and GIS to estimate movements, habitat use, and threats for adult Atlantic sturgeon in the ocean. American Fisheries Society 138th Annual Ottawa, Canada, August 17–21, 2008.

Fadaee, B., Pourkazemi, M., Tavakoli, M., Joushideh, H., Khoshghalb, M. R. B., Hosseini, M. R., and Abdulhay, H., 2006, Tagging and tracking juvenile sturgeons in shallow waters of the Caspian Sea (less than 10m depth) using CWT (Coded Wire Tags) and barbel incision. *J. Appl. Ichthyol.* **22(Suppl. 1)**:160–165.

Finney, S. T., Isely, J. J., and Cooke, D. W., 2006, Upstream migration of two pre-spawning shortnose sturgeon passed upstream of Pinopolis dam, Cooper River, South Carolina. *Southeast. Nat.* **5(2)**:369.

Grunwald, C., Maceda, L., Waldman, J., Stabile, J., and Wirgin, I., 2007, Conservation of Atlantic sturgeon *Acipenser oxyrinchus oxyrinchus*: delineation of stock structure and distinct population segments. *Conserv. Genet.* on-line first.

Hatin, D., Munro, J., Caron, F., and Simons, R. D., 2007, Movements, home range size, and habitat use and selection of early juvenile Atlantic sturgeon in the St. Lawrence estuarine transition zone. *Am. Fish. Soc. Symp.* **56**:129–155.

Heise, R. J., Slach, W. T., Ross, S. T., and Dugo, M. A., 2005, Gulf sturgeon summer habitat use and fall migration in the Pascagoula River, Mississippi, USA. *J. Appl. Ichthyol.* **21(6)**:461–468.

Hightower, J. E., Zehfuss, K. P., Fox, D. A., and Parauka, F. M., 2002, Summer habitat use by Gulf sturgeon in the Choctawhatchee River. *J. Appl. Ichthyol.* **18(4–6)**:595–600.

Israel, J. A., Blumberg, M., Cordes, J., and May, B., 2004, Geographic patterns of genetic differentiation among western U.S. collections of North American green sturgeon (*Acipenser medirostris*). *North Am. J. Fish. Manage.* **24**:922–931.

Khodorevskaya, R. P. and Krasikov, Y. V., 1999, Sturgeon abundance and distribution in the Caspian Sea. *J. Appl. Ichthyol.* **13**:106–113.

Khodorevskaya, R. P., Krasikov, E. V., Fedin, A. A., Fedorov, V. A., and Shvedov, V. V., 2002, Abundance and distribution of the beluga *Huso huo* in the Caspian Sea. *Voprosy Ikthiologii* **42(1)**:56–63.

King, T. L., Lubinski, B. A., and Spidle, A. P., 2001, Microsatellite DNA variation in Atlantic sturgeon (*Acipenser oxyrinchus oxyrinchus*) and cross-species amplification in the Acipenseridae. *Conserv. Genet.* **2**:103–119.

Kornienko, I. V., Voinova, N. V., Chistyakov, V. A., Timoshkina, N. N., Vechkanov, E. M., and Petrov, A. V., 2003, Polymorphism of primary sequence of a gene segment in the cytochrome b of mitochondrial DNA in the Azov population of *Acipenser gueldenstaedtii*. *Voprosy ikhtiologii* **43(1)**:73–77.

Kynard, B., Suciu, R., and Horgan, M., 2002, Migration and habitats of diadromous Danube River sturgeons in Romania: 1998-2000. *J. Appl. Ichthyol.* **18(4–6)**:529–535.

Laney, R. W., Hightower, J. E., Versak, B. R., Mangold, M. F., Cole, W. W. Jr., and Winslow, S. E., 2007, Distribution, habitat use, and size of Atlantic Sturgeon captured during cooperative winter tagging cruises, 1988-2006. *Am. Fish. Soc. Symp.* **56**:167–182.

Levin, A. V., 1997, The distribution and migration of sturgeons in the Caspian Sea. Sturgeon stocks and caviar trade workshop. in: *Sturgeon Stocks and Caviar Trade Workshop, Proceedings of a Workshop Held 9-10 October in Bonn, Germany*. IUCN Occasional Paper No 17. pp. 13–19.

Lindley, S. T., Moser, M. L., Erickson, D. L., Belchik, M., Rechisky, E. L., Kelly, J. T., Heublein, J., and Klimley, P., 2008, Marine migration of North American green sturgeon. *T. Am Fish. Soc.* **137**:182–194.

Mamina, K. N., 1995, *Analysis of reproduction of major exploited fish populations in Ural – Caspian Basin*. Doctor of Philosophy dissertation. Moscow State University. Moscow, Russia. In Russian.

Mitrofanov,V. P., Dukravec, G. M., and Peseridi, N. E., 1986, *Fishes of Kazakhstan* (V1) Nayka, Alma Ata, Kazakhstan. 272 p.

Paragamian, V. L. and Duehr, J. P., 2005, Variations in vertical location of Kootenai River white sturgeon during the prespawn and spawning periods. *T. Am Fish. Soc.* **134(1)**:261–266.

Peterson, D. L., Bain, M. B., and Haley, N., 2000, Evidence of declining recruitment of Atlantic sturgeon in the Hudson River. *N. Am. J. Fish. Manage.* **20(1)**:231–238.

Pikitch, E., Doukakis, P., Lauck, L., Chakrabarty, P., and Erickson, D. L., 2005, Sturgeon and paddlefish fisheries of the world: conservation strategies for the future. *Fish Fish.* **6**:233–265.

Pourkazemi, M., 2006, Caspian Sea sturgeon conservation and fisheries: past present and future. *J. Appl. Ichthyol.* **27(Suppl. 1)**:12–16.

Pourkazemi, M., Skibinski, D. O. F., and Beardmore, J. A., 1999, Application of mtDNA d-loop region for the study of Russian sturgeon population structure from Iranian coastline of the Caspian Sea. *J. Appl. Ichthyol.* **15 (4–5)**:23–28.

Rezvani Gilkolaei, S., 2000, Study of mtDNA variation of Russian sturgeon population from the south Caspian Sea using RFLP analysis of PCR amplified ND5/6 gene regions. *Iran. J. Fish. Sci.* **2(1)**:13–36.

Rezvani Gilkolaei, S. and Skibinski, D. O. F., 1999, Polymerase Chain Reaction (PCR) and direct sequencing of mtDNA from the ND5/6 gene region in Persian sturgeon *Acipenser persicus* from the southern Caspian Sea. *Iran. J. Fish. Sci.* **1(1)**:23–34.

Rodzen, J. A. and May, B., 2002, Inheritance of microsatellite loci in the white sturgeon (*Acipenser transmontanus*). Genome **45(6)**:1064–1076.

Secor, D. H., Arefjev, V., Nikolaev, A., and Sharov, A., 2000, Restoration of sturgeons: lessons from the Caspian Sea sturgeon ranching programme. *Fish Fish.* **1**:215–230.

Smith, T. I. J., McCord, J. W., Collins, M. R., and Post, W. C., 2002, Occurrence of stocked shortnose sturgeon *Acipenser brevirostrum* in non-target rivers. *J. Appl. Ichthyol.* **18(4-6)**:470–474.

Waldman, J. R., Nolan, K., Hart, J., and Wirgin, I. I., 1996, Genetic differentiation of three key anadromous fish populations of the Hudson River. *Estuaries* **19**:759–768.

Waldman, J. R., Grunwald, C., Stabile, J., and Wirgin, I., 2002, Impacts of life history and biogeography on the genetic stock structure of Atlantic sturgeon *Acipenser oxyrinchus oxyrinchus*, Gulf sturgeon *A. oxyrinchus desotoi*, and shortnose sturgeon A. brevirostrum. *J. Appl. Ichthyol.* **18**:509–518.

Welsh, A. and May, B., 2006, Development and standardization of disomic microsatellite markers for lake sturgeon genetic studies. *J. Appl. Ichthyol.* **22(5)**:337–344.

Whitlock, R. and McAllister, M., 2005, Developing a Bayesian mark-recapture modeling framework to inform stock recovery and rebuilding strategies for long-lived, anadromous species. In: *Theme Session on Rebuilding Programmes for Threatened Fish Populations (W)*. ICES Council Meeting documents no. 2005.

Wilson, S. G., Lutcavage, M. E., Brill, R. W., Genovese, M. P., Cooper, A. B., and Everly, A. W., 2005, Movements of bluefin tuna (*Thunnus thynnus*) in the northwestern Atlantic ocean recorded by pop-up satellite archival tags. *Mar. Biol.* **146**:409–423.

Wirgin, I., Waldman, J. R., Rosko, J., Gross, R., Collins, M. R., Rogers, S. G., and Stabile, J., 2000, Genetic structure of Atlantic sturgeon populations based on mtDNA control region sequences. *T. Am Fish. Soc.* **129**:476–486.

Wirgin, I., Waldman, J. R., Stabile, J., Lubinski, B. and King, T., 2002, Comparison of mitochondrial DNA control region sequence and microsatellite DNA analyses in estimating population structure and gene flow rates in Atlantic sturgeon *Acipenser oxyrinchus*. *J. Appl. Ichthyol.* **18**:313–319.

Wirgin, I., Grunwald, C., Carlson, E., Stabile, J., Peterson, D. L., and Waldman, J., 2005, Range-wide population structure of shortnose sturgeon *Acipenser brevirostrum* based on sequence analysis of the mitochondrial DNA control region. *Estuaries* **(28)3**:406–421.

Yang, D., Wei, Q., Wang, K., Chen, X., and Zhu, Y., 2005, Downstream migration of tag-released juvenile Chinese sturgeon (*Acipenser sinensis*) in the Yangtze River. *Acta Hydrobiol. Sin./Shuisheng Shengwu Xuebao* **29(1)**:26–30.

Zhang, S., Wang, D., and Zhang, Y., 2003, Mitochondrial DNA variation, effective female population size and population history of the endangered Chinese sturgeon, *Acipenser sinensis. Conserv. Genet.* **4(6)**:673–683.

Zhu, B., Zhao, N., Shao, Z., Lek, S., and Chang, J., 2006, Genetic population structure of Chinese sturgeon (*Acipenser sinensis*) in the Yangtze River revealed by artificial neural network *J. Appl. Ichthyol.* **22(Suppl. 1)**:82–88.

□□□H □□□ □□□□R□□□□ □□ □H□ □R□□□□R□ R□□□□□ □□□ □H□R □□R□□□□□ □□□□□□□□

□□□R□□□□□□*

Orenburg State Agrarian University, Orenburg, Russia

□□stract The problem of fish parasites threatening the food fish populations and, consequently, human health is an important aspect of water related environmental security. Analysis of the research results shows alarming parasitological situation with food fish populations in the Orenburg reservoirs. The fish parasites with various epidemiological values were found. Based on the research conducted for the last five years, it can be concluded that the fish parasitic fauna in the regional ponds is not fully investigated. Further research is needed to collect the data and plan rehabilitation programs.

□ey□ords□ Fish parasites, Orenburg oblast, protozoa, helminthic disease

The commercial catch in fishery reservoirs in the Orenburg region is based on 10–12 fish types: bream, pike perch, sazan, catfish, ide, pike, perch, roach, and crucian. There are parasites in the Orenburg region with various epidemiological values. Among protozoa the parameter of detection of trichodinosis is most stable. On the basis of the five years' data helminthic diseases of 11 kinds of activators have been recorded on the territories of the region. In the Orenburg region the epidemiological situation concerning helminthic diseases is adverse. The condition of fish stocks in water basins raises concerns. Research shows the complexity of the parasitological condition of the reservoirs in the Orenburg region.

Regional development of the region and its current social and economic conditions require more exact quantitative estimations of biological resources, in particular, the fish resources. In many parts of the region, fish production is an important source of income. Given the development of market relations, the fish market is currently oversaturated with the fish brought from various regions of the country. Therefore, a problem of fish parasites posing a threat to both food fish population and human health is

* A. Gryzunov, Orenburg State Agrarian University, Cheliuskincev, 18, Orenburg, 460795, Russia; gryzunov.orenburg@gmail.com

more acute than it has ever been. Until now, inadequate attention has been paid to addressing these questions. There is no complete data on fish contamination in different reservoirs of the region: current state, features and specifics have not been investigated yet.

Identifying specific structure of parasites, studying their biology and the ways of circulation, revealing life cycles in different hydrobiological conditions is of both theoretical and practical importance. The main purpose of the research is to study conditions of environmental veterinary and parasitological situation and to examine the current conditions of the fishing industry and its accompanying parasitic fauna in the reservoirs of Orenburg region.

Characteristics of the fish producing reservoirs

Fishery in the Orenburg region is based on the stock of the rivers, water basins, lakes and ponds. In the Orenburg region there are more than 617 rivers with the length over 10 km. Most of the rivers belong to the basin of the Ural river; others belong to the basins of the Volga and the Tobol. Majority of the rivers are shallow; only 19 of them have the average annual charge of more than 1 m^3/s. The stock includes 29 rivers with the length ranging from 50 to 100 km in length, 15 rivers – from 100 to 200 km, 9 rivers – over 200 km long. The largest river in the region is the Ural. With the overall length of 2,428 km, its length in the region constitutes more than 1,000 km. The primary refill source of the rivers of the Ural basin is snow cover, which makes up more than 80% of the annual drain. The refill occurs exclusively during the spring snow melting. The extent of rain refill is insignificant and prevails mainly in the basin of the Sakmara (11% of the annual drain). As a rule rains do not give much drain in the other river basins with an exception of especially intensive downpours.

The total stock of lakes used in fishery of the Orenburg region accounts for 22,000 ha, while the commercial catch covers over 2,000 ha. It consists mainly of the inundated lakes of the Ural and Volga basins. In the east of the region, there are lakes belonging to the Ural-Tobol watershed. The largest are the Shelkar, the Yega, the Kara (1,000 ha), Zhetykol (600 ha), and the Kairankol (950 ha). In spite of the fact that during the last years the eastern lakes have been periodically neglected by main fishery companies due to their remote location, they have gained the increasing trade value for the Society of Hunters and Fishermen of the Svetliy region, individual businessmen and organizations, which have made a contract with Federal State Agency "Kamuralrybvod". Other lakes are located on the banks of the largest rivers – the Ural and its inflows.

The overall area of the water basins is 35,546 ha, out of which the largest one is the Iriklinskoe reservoir, which surface is 26,000 ha, and is situated in the east steppe area of the Orenburg region. The fish catch in the Iriklinskoe reservoir constitutes 60–90% of the total regional catch. Water inflows to reservoir originate from the unregulated runoff from the Ural catchment area (30,472 km), discharges from the Magnitogorsk dam and different water sources below it. The share of the reservoir areas with depths up to 10 m makes 44%. It is the most productive and accessible fishing zone. The Sorochinskoye reservoir on the river Samara (the Volga basin) and the □hernovskoye reservoir are entirely used for the sport and amateur fishery. The □lshanskoye reservoir was also converted into the facility for sport and recreational fishery.

□onds in the Orenburg region are used as fish farms. The objects of cultivation are carps, sometimes grass-feeding fish.

□ommercial catch in the fish breeding reservoirs is based on 10–12 fish types: brean, pike perch, sazan, catfish, ide, pike, perch, roach, crucian. (Table 1).

Table 1. Fish catch in the Orenburg reservoirs

Fish types	Years										
	1995	1996	1997	1998	1999	2000	2001	2002	2003	2004	2005
□ike perch	30.74	12.28	9.48	32.71	4. 914	13.33	25	26.5	24.2	16.01	7.787
□atfish	0.095	0.016	0.50	0.311	0.115	0.7	2.5	1.3	–	0.3	0.206
Sazan	0.534	0.53	0.163	4.74	4.914	19.4	10.6	10.3	4	6.51	0.677
□ike	5.48	1.77	0.65	35.55	4.914	25.8	12.5	25.2	14.3	15.86	2.666
□ream	2.73	2.704	7.15	22.75	4.914	11.9	79	44.2	16.7	20.61	5.925
Ide	11.39	15.64	20.05	28.18	5.99	13.7	19	22.7	30	23.76	8.162
Asp	–	–	–	–	–	0.33	0.4	–	–	2.08	0.198
□rucian	20	17.83	11.77	63.48	22.17	78.6	66	98.9	83.93	93.19	33.028
□erch	25.97	71.53	60.36	62.923	66.78	212.7	161	113.6	104.89	111.29	152.819
□oach	35.76	46.86	30.55	30.663	30.26	54.3	31	29.8	20.9	21.8	22.104
Sopa	1.352	0.96	1.01	9.27	0.81	1.2	0.4	0.6	1.05	0.16	–
□ustera	0.34	0.82	0.875	7.26	1.02	1.7	0.3	4.2	1.4	0.18	1.029
Other fine	0.317	1.24	8.59	4.497	1.7	1.9	0.1	–	–	–	–
Whitefish	12.18	5.24	8.37	8.5	0.81	13.2	0.2	1.0	8.3	1.7	0.233
Vendace	142.3	88.78	285.72	21.9	303.58	16.6	0.4	8.4	22.2	17.5	13.76
In total	289.188	266.75	445.283	332.734	453.09	443.28	408.4	386.7	331.87	328.87	248.594

The data on the fish production in 2005 are as follows: carp – 200 t, trout – 4 t, White Amur – 0.1 t, □ester – 2 t. The fish catch in 2005 was 248,594 t (OrAdm 2006). The primary tool used in regional fishery are nets.

□arasito□ogica□conditions

According to the report of the Orenburg regional veterinary laboratory for 2001–2005, diseases are classified into two basic groups – *protozoan* and *helminthic*. The protozoa are represented by the *ichthyophthiriosis* (*infusorians Ichtyophthirius multifiliis*), *trichodinosis* (*infusorians of the Urceolariidae*), *myxosporidiosis*. Among the helminthes are *bothryocephalosis* (*cestodes Bothriocephalus acheilognathi*), *gyrodactylosis* (genus *Gyrodactylus*, families *Gyrodactylidae*), *dactylogyrosis* (genus *Dactylogyrus*), *diplostomosis* (*Trematodoses* families *Diplostomatidae*), *khawiosiset* (*Khawia sinensis*), *ligulidoses* (genus *Ligula*), *opisthorchosis* (*Opisthorchis felineus*), *postodiplostomosis* (*trematodes Posthodiplostomum cuticola*), *tetracotilosis* (trematodes families Strigeidae), *triaenophorosis* (*cestodes* sorts *Triaenophorus*, families *Triaenophoridae*), *philometroidosis* (*Philometroides lusiana*).

Among the protozoa the parameter of detection of trichodinosis is most stable, the geography of the fish infection varies much and includes the Krasnogvardeyskiy, Sol-Iletskiy, Sakmarskiy and Orenburg regions. The fact that all types of cultivated fish are vulnerable to this disease is troublesome since it can be a source of their destruction (□auer et al. 1981). The reservoirs of the Orenburg region also contain *ichthyophthiriosis*.

The five years□data show that there are 11 types of activators of helminthic diseases in the Orenburg region. Apart from that, the parasites with various epidemiological values are found in the fish of the reservoirs of Orenburg region.

The results of the research of Orenburg region have revealed that the epidemiological situation concerning helminthic disease – *opisthorchosis* is rather adverse, and its activators are worms: *trematode* – *Opisthorchis felineus*. As for the average number and frequency of detection of helminthes in the fish in the investigated reservoirs the dominating one is opisthorchosis, which was found in 2003, 2004 and 2005. The disease was detected in the fish sold on the markets of Orenburg and Orsk and fish caught in the Ural. The strong tendency of increased occurrence of this disease (from 0.15% in 2001 up to 4.32% in 2005) (Table 2) is observed. The given tendency assumes favorable conditions for existence and duplication of mollusks (intermediate helminthes carriers), and for pollution of the reservoirs (Ozeretskovskaja et al. 1985).

Table 2. Fish parasitic diseases

Disease	Years				
	2001	2002	2003	2004	2005
Fish protozoa					
Amount of units of the material (fish, patmaterial)	–	241	632	690	497
Ichthyophthiriosis	–	–	3 (0.47%)	2 (0.29%)	–
Myxosporidiosis	–	–	2 (0.31%)	–	–
Trichodinosis	5 (-)	5 (2.07%)	31 (4.9%)	1 (0.14%)	–
Fish Helminthes					
Amount of units of the material (fish, patmaterial)	1994	1379	2060	1992	1967
Bothryocephalosis	20 (1.0%)	–	1 (0.05%)	–	–
Gyrodactylosis	–	11 (0.8%)	10 (0.48%)	6 (0.3%)	–
Dactylogyrosis	–	5 (0.36%)	3 (0.15%)	3 (0.15%)	–
Diplostomosis	–	2 (0.15%)	–	–	–
Khawiosiset	5 (0.25%)	14 (1.02%)	6 (0.29%)	–	5 (0.25%)
Ligulidosis	1 (0.05%)	11 (0.8%)	21 (1.02%)	1 (0.05%)	–
Opisthorchosis	3 (0.15%)	–	9 (0.44%)	23 (1.15%)	85 (4.32%)
Postodiplostomosis	–	–	–	–	52 (2.64%)
Tetracotilosis	4 (0.2%)	–	–	7 (0.35%)	–
Triaenophorosis	–	–	–	–	2 (0.1%)
Philometroidosis	–	–	1 (0.05%)	2 (0.1%)	13 (0.66%)
Other helminthes	85 (4.26%)	111 (8.05%)	124 (6.01%)	111 (5.57%)	151 (7.7%)
Total	118	154	175	153	308

Gyrodactylosis is regularly found in the Sakmarskiy region (2002, 2004). This disease was also found in Siberia and in the Ural area (Kashkovskij et al. 1974), which is confirmed by our data.

□igulidosis and khawiosiset are widely circulating at rather low parameters of the invasion percent. The least spread helminthes on the territory of our region is *diplostomosis* found out in 2002 in the Sakmarskiy region.

Conc□sion

□ondition of the fish stocks in water basins is alarming. The mode of work of water basins is determined by the rules of maintenance in which the interests of fish management are not taken into account. The basic purpose of water basins regulates their hydrological mode: deep working during wintertime, accumulation of water during the spring period and constant expenditure for needs of the population. It leads to the reduction of spawning places, reduction of feeding reserve, deterioration of epizootic conditions (OrAdm 2005).

□espite of the significant amount of parasites, not all of them have identical epizootologic values. The majority of species are found relatively rare. The research analysis testifies complexity of the parasitological condition of the reservoirs of the Orenburg region. Such parasitic fish diseases as *dactylogyrosis*, *diplostomosis*, *postodiplostomosis*, *ichthyophthiriosis* harns young fish. *Gyrodactylosis*, *philometroidosis*, *trichodinosis* negatively affect the rate of fish growth, i.e. their biomass (□auer et al. 1981). This research as well as the previous findings of other researchers during the last five years suggests that the parasitic fauna of the fish is not fully investigated. The further research is required to compile the data on parasitic fauna in the regional reservoirs and to study their interlinkages with geomorphological, hydrobiological reservoir features as well as their impact on the fish population.

□eferences

□auer O.□., Musselius V.A., Strelkov □A., 1981, □ond fish diseases, M.: Food-Industry 320 p.

Kashkovskij V.V., □azmashkin □.A., Skripchenko □.□. 1974. Fish diseases and parasites in fish-breeding economy of Siberia and Ural. Sverdlovsk. 159 p.

OrAdm, 2005. □eport on state of the environment in the Orenburg region in 2004. Administration of the Orenbug □egion. Orenburg. 179 p.

OrAdm, 2006. □eport on state of the environment in the Orenburg region in 2005. Administration of the Orenbug □egion. Orenburg. 181 p.

Ozeretskovskaja □.□., Zalnova □.S., Tumolskaya □.I., 1985. The clinics and treatment of helminthes, □., 95 p.

2.3 RESTORATION ACTIVITIES

ESTABLISHMENT OF THE INTERNATIONAL URAL STURGEON PARK TO SECURE STURGEON CONSERVATION AND TO FACILITATE SUSTAINABLE INTEGRATED WATER MANAGEMENT

VIKTOR LAGUTOV*

Central European University, Hungary

VLADIMIR LAGUTOV

DonEco Research and Consulting, Russia

Abstract The only free-flowing river in the Caspian basin, the Ural River, is a unique ecosystem with a preserved natural hydrological regime and the last remaining in the Caspian basin unaltered sturgeon spawning habitats. To secure its further preservation the river basin ecosystem and human activities in the region must be managed in an integrated sustainable manner. Though there is now an international consensus on the need for an integrated approach to sustainable river basin management, there is no standard definition of the term "sustainable" or consensus on how to reach this state. Sustainable development of watersheds should consider three main components: economic, social and environmental, which can hardly be reached in real-life watershed management. Using sturgeon species as a natural indicator and an incentive for transboundary IWRM cooperation in the Ural river basin is suggested. To secure basin IWRM and sturgeon stock restoration the Ural River Sturgeon International Park should be established. The Ural River Basin Project, which aims at the creation of such a Park, is described in this paper. Activities towards successful integrated water management in the Ural Park will not only work towards sustainable watershed management, but also secure preservation and restoration of sturgeon. Local communities (Cossacks) involvement in sturgeon conservation and water management also resolves social and economic problems by restoration of the traditional life style.

Keywords: Watershed, river basin, integrated water resource management, indicator species, integrated environmental assessment, community-based environmental protection, Ural river, Cossacks, integrated modelling, uchug, sturgeon, beluga

* Viktor Lagutov, Central European University Nador ut.9, H1051, Budapest, Hungary; lagutov@uralbas.ru

Background

Rivers and their associated ecosystems and biodiversity provide the basis of life for a large portion of the world's population. Fresh water is expected to become the most limiting resource in many parts of the world in the near future. The world's freshwater resources are under increasing pressure. Population growth, improving standards of living, blooming economic activities and the degradation of aquatic ecosystems lead to increased competition for and conflicts over the limited freshwater resources. Water resources mismanagement results not only in breakdown of economic activities, but also in biodiversity loss, habitat degradation, social and political tensions.

The need for holistic cross-sectoral approach to water resources management is increasingly recognized and has resulted in a drastic increase in the number of watershed management programs worldwide. Depending on institutional needs and regional priorities various concepts of Integrated Water Resources Management (IWRM) have been developed recently. In brief IWRM is a process which can assist countries in their efforts to deal with water issues in a cost-effective and sustainable way.

□asin cooperation is often complicated by the high number of parties in the water management scheme and the complex nature of administrative and national borders. The Ural river basin, shared by □ussia and Kazakhstan (Figure 1), is no exception in this regard. The system of integrated river basin management that once existed in the Soviet Union collapsed in the early 90s and drastic deterioration of environmental conditions of all transboundary watercourses can be observed since then.

The Soviet integrated water management system was an effective tool in water management; however the priorities in water usage were given to the development of growing regional agriculture and industries without any attention being paid to the needs of the environmental flows and ecosystems. The rapid degradation of the ecosystem and its biodiversity (including decline in sturgeon stock) coincides with the collapse of the Soviet Union and the disintegration of the united basin water management system. The growing concern over environmental aspects of watershed management has been difficult to feed into practical water management plans due to, among other reasons, lack of transboundary cooperation. Many countries of the former Soviet Union do not have even agreements on transboundary water management. Though some agreements do exist, they are of a superficial character and not fully implemented. These documents may not be effective tools to tackle the issues addressed. The current

Figure 1. The geographic position of the internationally shared Ural river basin

state of IWRM in the region can be characterized by a lack of cooperation among countries and a shortage of incentives for such cooperation (UNECE 2003). Paradoxically, the general attitude towards transboundary water cooperation is positive and there is a strong need to stimulate it and present the best management practices.

Although Russia and Kazakhstan are both remnants of the Soviet Union and at that time had a single indivisible water management system, at present official transboundary cooperation on water management and regional environmental issues is almost negligible. However, the cultural and personal links are still very strong and the need for cooperation is well understood by stakeholders in both countries.

From the IWRM point of view the basin of the Ural River is a unique ecosystem. All components of traditional sustainable management – social, economic and environmental – can be harmonically linked here and considered jointly. It provides a perfect case study to develop, test and put into practice a sustainable watershed management strategy taking into account all the principles of IWRM.

While there is plenty of experience and knowledge in transboundary river management accumulated in European countries as well as the former Soviet Union, this knowledge cannot be simply copied to the Ural river basin given its different institutional and regional specifications. Careful evaluation of best practices and consideration of regional specifics is needed to develop a sound sustainable basin development strategy.

IRWM and sturgeon restoration

IWRM problems and obstacles for its success

The issue of sustainable watershed management and development are widely discussed. Though many articles and handbooks have been written and numerous attempts have been made to put this concept into practice, there is no uniform terminology accepted by all stakeholders or consensus on the best way to achieve sustainability in water resource use. There are numerous versions of IWRM approaches and guidelines.

Nevertheless, some fundamental principles underlying best management practices are common for most approaches. The IWRM principles proclaimed at Rio-92 are most commonly used. According to this approach the six basic principles of Integrated Water Management are:

1. The river basin is the most appropriate administrative unit for water management.
2. Water resources and the land which forms the river basin area must be integrated, in other words, planned and managed together.
3. Social, economic and environmental factors must be integrated within water resources planning and management.
4. Surface water and groundwater and the ecosystems through which they flow must be integrated within water resources planning and management.
5. Public participation is necessary_for effective water resources decision making.
6. Transparency and accountability in water management decision making are necessary features of sound water resources planning and management.

This approach was incorporated in the European Union Water Framework Directive (EU WFD) adopted by EU Member Countries in 2000 and in National Water Codes in many countries.

Though there is now an international consensus on the need for an integrated approach to sustainable river basin management, implementation of these nice principles is problematic.

The concept of IWRM is closely linked to the idea of sustainable development (SD). The generally accepted definition of sustainable development defines it as "development which meets the needs of the present, without compromising the ability of future generations to meet their own needs".

As repeatedly indicated by many authors, both concepts, SD and IWRM, have ambiguities in definitions and practical implementation (Anthony et al. 2003; Jewitt 2002; Jonker 2002). Among them the following can be mentioned: (1) the standard definition assumes a common understanding of what development means; (2) it assumes the present generation knows what the needs of future generations will be; (3) it does not explicitly link society and resources, the two elements in development; (4) it is impossible to measure at what stage of development future generations are being compromised; (5) it does not seem to consider the different time spans between human lifecycles and natural cycles.

With regards to these considerations a better definition of sustainable development could be "the improvement of people's livelihoods without disrupting the natural cycles". Based on this approach a more appropriate definition of IWRM would be "managing people's activities in a river basin in a manner that promotes sustainable development (improves livelihoods without disrupting the water cycle)". Unfortunately, the traditional approach still prevails in water management.

As a rule researchers and managers continue to address IWRM issues from a narrow, sectoral perspective. Many watershed management projects based on these principles not only indicate no success so far, but also often worsen the environmental situation in watersheds.

Apart from these conceptual problems there are a number of obstacles in everyday water management and practical implementation of IWRM principles.

The first commonly accepted principle recognizes a river basin as the most appropriate unit for considering the management of water resources. Nowadays this principle is mentioned in almost every water management-related directive or policy recommendation, though rarely duly realized even in national environmental management practices. For example, the new Russian Water Code adopted in 2006 proclaims the need for river basin-wide water management strategies, but stipulates national water management depending on the existing administrative territorial division (RF 2007).

Another undisputable point in the theory of IWRM is that sustainable development of watersheds should consider three main components: economic, social and environmental. However, the history of human communities' development in river basins shows that this is hard to achieve. Some components are often neglected in favour of others.

Freshwater is a limited resource, which is very much affected by uncontrolled weather conditions and often even the best management strategies cannot provide enough water for all users. This consideration leads to the necessity to prioritize water users in conditions of both scarce and available water resources. Traditionally the first priority in water use is given to economic development at the cost of environmental needs. However, it is clear that such an approach cannot be sustainable even in case of abundant water resources taking into account the constantly growing economic needs and related anthropogenic impacts.

The very definition of “sustainable” applied for water use is a vague concept, as has been repeatedly indicated by some authors (Hedelin 2007; Lagutov 1995, 1997; RAMSAR 2002). Despite numerous regulations and projects the attempts to synchronize understanding of the IWRM concept by different stakeholders and to introduce integrated water management into management practices are often not very successful. Not only different stakeholders and water users, but also different scientists define this concept in various, often contradictory, ways. This creates problems for a participatory approach, one of the pillars of sustainable watershed management, which implies transparency and participation in decision-making for all involved stakeholders and water users. Apart from that, particular essential ecological water services such as biodiversity needs are often not represented by any stakeholders, organizations or communities who participate in the decision-making process of IWRM. Hence, these needs are often neglected even in case of participatory decision-making.

Another basic yet controversial IWRM principle is the introduction of the economic analysis of water use (EU 2000). Though seeming to be a good idea, it often cannot be duly implemented. Assessment of a certain species’ extinction in monetary terms or, even more, economic comparison of such a loss to, for example, electricity generation, is hardly possible.

In addition the integrated watershed management is often complicated by the transboundary nature of the river basins (Jansky et al. 2004; Kauffman 2002; Margesson 1997; van Ast 2000). IWRM implies the need to manage transboundary water resources jointly, which can rarely be achieved. The interests of a particular country and its willingness to participate in the process directly depend on its upstream or downstream position along the river stream. For instance, EU Member States adopted the Water Framework Directive (WFD) in 2000 and are rapidly moving towards unifying management systems and standards (EU 2000). According to WFD, Member States are obliged to protect, enhance and restore all surface waters with the aim of achieving good ecological status by 2015. However, by 2007 the pressing issue of transboundary water management for EU members not only lacks enforcement, but even the principles have not been agreed on yet (UNECE 2007). The natural hydrological cycle is perceived by managers and decision-makers as a uniflow river stream. If any damage to the river ecosystem (pollution, hydrological cycle disruption, etc.) is made by an upstream country and downstream countries are bearing ecological or economic losses it seems that no harm is caused back upstream. The case of the cyanide spill at Baia Mare (Romania) in March 2000 revealed the problems of integrated transboundary water management and political implications very well (UNEP 2004).

New approaches should be sought and applied to integrated water resources management to make it an effective tool in practical environmental management. These approaches should be based on an ecosystem's sustainability, e.g. water cycles, and non-disruptive character of human activities with regards to ecosystem functions. An ecosystems approach to IWRM focusing on the role of the hydrological cycle is under discussion in the scientific literature (Jewitt 2002; RAMSAR 2002).

IWRM indicators

Careful selection of appropriate indicators in altered watersheds is an essential part of sound policy and decision-making in IWRM. On the one hand, these indicators should integrate the long-term temporal and spatial basin-wide environmental characteristics of a watershed. On the other, it should ideally be possible to assess the socio-economic activities in a watershed using this indicator. However, indicators in general and integrated indicators in particular are still not a well elaborated aspect of IWRM (Chaves and Alipaz 2007; He et al. 2000).

There are many indicators and indices suggested to evaluate the progress in a particular aspect of the IWRM process. For example, the number of published articles or sent messages to stakeholders are suggested as indicators of public participation or awareness raising in IWRM (Hedelin 2007).

Another case study for indicators' usage in IWRM can be drawn from WFD, which aims at "achieving good status of surface water" (EU 2000). Surface water is defined as of good ecological quality if there is only slight variation from the ecosystem with minimum anthropogenic impact. A long list of different indicators to be selected from is suggested for appropriate authorities, who can choose several indicators to work with and set up their own standards. The indicators are isolated and treated separately, which by itself cannot result in sound policy (Chaves and Alipaz 2007). Most of these indicators and, correspondingly, activities within WFD concentrate on water quality. At the same time other river-floodplain system characteristics (i.e. habitats fragmentation), economic or social aspects are either not taken into consideration or inadequately considered.

In comparison to isolated indicators of the physical environment, economic or social aspects of the IWRM process the ecological and biodiversity indicators are usually either not taken into account or little attention is paid to them in water management practices. At the same time ecological aspects, in particular biodiversity conservation, are an essential part of the SD process and they should be applied to IWRM to consider watershed development as sustainable. The concepts of "key-", "indicator-" or "keystone-" species for sustainable development and IWRM have been widely discussed in last 15 years (AzovBas 2002; Russian State Duma 1995; Lagutov 1995, 1997; WWF 2002a, b).

Moreover, given the holistic, "integrated" nature of IWRM it is essential to introduce some river basin-wide single natural indicator of sustainable watershed development which can bring together different sectors and stakeholders concerned with IWRM, allowing the interests of various water users (including ecosystem services) to be taken into account. This indicator should also encourage involvement of different disciplines related to water management as well as incorporate concerns of ecological, socio-economic and policy aspects of sustainable development.

Such integrated natural indicators of sustainable watershed management are sorely lacking in practical environmental activities and the need for this indicator has been mentioned by many authors (Chaves and Alipaz 2007; Jewitt 2002; van Delden et al. 2007).

Sturgeon population as an incentive for transboundary integrated water management

One of the biggest obstacles to transboundary IWRM in Central Asia in general and in the Ural river basin in particular is the lack of incentives for cooperation. This statement is especially true in cases of upstream-downstream watershed division as in the case of the Ural river basin. The selection of these incentives is always region-specific and depends on the current state of international affairs and environmental conditions.

Being a unique ecosystem the Ural river basin provides an encouraging incentive for transboundary IWRM through preservation of the sturgeon species.

There is no need to describe the importance of sturgeon conservation and worldwide concern over its fate. The importance of this flagship species' preservation is acknowledged by many international Conventions and Agreements (CITES 2004; FAO 2007; TRAFFIC 2003, 2007; WWF 2002a). The reason for such an interest in this species' preservation is its high commercial value. Sturgeon caviar is synonymous with luxury and wealth worldwide.

Sturgeon population as an indicator of the sustainability of watershed management

Apart from its high economic value, sturgeon is a perfect indicator (an umbrella) species for the river basin it inhabits (AzovBas 2002; Lagutov 1995, 1996, 1997, 1999; Uralbas 2007). The presence and well being of the sturgeon population in a river network indicates the "good quality" of a river ecosystem's health. Sturgeons also represents regional economic development and social structure, as poaching and illegal fishing which reduce sturgeon populations develop in areas with a poor unemployed population.

Securing natural sturgeon reproduction, protection and sustainable management of sturgeon stock is directly linked to integrated water resources management in the river basin and sustainable watershed development. These activities influence each other and should be considered only in an integrated manner.

Preserving sturgeon in the region would not only be of pure environmental benefit, but would also greatly contribute to economic and social stability in the region as well as food and water security. Thus, the measures aimed at preservation and sustainable management of the Ural sturgeon population can bring together environmental and socio-economic aspects of sustainable development and underpin the strategies for sustainable watershed development.

In this way, sturgeon meets the requirements for integrated IRWM indicator discussed above.

It should be noted that fish have been used as indicators for solely ecological status assessment for about 20 years as one of many indicators along with phytoplankton and amphibians (Hughes and Oberdorff 1999). For example, fish populations are one of many ecosystem health indicators in the EU's WFD. To date, however, even EU Member States have not yet included fish in their routine monitoring programs. Sturgeon is one of the suggested indicators for biodiversity abundance. However, use of this indicator for European rivers is a matter for the very distant future, if at all, since it is totally extinct from every European river without hopes for restoration due to habitat loss and damming. The only exception is some land-locked freshwater sturgeon subspecies of little ecological and economic value, which cannot be used as an indicator in the same way as other sturgeon species (e.g. sterlet in the Danube). WWF's European Freshwater Programme also considers Sturgeons as habitat Flagship species, Species of special concern and Indicator species (WWF 2002a, b).

Ural basin project

Ural sturgeon international park

The Ural River Basin Project was launched to facilitate the sturgeon restoration and sustainable watershed management in the Ural River Basin in 2007. The Project is a joint initiative by Central European University, the

Russian Environmental NGO "Green Don", local communities and a number of Russian and Kazakhstan NGOs and environmental state agencies.

The underlying idea of the Project is the concept of sustainable basin development by securing natural reproduction of migratory sturgeon species. In order to assure the implementation of this idea an international Ural Sturgeon Park should be established, spreading through the full extent of sturgeon migration routes and habitats, from the spawning grounds in the river upper branches to the river mouth. The Park borders should be drawn based on the 100 year flood level to secure undisturbed ecosystem functioning under possible extreme conditions. This approach differs from the utilized now approach to the creation of small patches of reserves through the river stream network (Bolshov 2000; RK 2002). The entire extent of the sturgeon migration routes and habitats in the Ural river should be equally protected.

The population density in the considered areas is very low, the industries and agriculture influence is minimal.[1] Such a Park should also have features of a Biosphere Reserve and Ethno-Natural Protected Area. Integrated water management and community-based management of sturgeon stocks can be the basis for sustainable basin development. In this way the Project aims not only to preserve this flagship species, but also to solve social and economic problems by restoration of the traditional life style of local communities.

The productive sturgeon spawning habitats are located in the upper branches of the Ural river in the Russian part of the watershed,[2] while the most of the migration paths are in Kazakhstan. To achieve this, close cooperation and agreement should be established not only on communities' level, but mainly on the level of local and regional authorities of Russia and Republic of Kazakhstan. The proposal for the creation of such an International Park should be developed in collaboration with all the interested parties and a cross-sectoral feasibility study should be carried out in cooperation with national and international agencies. The final proposal should take into account the interests of all stakeholders with priority given to sturgeon conservation to secure regional sustainable development.

High economic and social values of sturgeon allow the combination of both ecological and socio-economic aspects of sustainable development. Investment in IWRM and sturgeon conservation can be largely repaid later by "sustainable extraction" of sturgeon upon stock restoration.

While the establishment of a Ural Park seems to be long-term distant goal, other activities have been carried out in the framework of the Project. Public awareness raising has been approached through a number of regular publications in regional and local mass-media. The website of the Ural

[1] See paper on the Ural river in this volume.

[2] See paper on the Ural sturgeon in this volume.

Basin Project was launched at the beginning of 2007. A number of research projects on the river ecosystem and riverine biodiversity has being also undertaken in cooperation with regional organizations (i.e. GIS databases creation, river ecosystem and sturgeon population modelling).

The areas under the scope of the Project include different environmental disciplines and anthropogenic activities related to the well-being of the sturgeon population, taking into account its triple function in the river ecosystem (as indicator species, flagship species and species of special concern). By adopting this holistic, integrated approach the Project will be a focal point for specialists on water quality, fishery, international and national environmental law, as well as sturgeon experts.

One of the Project's goals is to develop a network of specialists involved in different aspects of integrated water management and sturgeon conservation. Such a network should unite not only different scientists (biologists, hydrologists, economists, chemists, lawyers, etc.), but also water users (industry, agriculture, local communities, etc.) to provide integrated interdisciplinary analysis of watershed-related problems and develop sound recommendations for decision-makers. Managerial insight and opinion should also be taken into account through their involvement and feedback to the developed recommendations supplied to them. Figure 2 displays the idea of this network and the role of the Ural Basin Project in it.

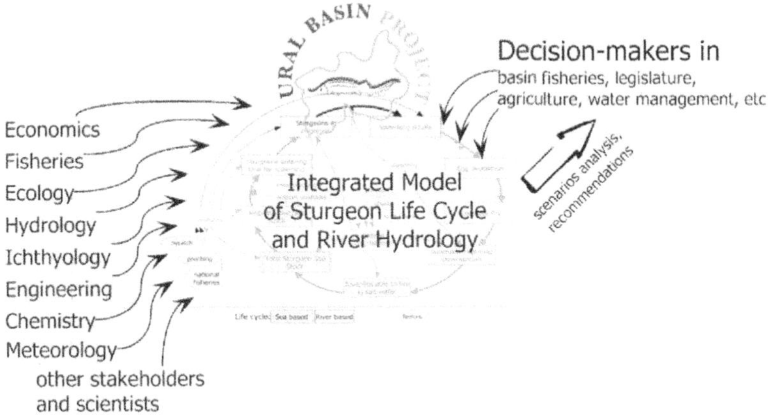

Figure 2. The network of specialists in integrated watershed management of the Ural river and role of Ural Basin Project

The cooperation with educational institutes aimed at the review of current environmental-related courses is carried out as a part of the project. In particular, it is planned to include in syllabi discussions of transboundary environmental management and nature protection and to introduce to institutions and schools of the Ural Basin experimental training courses for officers of environmental agencies and state services.

The First Ural River Basin International Workshop "Rescue of Sturgeon Species by means of Transboundary Integrated Water Management in the Ural River Basin" was held in Orenburg (Russia) on June 13–16, 2007 within the framework of the Ural Basin Project. Organized by the Research and Consulting Center DonEco and Central European University, the Workshop was also conducted with active involvement and assistance by the Russian Federal Agency for Environmental Inspections. The workshop was attended by more then 60 experts, researchers and practitioners from Governmental Environmental Agencies, NGO and business representatives from both basin countries (Russia and Kazakhstan), and representatives from relevant international organizations such as the Food and Agriculture Organization of the United Nations, the Secretariat of Wetland Convention (RAMSAR), the International Association on Danube Research, and many others covering the whole spectrum of Ural Basin management stakeholders. More information on this and other Project activities can be found on the Project's website at http://uralbas.ru.

Public participation in IWRM and biodiversity protection (Ural Cossacks communities)

As stated above, public participation is one of the essential principles of IWRM and sustainable watershed development. Any nature protection activities are ineffective if they are not supported by local communities. Moreover, some authors mention the rights of local communities over water and water ecosystem related resources as an important factor contributing to sustainable basin development (Karpov 1911; Kgarebe 2002).

In many cases practical implementation of these requirements are hardly possible or has a limited, formal character, since often local communities have no incentives to participate in these activities. This is well illustrated by conservation of sturgeons with high market value, which makes this species a subject for poaching. So, poaching and illegal fishing are widespread regional threats in the Ural region nowadays.

However, a high level of public participation can be easily achieved in the Ural watershed. Active cooperation of local communities with regional authorities might be possible thanks to the peculiarities of regional identity. This area is historically populated by Ural Cossack communities, a self-governing paramilitary ethnic group. Cossack troops were traditionally involved in various State services in Russian Empire. They were either protecting Russia's borders in their areas or serving as combatants during military campaigns. In exchange for military service they enjoyed exclusive rights to control natural resources on their territory (e.g. fish and water) and paid no taxes (Borodin 1901; Brockhaus and Efron 1898; Semple 1907; Von Harthausen 1972).

The Self-governing Lands of Cossack Communities in the Russian Empire were historically located in the river basins (Don, Volga, Cuban, Terek, Ural, Amur, etc.) in the frontier areas. Cossacks were living in small villages (*stanitcas*) throughout the river floodplains, relying on fishing, hunting and small scale farming as food sources. Any industrial or agricultural activity on their Lands had to be confirmed at Cossack gatherings ("Cossack Circles").

The Ural Cossacks, one of the oldest Cossack communities in Russia, controlled the entire territory and resources of lower Ural basin and adjacent sea area. Their historical settlements are stretched in a line along the bank of the Ural river for more then 450 km. As can be seen from Figure 3, a reprint of an old Russian map from the beginning of the 20th century, the

Figure 3. The territory of Ural Cossacks Land before 1917 fully covered the sturgeon habitats from the river mouth to sturgeon spawning grounds upstream

Ural Cossacks' Land closely matches the shape of the Ural basin, covering all sturgeon habitats.

The traditional life style of the Ural Cossacks directly relates to the problems of sustainable water management in a river basin. Living in harsh environmental conditions characterized by low soil fertility they had to fully rely only on the river ecosystem to support their communities.

Consequently, all the aspects of water usage and fishery were very carefully described, regulated and enforced (Borodin 1901; Dal 1961). There were fishery and water laws. Out of two elected commanders (*atamans*) one was a military commander, while the other one was solely responsible for fishery. Special mounted troops guarded the river streams during spawning migrations.

Baron August Von Harthausen in his book "Studies On The Interior Of Russia", first published in German in 1847, described the Ural Cossacks as follows:

> they do not farm the land at all... and live principally from fishing... Fishing is precisely regulated. It is limited to specific times in the winter, spring, and autumn. Whoever dares to catch a fish out of season loses his share for that year. Even if the Cossack happens to find a sturgeon which has been tossed onto the land, he will carefully throw it back into the water rather than take it home....(Von Harthausen 1972)

The characteristic feature of the Ural fishery was *uchug*, the temporal metallic or wooden fence constructed through the river stream near the city of Uralsk. The fence prevented spawning sturgeons from autumn race from going upstream out of Ural Cossacks territory. Sturgeons had to hibernate in the wintering habitats downstream of *uchug*.

Any sturgeon fishing was limited to several days during the year. In winter it was aimed at hibernating sturgeons below *uchug*. During these days Cossacks were gathering at the bank of the Ural. Following the signal of the fishing ataman they rushed to the ice covered Ural, cut holes through ice and equipped with special spears were taking sturgeons out of the wintering holes, river depressions were sturgeons hibernated during the winter (Dal 1961). It should be noted that any fish caught during the first day was sent as a gift to the Emperor (*tcarskij kusok*).

The fishing in the spring and autumn was also precisely defined and organized by fishing atamans. During several days all Cossacks were fishing as a group only in predefined segments of the river after the ataman's signal. Any fishing in the river or sea in summer was prohibited.

The catch in the sea was carried out with *okhans*, nets with coarse, more then 0.5 m, mesh (Malecha 2002). In the river fishing with coarse-meshed nets was allowed only in strictly limited days upstream of uchug (around 500 km from the delta).

Every spring during the vernal spawning migrations it was dismounted to secure natural sturgeon reproduction.

Only Cossacks were allowed to fish in the Ural river. Merchants, always coming from other regions, had to wait at the river bank and to buy the fish from the Cossacks.

Until Russian Civil War in 1917 when Cossacks were deprived of their privileges the entire water course of the lower and middle Ural was used exclusively for fishery (Brockhaus and Efron 1898). No any other kind of activity was allowed, including navigation. Ferriage through the Ural was allowed only in a couple of places through the entire territory in order not to frighten the fish.

Sturgeon and river worshipping by Ural Cossacks was reflected on their coat of arms: sturgeon and water were the only items depicted on it in addition to their weapons.

It should be stressed that sturgeon meat and caviar was not considered as a kind of luxury, but rather as ordinary people food (Borodin 1901; Dal 1961). So, the precise regulation and careful protection of sturgeon stock was not an attempt to maximize the profit, but to secure sustainable, as we call it now, consumption of the food resource. The harvested fish was mainly consumed by local population. The fishing targeted only beluga, ship and Russian sturgeon. Other sturgeon species considered to be of no importance. The annual export of caviar by the Ural Cossacks for these three species was about 130 t, meat – more then 1 thousand tons (Brockhaus and Efron 1898).[3]

Unfortunately, this interesting experience of sustainable river-related management is not adequately reported in Soviet, and correspondingly, foreign literature, due to the persecution of Cossacks by the Soviet regime during the 20th century.

Thanks to such an environmentally-friendly attitude and rational prudent fishing the sturgeon stock in the Ural river lasts much longer then in other Caspian regions. It was the main source for record-breaking catch in the second-half of 20th century, when Ural catch along substituted more then a half of all Caspian catch in f.USSR.[4] Such an attitude to natural resources by every member of local communities is a rare phenomenon. Partial explanation to it is the community exclusive rights over the resource and its utilization.

Cossacks were, probably, the most severely persecuted ethnic group after the Bolshevik Revolution in 1917. Revival of the Cossack movement

---

[3] For comparison the CITES sturgeon caviar quota for Kazakhstan in 2007 is 5 t, which cannot be reached even under the condition of lack of local consumption and producers deficit for hatchery-based restocking programs.

[4] See paper on Ural sturgeon population dynamics an catch in this volume.

is a widespread phenomenon through the whole territory of Russia nowadays. Recovering from repressions they are actively looking for their place in modern Society and possibilities of State Service, demanding changes in legislation and society structure (BBC 2007). For example, often regional Cossack organizations serve as voluntary mounted police in cities in their region.

While it might be impossible to fully restore Cossacks' former rights and privileges, this potential and these grass-roots initiatives should be utilized. The idea of using reviving Cossack groups for environmental protection has been actively promoted in the last decade by some Cossack and NGO leaders (AzovBas 2002; Russian State Duma 1995; Lagutov 1995; Uralbas 2007). The involvement of local communities in nature protection activities (e.g. establishment of ethno-natural protected territories) in the Ural river basin may not only protect this species and ecosystem of worldwide concern, but also stabilize the social and economic situation in the region by providing employment. In this case, Cossack groups can be effectively used for guarding the protected areas to prevent poaching and serving as rangers.

The consultations with the local communities (main regional Cossacks organizations) in the region have shown their interest and full support for this initiative. Moreover, local Cossacks communities are already reported to oppose poachers and guard the territory next to their *stanitsas* (villages) independently and in cooperation with State Environmental Inspection Agencies. The first joint anti-poaching patrols involving state fishery inspectors and voluntary-based Cossack troops were successfully carried out in the Orenburg region (Uralbas 2007).

Legal and institutional framework

The existing legislative base including, but not limited to, National Constitutions, International Conventions, ratified by both countries, national laws and a number of bilateral agreements creates an adequate legal framework to initiate transboundary watershed management cooperation. However, further improvements and amendments are needed for its successful realization.

The new RF Water Code, a framework national law regulating the protection and use of water resources, was adopted in Russia in June 2006, and came into force on 1 January 2007. The Code is mostly based on existing national water legislation. It also incorporates a number of contemporary approaches to water management presented in IWRM and, as an example of a well elaborated water Directive, the European Union Water Framework Directive. In particular, one of its innovations is the introduction of a basin management approach to water management practice, institutional coordination based on a basin approach and the creation of basin councils. It also envisages comprehensive basin management schemes that are to be developed for the purposes of integrated water management.

However, the EU WFD urges "management of a river basin as a single system of water management" and suggests that the usual administrative boundaries should no longer be applied to water basin management. At the same time, the territorial unit for water management (basin "okriug") in the Russian Water Code is based on the existing administrative structure in Russia. The Water Code simply coordinates water policies between the federation, the 89 federal subjects and the municipalities by defining the scope of their competences within the traditional administrative borders. It also aims at coordination between multiple stakeholders and water-users.

It should also be noticed that a basin management approach was also foreseen by the 1996 Water Code of the Russian Federation, though without any practical implementation. The legal framework for integrated river basin management was also developed and adopted in the Soviet Union in the 1960s.

By contrast, Kazakhstan is a few steps ahead in implementing the basin water management principles. The new edition of its Water Code, incorporating the principles of river basin management, was adopted in 2003. The Basin Councils for most of the river basins within the territory of Kazakhstan are already established and functioning. A number of internationally-funded projects on IRWM are undergoing. However, according to UNDP reviews the situation with water resources management in Kazakhstan "is best described as being fragmented, underfunded and poorly governed" and there is still a long way to go to implement IWRM principles.

Transboundary aspects of watershed management have received very little attention in either country so far. Though both countries have ratified a number of international and bilateral conventions and agreements on transboundary water issues and pollution, they have not been enforced yet.

The preservation of the ecosystem and sustainable watershed management of the Ural River depends not only upon efficient cooperation by both basin countries, Russia and Kazakhstan, but also on active involvement of international institutions. The latest trends show that international organizations and donors are increasingly interested and willing to fund and participate in transboundary water management projects and threatened biodiversity conservation (in particular sturgeon and other flagship species) (CITES 2003; Raymakers and Hoover 2002; Turnock 2001; UNECE 2006; WWF 2003).

Both countries are parties to international conventions and agreements on various aspects of water protection, biodiversity conservation and sustainable development, though this is not explicitly reflected in water management practices.

Taking into account the above considerations and the development stage for water regulations and methodologies in both basin countries, the case of the Ural River Basin is ideal for a pilot study on drafting and implementation of transboundary basin management directives aimed at sustainable watershed development in post-soviet countries.

Integrated modelling

As has been indicated by many authors, one of the most convenient and useful ways to approach the highly interdisciplinary nature of environmental assessment and management in river basins is to use techniques and tools of Integrated Assessment and Modelling (Harris 2002; Jakeman and Letcher 2003; Jansky et al. 2004; Janssen and Goldsworthy 1996; Krysanova et al. 2007; Lagutov 1997; Letcher et al. 2007; Parker et al. 2002; Shen et al. 2005).

Thus, special attention within the project activities is paid to the modelling of sturgeon population and water management issues. Such a model has proven to be a very useful tool in integrated water-resources management in a river basin and analysis of sturgeon protection activities (Lagutov 1996, 1997, 2003). Moreover, some authors state that sustainable river basin management is only possible by means of applying catchment models to evaluate management alternatives (Fohrer 2005; Refsgaard et al. 2005).

Given the wide acceptance of the need for integrated modelling it may be somewhat surprising that there is no understanding of the ways in which integrated modelling can be useful in environmental assessment and management.

Traditionally the role of modelling is perceived by environmental managers and practitioners as a way to produce end-user tools for predictions and analysis of the consequences of management strategies for decision-makers (Giupponi 2007; Mysiak et al. 2005; Scoccimarro et al. 1999). These tools, called "Decision Support Systems" (DSS), are computer-based programs with easy-to-use interfaces allowing managers and practitioners to take into account expert opinion in some areas. Despite their popularity, the success of DSS development is uncertain and many computerized decision-support tools have failed when dealing with complex and unstructured problems (Larocque et al. 2006). The level of uncertainty in environmental models and associated socio-economic subsystems is very high, and often it is impossible to forecast the behaviour of a certain ecosystem and/or related management strategies. Limitations and disadvantages as well as advantages and benefits of using DSS should be well understood by end users.

Undoubtedly, DSS is still a very important function of environmental modelling, but sometimes it is not the main goal of the modelling efforts.

Another important function of integrated modelling is serving as a framework for the organization of existing multi-disciplinary knowledge, to identify gaps in knowledge and to bring scientists, stakeholders and decision-makers together (Keyl and Wolff 2007; Parker et al. 2002; Suter and Glenn 1999).

Development of a conceptual watershed model can be used as a tool to facilitate debates and consultations among stakeholders and scientists, thus to enhance the participatory process (Lanini et al. 2004; Sendzimir et al. 2007). The integrated modelling should be perceived not as a finished product but as a tool for problem exploration and communication of results.

In accordance with Projects' underlying principles and priorities, the river hydrological modelling is combined with sturgeon population models. This should introduce a long term perspective to water management strategies.

Process-based spatial model simulating river hydrology and upstream-downstream migration of sturgeon populations has being developed. Corresponding human activities (i.e. water intakes, pollution, fishery, etc.) as well as environmental conditions are also simulated. Based on the input control parameters and calculated river characteristics, Habitat Suitability Indices are generated along the river stream. Using these indices the possible locations of spawning, wintering and feeding grounds are to be identified.

Modelling efforts of hydrological and population processes should be supported by reliable data on various parameters (Thorsten et al. 2004; Vidal et al. 2007) A single river ecosystem-related monitoring system which was established in the Soviet Union has collapsed in both countries in the 1990s. Though each basin country is now trying to develop a monitoring system independently, a lot of information is still not available or biased (e.g. data on sturgeon catches). Using an integrated modelling approach, missing data can be substituted by expert opinion (Liu et al. 2007).

To support modelling efforts and to collect data available on the Ural river basin GIS databases of the Ural River ecosystem are being developed. There are a number of techniques which can be used for linking GIS and environmental modelling (Aspinall and Pearson 2000; Pullar and Springer 2000).

Discussion

Sturgeon species can be considered a perfect natural bioindicator of a river basin's health and sustainability of integrated watershed management.

These two issues are closely interconnected and should not be approached independently.

At the same time sturgeon species allows to consider all aspects of sustainable development: ecological, social and economic. The activities towards its conservation and restoration bring together all stakeholders in integrated watershed management. In this way the general and vague concept of Sustainable Development gets the definite realistic definition and mechanisms for practical implementation. It evolves into the new Concept of Basin Sustainable Development (AzovBas 2002; Lagutov 1995, 1999) which can be implemented and enforced in practical IWRM process.

The Ural river ecosystem with affiliated traditional Cossacks life style is a unique natural and cultural phenomenon which should be protected internationally. The establishment of International Ural Sturgeon Park with active involvement of reviving Cossacks communities and international monitoring seems to be the only possibility to secure Caspian sturgeon conservation and rehabilitation. Its conservation will also assist in the Region's sustainable economic and social development.

The need for transboundary cooperation is well understood in both basin countries. The existing legal national frameworks allow such cooperation, though the practical steps in joint watershed management and transboundary biodiversity conservation is badly needed.

In case of successful rehabilitation the Ural sturgeon population can serve as a gene pool for the sturgeon restoration programs worldwide.

References

Anthony, A.J., Alsharhan, A.S., et al. 2003. Integrated water resources management is more a political than a technical challenge. In *Developments in Water Science.* ed., 9–23. Elsevier, Amsterdam.

Aspinall, R. and Pearson, D. 2000. Integrated geographical assessment of environmental condition in water catchments: linking landscape ecology, environmental modelling and GIS. *Journal of Environmental Management.* 59. (4): 299–319.

AzovBas. 2002. Proceedings. Azov Sea Basin Workshop Novocherkassk, Russia, Green Don.

BBC. 2007. Russia's Cossacks rise again. [on-line] BBC News. http://news.bbc.co.uk/2/hi/europe/6937562.stm [cited 10 November 2007].

Bolshov, A. 2000. Otcenka suchshestvuiushih nacionalniuh okhraniaemiuh territorij i ih statusa zaschity, osobenno transgranichniuh i migriruiuschih vidov. Kazakhstan. Biodiversity CRTC, Second Regional Caspian Workshop, Alma-aty, Caspian Environment Program.

Borodin, N.A. 1901. *Uralskie Kazaki i ih rybolovstva.* St. Petersburg, Izdanie "Vestnik Kazachih Voisk".

Brockhaus, F.A. and Efron, I.A. 1898. *Brockhaus and Efron Encyclopedic Dictionary.* St. Petersburg: Brockhaus & Efron Publishing House.

Chaves, H.M.L. and Alipaz, S. 2007. An integrated indicator based on basin hydrology, environment, life, and policy: the watershed sustainability index. *Water Resource Management.* 21. 883–895.

CITES. 2003. Sturgeons of Romania and CITES. [on-line] Black Sea Sturgeon Management Action Group. 30/06/03. http://rosturgeons.danubedelta.org/ [cited May 1 2004].

CITES. 2004. A brief history of sturgeons and CITES. [on-line] Steering Committee. Convention on International Trade in Endangered Species of Wild Fauna and Flora. http://www.cites.org/eng/prog/Sturgeon/history.shtml [cited April 2004].

Dal, V.I. 1961. Uralskij Kazak (1842). In *Dal, V.I. Novels and essays.* ed., Moscow: Pravda.

EU. 2000. European Union Water Framework Directive. European Union.

FAO. 2007. *The State of World Fisheries and Aquaculture 2006.* Rome: FAO Fisheries Department.

Fohrer, N. 2005. Advances in sustainable river basin management. *Ecological Modelling.* 187. (1): 1.

Giupponi, C. 2007. Decision Support Systems for implementing the European Water Framework Directive: the MULINO approach. *Environmental Modelling & Software.* 22. (2): 248–258.

Harris, G. 2002. Integrated assessment and modelling: an essential way of doing science. *Environmental Modelling & Software.* 17. (3): 201.

He, C., Malcolm, S.B., et al. 2000. A conceptual framework for integrating hydrological and biological indicators into watershed management. *Landscape and Urban Planning.* 49. (1–2): 25.

Hedelin, B. 2007. Criteria for the assessment of sustainable water management. *Environmental Management.* 39. 151.

Hughes, R.M. and Oberdorff, T. 1999. Applications of IBI concepts and metrics to waters outside the United States and Canada. In: Simon T.P. (Ed.) Assessing the sustainability and biological integrity of water resources using fish communities. London, CRC Press, pp. 79–96.

Jakeman, A.J. and Letcher, R.A. 2003. Integrated assessment and modelling: features, principles and examples for catchment management. *Environmental Modelling & Software.* 18. (6): 491–501.

Jansky, L., Pachova, N.I., et al. 2004. The Danube: a case study of sharing international waters. *Global Environmental Change Part A.* 14. (Supplement 1): 39–49.

Janssen, W. and Goldsworthy, P. 1996. Multidisciplinary research for natural resource management: conceptual and practical implications. *Agricultural Systems.* 51. (3): 259.

Jewitt, G. 2002. Can integrated water resources management sustain the provision of ecosystem goods and services? *Physics and Chemistry of the Earth, Parts A/B/C.* 27. (11–22): 887–895.

Jonker, L. 2002. Integrated water resources management: theory, practice, cases. *Physics and Chemistry of the Earth, Parts A/B/C.* 27. (11–22): 719–720.

Karpov, A.B. 1911. *Uralci. Istoricheskij ocherk.* Uralsk: Army Publishing House.

Kauffman, G.J. 2002. What if... the United States of America were based on watersheds? *Water Policy.* 4. (1): 57–68.

Keyl, F. and Wolff, M. 2007. Environmental variability and fisheries: what can models do? *Reviews in Fish Biology and Fisheries.* doi: 10.1007/s11160-007-9075-5.

Kgarebe, B.V. 2002. Water resources management: the challenge of integration. *Physics and Chemistry of the Earth, Parts A/B/C.* 27. (11–22): 865.

Krysanova, V., Hattermann, F., et al. 2007. Implications of complexity and uncertainty for integrated modelling and impact assessment in river basins. *Environmental Modelling & Software.* 22. (5): 701.

Lagutov, V. 1995. *Mekhanizm unichtozheniya riubniuh zapasov Yuga Rossii i puti ih spaseniya* [The mechanism of extermination of fish stocks in South Russia and possibilities of their restoration]. Novocherkassk: Donskaia Rech.

Lagutov, V. 1996. Imitacionnoe modelirovanie zhiznennogo tsikla osetroviuh v basseine reki Don. [Simulation of Sturgeon Life Cycle in the Don River Basin]. MSc Thesis, Department of Applied Mathematics, South Russia State Technical University, Novocherkassk.

Lagutov, V. 1997. Simulation Model of Sturgeon Life Cycle as a Tool in Sustainable Water Management. MSc Thesis, Department of Environmental Sciences, Central European University, Budapest.

Lagutov, V. 1999. New Environmental Policy. Sustainable development: from myth to reality. 13th Global Biodiversity Forum, San Jose, Costa Rica.

Lagutov, V. 2003. *Modelling of radionuclides washout with surface water runoffs from radioactive waste storage site.* Luxemburg, International Institute for Applied Systems Analysis.

Lanini, S., Courtois, N., et al. 2004. Socio-hydrosystem modelling for integrated water-resources management–the Herault catchment case study, southern France. *Environmental Modelling & Software.* 19. (11): 1011.

Larocque, G.R., Mauriello, D.A., et al. 2006. Ecological Models as Decision Tools in the 21st Century: Proceedings of a conference organized by the International Society for Ecological Modelling (ISEM) in Quebec, Canada, August 22–24, 2004. *Ecological Modelling.* 199. (3): 217.

Letcher, R.A., Croke, B.F.W., et al. 2007. Integrated assessment modelling for water resource allocation and management: a generalised conceptual framework. *Environmental Modelling & Software.* 22. (5): 733.

Liu, Y., Guo, H., et al. 2007. An optimization method based on scenario analysis for watershed management under uncertainty. *Environmental Management.* 39. 678.

Malecha, N.I. 2002. *Slovar govorov uralskih (jaickih) kazakov.* Orenburg: Orenburg Publishing.

Margesson, R. 1997. Environment and international water management: dealing with the problems of the Danube Delta. *Environmental Impact Assessment Review.* 17. (3): 145–162.

Mysiak, J., Giupponi, C., et al. 2005. Towards the development of a decision support system for water resource management. *Environmental Modelling & Software.* 20. (2): 203.

Parker, P., Letcher, R., et al. 2002. Progress in integrated assessment and modelling. *Environmental Modelling & Software.* 17. (3): 209.

Pullar, D. and Springer, D. 2000. Towards integrating GIS and catchment models. *Environmental Modelling and Software.* 15. (5): 451–459.

□AMSA□. 2002. Sustainable Management of Water □esources: The □eed for a Holistic □cosystem Approach □unning out of Freshwater or Maintaining Freshwater through a □cosystem □ased Approach - An □asy □hoice. 8th Meeting of the □onference of the □ontracting □arties to the □onvention on Wetlands (□amsar, Iran, 1971), Valencia, Spain, The □amsar □onvention on Wetlands.

Running out of Freshwater or Maintaining Freshwater Through a Ecosystem Based Approach - An Easy Choice. 8th Meeting of the Conference of the Contracting Parties to the Convention on Wetlands (Ramsar, Iran, 1971), Valencia, Spain, The Ramsar Convention on Wetlands, 2005.

Raymakers, C. and Hoover, C. 2002. Acipenseriformes: CITES implementation from Range States to consumer countries. *Journal of Applied Ichthyology.* 18. (4–6): 629–638.

Refsgaard, J.C., Nilsson, B., et al. 2005. Harmonised techniques and representative river basin data for assessment and use of uncertainty information in integrated water management (HarmoniRiB). *Environmental Science & Policy.* 8. (3): 267.

RF. 2007. ВОДНЫЙ КОДЕКС РОССИЙСКОЙ ФЕДЕРАЦИИ. (Water Code of the Russian Federation). Legal Information Consortium «Kodeks» (In Russian).

RK. 2002. National Action Plan for rehabilitation of the Caspian Sea environment for 2003-2012. Ministry of Environmental Protection, Republic of Kazakhstan.

Russian State Duma. 1995. Committee on Ecology, Russian State Duma. Hearing on "Improvement of fishery policies in the Azov Sea Basin". Report by Dr. V. Lagutov.

Scoccimarro, M., Walker, A., et al. 1999. A framework for integrated catchment assessment in northern Thailand. *Environmental Modelling and Software.* 14. (6): 567–577.

Semple, E.C. 1907. Geographical boundaries. II. *Bulletin of the American Geographical Society.* 39. (8): 449–463.

Sendzimir, J., Magnuszewski, P., et al. 2007. Anticipatory modeling of biocomplexity in the Tisza River Basin: first steps to establish a participatory adaptive framework. *Environmental Modelling & Software.* 22. (5): 599.

Shen, J., Parker, A., et al. 2005. A new approach for a Windows-based watershed modeling system based on a database-supporting architecture. *Environmental Modelling & Software.* 20. (9): 1127.

Suter, I. and Glenn, W. 1999. Developing conceptual models for complex ecological risk assessments. *Human and Ecological Risk Assessment.* 5. (2): 375–396.

Thorsten, W., Eloy, R., et al. 2004. Dealing with uncertainty in spatially explicit population models. *Biodiversity and Conservation.* 13. (1): 53.

TRAFFIC. 2003. Bleak outlook for world caviar trade: North America sturgeon, paddlefish stocks at risk. [on-line] 22 May 2003. http://www.traffic.org/news/press-releases/bleakoutlook.html [cited 2004 April 20].

TRAFFIC. 2007. Stamp of approval for caviar. [on-line] TRAFFIC. http://www.traffic.org/home/2007/6/29/stamp-of-approval-for-caviar.html [cited 2007].

Turnock, D. 2001. Cross-border conservation in East Central Europe: the Danube-Carpathian complex and the contribution of the World Wide Fund for Nature. *GeoJournal.* 55. (2–4): 655–681.

UNECE. 2003. *Transboundary Water Cooperation in the newly independent States (NIS).* Moscow-Geneva United Nations Economic Commission for Europe.

UNECE. 2006. *Preliminary Assessment of the Status of Transboundary Rivers Discharging Into the Caspian Sea and Their Major Transboundary Tributaries.* United Nations Economic Commission for Europe.

UNECE. 2007. Transboundary Accidental Water Pollution, Liability and Compensation: Challenges and Opportunities, Budapest, Hungary, United Nations Economic Commission for Europe.

UNEP. 2004. Learning from Baia Mare. [on-line] The Environment Times. http://www.environmenttimes.net/article.cfm?pageID=112 [cited 10 December 2007].

Uralbas. 2007. Proceedings. First Ural River International Workshop Orenburg, Russia.

van Ast, J.A. 2000. Interactive management of international river basins; experiences in Northern America and Western Europe. *Physics and Chemistry of the Earth, Part B: Hydrology, Oceans and Atmosphere.* 25. (3): 325–328.

van Delden, H., Luja, P., et al. 2007. Integration of multi-scale dynamic spatial models of socio-economic and physical processes for river basin management. *Environmental Modelling & Software.* 22. (2): 223.

Vidal, J.-P., Moisan, S., et al. 2007. River model calibration, from guidelines to operational support tools. *Environmental Modelling & Software.* 22. (11): 1628–1640.

Von Harthausen, A. 1972. *Studies on the Interior of Russia.* Chicago/London: University of Chicago Press.

WWF. 2002a. WWF Factsheet: Sturgeon. The Materials of 12th Meeting of the Conference of the Parties to CITES, Santiago, 3–15 November 2002.

WWF. 2002b. *Sturgeon 2020. Initiative for Providing Habitats for Sturgeons.* Program despcription and schedule, Budapest, WWF.

WWF. 2003. Global 200. Blueprint for a living planet. [on-line] World Wildlife Fund. http: //www.panda.org/about_wwf/where_we_work/ecoregions/global200/pages/regions/region159.htm [cited April 2004].

RESULTS OF THE FIRST INTERNATIONAL URAL RIVER BASIN WORKSHOP (NATO-ARW)

Rescue of Sturgeon Species by Means of Transboundary Integrated Watershed Management in the Ural River Basin

Resolution adopted by the First International Ural River Basin Workshop Orenburg, Russia, 13–16 June 2007

The First Ural River Basin International Workshop "Rescue of Sturgeon Species by means of Transboundary Integrated Water Management in the Ural River Basin" (NATO-ARW) was held in Orenburg on June 13–16 within the Ural Basin Project framework. Organized by Research and Consulting Center DonEco and Central European University the Workshop was co-sponsored by the Security Through Science Programme (NATO-ARW) and the Caspian Environmental Program. The Workshop was conducted under the auspices of Orenburg Regional Government and Parliament with active involvement and assistance by the Russian Federal Agency for Environmental Inspections. The Project's ultimate goal is the establishment of a Ural Sturgeon Park to facilitate conservation and restoration of the sturgeon population in the Caspian Sea, rehabilitation of the Ural River ecosystem, and sustainable development of the Ural Basin.

The workshop was attended by more then 60 experts, researchers and practitioners from Governmental Environmental Agencies, NGO and business representatives from both basin countries (Russia and Kazakhstan), and representatives from relevant international organizations such as the Food and Agriculture Organization of the United Nations, the Secretariat of Wetland Convention (RAMSAR), the International Association on Danube Research, and many others covering the whole spectrum of Ural Basin management stakeholders.

Drawing on recommendations of previous workshops and on the results and materials of other conferences and meetings dedicated to the problems of the Caspian Sea.

Workshop participants EMPHASIZE that:

1. The Ural Basin Ecosystem has world wide importance as it is uniquely positioned for conservation of the sturgeon species and has strong historical traditions in the rational use of living aquatic resources.
2. Being the only free-flowing river in the Caspian Basin, the Ural has the only available natural spawning grounds and undisturbed migration routes of sturgeon species. The Ural is the only river in the Caspian Sea catchment where sturgeon species can reproduce in the wild, which undoubtedly makes it the only remaining opportunity for the conservation of sturgeon species in the Caspian Basin and for successful restoration programs worldwide.
3. Sturgeon species can be considered a perfect natural bioindicator of the river basin's health. Their conservation will serve the Region's sustainable economic and social development.
4. Despite the unquestionable importance of the Ural's ecosystem and its crucial role in conserving the sturgeon population of the Caspian Sea, there have been limited practical steps undertaken and no transboundary cooperation developed to date. More urgent and radical measures are needed for the sustainable protection of the Ural River ecosystem and reinstitution of the rational use of natural resources.
5. The Ural-Caspian Basin conservation challenge requires an interdisciplinary holistic approach. The joint efforts of stakeholders and experts from different fields (lawyers, bio scientists, economists, politicians, members of local business community and many others) are extremely important for securing successful environmental protection programs.
6. Ecosystem and biological resources preservation depends upon efficient cooperation by both basin countries, Russia and Kazakhstan, and active involvement of international institutions.
7. The existing legislative base including, but not limited to, National Constitutions, International Conventions, ratified by both countries, national laws and a number of bilateral agreements creates an adequate legal framework for this kind of cooperation.
8. Both national and international experience of transboundary river basins management and biodiversity conservation should be incorporated in the planning of environmental protection activities in the Ural River Basin.

CONSIDERING the above and taking into account the following:

- Water streams and migratory species are national/federal property.
- Transboundary integrated water management and preservation of sturgeon species are priority issues on the international environmental agenda.
- Sturgeon species play an important role in the river ecosystem and river affiliated communities.

Workshop participants URGE the Presidents, Governments and Parliaments of Russia and Kazakhstan as well as international organizations to undertake the following steps:

1. Recognize the Ural River Basin as an ecosystem of worldwide importance, guarantee its maximum protection and the preservation of its natural ecosystem regimes and proclaim the Ural Basin a model territory and pilot project in transboundary integrated water management and biodiversity conservation.
2. Make the creation of an International Ural Sturgeon Park and sustainable development of the Ural River Basin one of the priority issues in regional cooperation between Russia and Kazakhstan.
3. Establish a Joint Russia-Kazakhstan Ural Basin Interparliamentary Commission to be responsible for restoration and sustainable usage of the biological and natural resources of the Ural River Basin.
4. Proceed with bilateral negotiations on the international legal status of the Ural Sturgeon Park initiated by the Research and Consulting Center DonECO in 2003.
5. Prepare and ratify a Ural Basin Convention dedicated to the preservation of Ural River biodiversity and community-based management and community rights for natural resources.
6. Draft, implement and enforce national laws on protection and sustainable usage of Ural Basin renewable resources in Russia and Kazakhstan, drawing on a basin approach to sustainable development and using the sturgeon as an indicator species.
7. Create an International Protected Area along the Ural River floodplain within the boundaries drawn by a 100-year flood, as defined by the concept of sustainable basin development under the auspices of the United Nations.
8. In cooperation with the United Nations develop and put into practice international technical standards and regulations for the environmental management of transboundary territories and water basins.
9. Encourage and facilitate the application of an interdisciplinary holistic approach to planning and management of environment-related activities.
10. Set up a joint transboundary Russia-Kazakhstan system of water quality monitoring in the river Ural using modern methods and technologies, including, but not limited to, remote sensing, satellite imagery, and biomonitoring.
11. Prohibit disposal of any untreated, heated and contaminated waters, sewage and waste waters with pollutant concentration higher than the maximum concentration limit for fish ponds to water streams of the Ural River Basin.
12. Monitor new industrial and agricultural projects in the basin and enforce the implementation of independent environmental impact assessment with mandatory consideration of impact on the river ecosystem and conservation of aquatic biodiversity.

13. Incorporate into the national legislation of the basin countries the strictest international environmental standards, principles and processes and the strictest rules for monitoring of those standards.
14. Ban fishing of sturgeon and other migratory species in the Ural-Caspian Basin until the level of recruitment from natural reproduction of these species is restored to a level within the limits recommended by the expert group of the Ural Basin Project.
15. Revise national and regional Red Lists to include sturgeon species of Caspian and Ural basins which are endangered according to the IUCN Red List.
16. Intensify measures to combat poaching and illegal fishing by introduction of severe punishments, better transboundary cooperation of fishery inspections of Russia and Kazakhstan and active involvement of local communities.
17. Strive against an unreasonably high "scientific" catch by banning any kind of sturgeon fishing, including fishing for scientific and artificial propagation purposes. Some permits, the use of which have to be closely and carefully monitored, can be given to particular institutions upon specific request for justified scientific, commercial and restocking purposes.
18. Improve methods used in assessment of sturgeon stock in the Caspian Basin by following international standards and applying worldwide recognized techniques.
19. Develop and equip dams and hydrological constructions with state-of-the-art (i.e. contactless) counting devices to monitor migrating specimens.
20. Base sturgeon re-stocking activities (i.e. establish new sturgeon hatcheries, release fingerlings) not in the mouth of the Ural river but further upstream near the spawning places close to the city of Uralsk and on the river Ilek.
21. Restore sturgeon migration routes in the regulated rivers of the Caspian Sea basin to ensure the natural restoration and sustainable usage of sturgeon by improving existing fish passes through dams, or equipping those dams without fish passes, using state-of-the-art fish passage designs, preferably nature-like design options where possible.
22. Secure regular flooding of river floodplains during the spawning periods.
23. Encourage fundamental and applied research on management issues of transboundary areas to secure sustainable development in the basin.
24. Review the education system, in particular state and business management to include in syllabi discussions of transboundary environmental management and nature protection and to introduce to institutions and schools of the Ural Basin experimental training courses for officers of environmental agencies and state services.
25. Assist in and facilitate regular Ural Basin Workshops to provide scientific grounding, public participation, stakeholder involvement and public support for the activities of the Ural Basin Project and the Ural Basin Interparliamentary Commission.
26. Seek cooperation and synergies with the FAO TCP/INT/3101 "Capacity building for the recovery and management of the sturgeon fisheries of the Caspian Sea".

Workshop Participants APPROVE

- Working Group activities regarding the development and implementation of the Ural Basin Project chaired by Dr. Vladimir Lagutov and including the representatives from Russia, Kazakhstan and international organizations.
- Inclusion of the experts who participated in the First Ural Basin Workshop in the Advisory Board of the Ural Basin Project.

To deliver the strategic goals outlined above and to ensure further development of the Ural Basin Project the workshop participants RECOMMEND the UBP Working Group to undertake the following steps within the framework of the Ural Basin Project:

1. Inform the governments of Russia and Kazakhstan, regional and local authorities and relevant international organizations about the Project's activities.
2. Publish the results and materials presented at the First International Ural Basin Workshop.
3. Include representatives of the cities of Uralsk and Atyrau (Guriev), both basin countries and representatives from all Caspian countries in the Project Working Group.
4. Approach relevant national and international institutions with an initiative on drafting and carrying out a feasibility study for the Ural Sturgeon Park proposal within the framework of the Ural Basin Project.
5. Propose using the case of the Ural River Basin to develop and test universal environmental basin legislation, originating from the concept of sustainable basin development.
6. Explore the existing legal framework in national and international legislation to support the creation of an International Sturgeon Park.
7. Summarize the current situation with water bioresources in the Ural River Basin in a position paper to be used as a background paper for fundraising and negotiations with international and national institutions on possible cooperation.
8. Identify stakeholders in transboundary integrated water management and biodiversity conservation in the Ural river, establish working relationships and encourage them to participate in the Ural Basin Project.
9. Carry out a preliminary analysis of the factors influencing the sturgeon population through its life cycle.
10. In accordance with the identified factors establish and develop a network of specialists and practitioners in sturgeon and water-related issues of the Ural basin using *inter alia* the Project website.
11. Develop a Ural basin GIS database for these factors.
12. Draft amendments to the Red Lists of Russia and Kazakhstan as well as regional Red Lists to include IUCN Red List endangered sturgeon species and push for adoption of these amendments.

13. Conduct feasibility studies and draft proposals on (i) restoration and rehabilitation of the regulated rivers of the Caspian Basin such as the Volga, Terek, Kura-Araks, etc., (ii) practical steps for using sturgeon species for bioindication.
14. Promote and popularize the idea of the priority of natural reproduction of sturgeons over artificial propagation and facilitate its adoption by national and international environmental institutions and agencies.
15. Encourage stakeholders and workshop participants to participate in development and information provision of the Ural Basin Project website (http://uralbas.ru).
16. Design and support an independent information campaign on the importance of Ural River habitats preservation and on Project activities in basin countries and within international environmental communities.
17. Prepare and hold a Second Ural Basin Workshop in Kazakhstan within the framework of the Ural Basin Project.
18. Intensify the consultations with international institutions to explore the possibilities of cooperation and financial support for the Ural Basin Project.

Participants of the First Ural Basin Workshop would like to express their gratitude to the initiators and Working Group of the Ural Basin Project for the preparation and organization of the Workshop; the NATO Security through Science Programme, the Caspian Environmental Program and Central European University for their financial support; the Research and Consultative Center DonECO and the Orenburg Regional Branch of the Russian Federal Agency for Environmental Inspections for technical assistance and active support in the organization of the Workshop; the Orenburg regional authorities for enabling the workshop to take place in Orenburg.

INDEX

V

W

GPSR Compliance

The European Union's (EU) General Product Safety Regulation (GPSR) is a set of rules that requires consumer products to be safe and our obligations to ensure this.

If you have any concerns about our products, you can contact us on ProductSafety@springernature.com

In case Publisher is established outside the EU, the EU authorized representative is:

Springer Nature Customer Service Center GmbH
Europaplatz 3
69115 Heidelberg, Germany

Batch number: 10755284

Printed by Printforce, the Netherlands